Signal Processing for Computer Vision

Signal Processing for Computer Vision

by

Gösta H. Granlund
Linköping University

and

Hans Knutsson
Linköping University

KLUWER ACADEMIC PUBLISHERS
DORDRECHT / BOSTON / LONDON

Library of Congress Cataloging-in-Publication Data

Granlund, Gosta H.
Signal processing for computer vision / by Gosta H. Granlund and Hans Knutsson.
p. cm.
Includes bibliogrphical references and index.
ISBN 0-7923-9530-1 (acid-free paper)
1. Computer vision. 2. Signal processing--Digital techniques.
I. Knutsson, Hans. II. Title.
TA1634.G73 1995
006.4'2--dc20 94-39290

ISBN 0-7923-9530-1

Published by Kluwer Academic Publishers,
P.O. Box 17, 3300 AA Dordrecht, The Netherlands.

Kluwer Academic Publishers incorporates
the publishing programmes of
D. Reidel, Martinus Nijhoff, Dr W. Junk and MTP Press.

Sold and distributed in the U.S.A. and Canada
by Kluwer Academic Publishers,
101 Philip Drive, Norwell, MA 02061, U.S.A.

In all other countries, sold and distributed
by Kluwer Academic Publishers Group,
P.O. Box 322, 3300 AH Dordrecht, The Netherlands.

Printed on acid-free paper

Printed in the Netherlands

CONTENTS

PREFACE ix

1 INTRODUCTION AND OVERVIEW 1

1.1 Hierarchical Computing Structures 3
1.2 Low level representation and operations 12
1.3 Description in terms of symmetry 20
1.4 Cascading of operations 24
1.5 Compatibility representation 25
1.6 Description of size and scale 27
1.7 Model based processing 33
1.8 Representation of volumes and time sequences 38
1.9 Classification and response generation 39

2 BIOLOGICAL VISION 41

2.1 Motivation 41
2.2 Overview of the visual system 42
2.3 Properties of neurons 43
2.4 The retina 49
2.5 Color vision 59
2.6 The visual pathways and the lateral geniculate nucleus 67
2.7 The primary visual cortex 69
2.8 Columnar organization and the layers of the cortex 75
2.9 Possible implementations of filters 80
2.10 Channel organization of features 83
2.11 The visual pathways beyond the primary visual cortex 89

3 LOW LEVEL OPERATIONS 97
3.1 Information representation 97
3.2 Complex valued convolution functions 102
3.3 Compact representation 105
3.4 Examples of processing 109
3.5 More on compatibility representation 112
3.6 General properties of a useful information representation 114

4 FOURIER TRANSFORMS 117
4.1 Introduction 119
4.2 Basics 122
4.3 Three aspects of the Fourier transform 143
4.4 Separability 153
4.5 Analytic signals 164
4.6 Examples of statistical calculations 174
4.7 Transform pairs 177
4.8 The Fourier transform and discrete functions 180
4.9 The discrete Fourier transform 192

5 KERNEL OPTIMIZATION 199
5.1 Spatial restrictions 199
5.2 Distance measure 200
5.3 Minimizing the distance 201
5.4 The weighting function 201
5.5 Optimization results 203
5.6 Kernel evaluation 209

6 ORIENTATION AND VELOCITY 219
6.1 Tensors - A short introduction 222
6.2 Representing orientation 224
6.3 Orientation estimation 230
6.4 Tensor Construction 239
6.5 Interpretation of the orientation tensor 249
6.6 Time sequences - velocity 253
6.7 Performance measures 257

7 LOCAL PHASE ESTIMATION 259

7.1 What is local phase? 259
7.2 Local Phase in Scale–space 266
7.3 Phase in higher dimensions 271
7.4 Applications using Local Phase Estimates 275

8 LOCAL FREQUENCY 279

8.1 Frequency estimation 281
8.2 Wide range frequency estimation 288
8.3 Experimental results 290

9 REPRESENTATION AND AVERAGING 297

9.1 Background 297
9.2 Invariance and equivariance 298
9.3 Signals and certainties 299
9.4 Vector and tensor representation of local orientation 301
9.5 Averaging 303
9.6 Normalized averaging 304

10 ADAPTIVE FILTERING 309

10.1 Adaptive filtering 311
10.2 Tensor controlled filter 313
10.3 The control tensor 313
10.4 Adaptive filter synthesis 320
10.5 Examples of filtering 333

11 VECTOR AND TENSOR FIELD FILTERING 343

11.1 Introduction 343
11.2 Orientation in vector and tensor fields 343
11.3 Rotational symmetries 350
11.4 Curvature estimation 361

12 CLASSIFICATION AND RESPONSE GENERATION 367

12.1 Information needed for classification 368
12.2 Linear discriminant functions 371

12.3 Minimum distance classification 378
12.4 Convolution operations 380
12.5 Generalized discriminant functions 385
12.6 Training of linear discriminant classifiers 389
12.7 Perceptrons and feedforward neural networks 390
12.8 Clustering and unsupervised classification 396

13 TEXTURE ANALYSIS 399
13.1 Introduction 399
13.2 Issues in texture analysis 401
13.3 Human aspects of texture perception 402
13.4 Feature-based texture analysis 405
13.5 Structural approach using certainty gating 412

REFERENCES 419

INDEX 433

PREFACE

Interest in computer vision has increased considerably as other disciplines of computation and control have developed to require better flexibility and precision in their controlling information. Despite considerable development over the past few decades, the vision field is still in its infancy. Many fundamental questions remain of how to handle and represent spatial information in an effective way.

The intent of writing this book has been to give an introduction to robust low-level methods for computer vision. The field has seen a variety of methods originating from different researchers over a few decades. This book is not a review of published material in the field, but mainly an account of methods developed and tested at the Computer Vision Laboratory (CVL) of Linköping University in Sweden. The book includes as well certain methods developed at other sites; methods that conform to the requirements of robustness and modularity that are deemed essential for vision systems.

Although this book is attributed to two principal authors, both the work described and its documentation is the collective effort of a number of people who have contributed in various ways over the years.

Chapter 1, written by Gösta Granlund, establishes the motivation and use of hierarchical operation structures to provide a systematic organization for the implementation of complicated models. The chapter gives an intuitive treatment of most aspects that are considered in the later chapters.

Chapter 2, written by Gösta Granlund and Jörgen Karlholm, with contributions by Carl-Johan Westelius and Carl-Fredrik Westin, gives an overview of important biological vision mechanisms. Although a great deal is known about neural processing of visual information, most essential questions about biological vision remain as yet unanswered. Nonetheless, the knowledge available has already provided useful guidance to the organization of effective machine vision systems.

Chapter 3, written by Gösta Granlund with contributions by Johan Wiklund, gives an introductory treatment of operations and representations for low-level features in multi-dimensional spaces. An important issue is how to combine contributions from several filters to provide robust statements in accordance with certain low-level models. This chapter gives an introduction to the problems of unambiguous mappings in multi-dimensional spaces.

These introductory chapters establish the framework and motivate a more detailed treatment, which begins in Chapter 4. This chapter, written by Klas Nordberg, is on Fourier methods, with a particular emphasis on definitions and theorems essential to the understanding of filtering procedures in multidimensional spaces. This is a central issue in computer vision.

Chapter 5, written by Hans Knutsson, presents a method for obtaining an optimal n-dimensional set of filter coefficients for any given frequency response. An optimality criterion is defined that enables different frequencies to be given individual weights. Appropriate forms of frequency weight functions are discussed and a number of optimization examples are given.

Chapter 6, written by Hans Knutsson, introduces the use of tensors in estimation of local structure and orientation. The tensor representation is shown to be crucial to unambiguous and continuous representation of local orientation in multiple dimensions. In addition to orientation the tensor representation also conveys the degree and type of local anisotropy. The orientation estimation approach is developed in detail for two, three and four dimensions and is shown to be extendable to higher dimensions. Examples and performance measures are given for processing of images, volumes and time sequences. The processing and tests in three dimensions are the work of Mats Andersson, Leif Haglund and Johan Wiklund.

Chapter 7, written by Carl-Johan Westelius deals with the concept of phase and phase representation in multiple dimensions. Phase is an important concept, which emerges in several contexts in vision. The chapter provides a detailed treatment of phase properties in various situations and deals with how to maintain continuity in phase representation. An example is given of how local phase in difference scales can be used for disparity estimation. The chapter contains original material on the representation of phase for signals in three dimensions and higher dimensions.

Chapter 8, written by Hans Knutsson with contributions by Carl-Fredrik Westin, deals with the estimation of local frequency and bandwidth. Local frequency is an important concept which provides an indication of the appropriate range of scales for subsequent analysis. A number of one-dimensional and two-dimensional examples of local frequency and bandwidth estimation are given.

Chapter 9, written by Carl-Fredrik Westin with contributions by Hans Knutsson, considers what the important properties are for an information representation to behave well in various transformations. There is an extended discussion on the necessity to separate between class membership and certainty of a signal.

Chapter 10, written by Hans Knutsson, presents a computationally efficient technique for adaptive filtering of n-dimensional signals. The approach is based on the local signal description given by the orientation tensor discussed in Chapter 6. The adaptive filter output is synthesized as a tensor-controlled weighted summation of shift-invariant filter outputs. Several examples of adaptive filtering in two and three dimensions are given. Most of the results in the example section were produced by Leif Haglund. The chapter contains original results on the extension of the techniques to n dimensions.

Chapter 11, written by Carl-Fredrik Westin discusses techniques for processing of higher order data such as vector and tensor fields. As abstraction implies a more complex descriptor, developing methods for processing of higher order data is an essential part of any hierarchical or layered approach to vision. The chapter focuses on models for extracting local symmetries and discontinuities in higher order fields.

Chapter 12, written by Gösta Granlund with contributions by Jörgen Karlholm, is not original, but presents methods for linear classification in the tradition of N. J. Nilsson as well as R. O. Duda and P. E. Hart. Part of the motivation for including this well-known material is to allow the vision structure to be brought to a logical conclusion in which feature properties are combined to form responses or class statements. Another motivation developed here is to display the similarity in structure between convolution operations and linear discriminant functions. This brings all operations for feature extraction and classification to the use of a common component, linear discriminants. This is also illustrated in the form of perceptrons, which allows a transition to the modern theory of neural networks.

Chapter 13, written by Morgan Ulvklo, deals with texture analysis, an important application of the methods described in earlier chapters. It introduces ideas from preattentive vision, which gives clues for the extraction of texture primitives. There is also a discussion on how to handle features whose significance varies with spatial position.

In addition to writing Chapter 4, Klas Nordberg has orchestrated the production of the book with its detail work on organization, layout and figures and he is responsible for making the LaTeX compilations run. Catharina Holmgren has contributed with entry of text and pictorial material, as well as proof-reading of the manuscript. Finally, the authors want to acknowledge contributions made by all other members, past and present, of the CVL over the years since the start of its activities in the late seventies.

Matt Fichtenbaum of GenRad, Inc., Concord Massachusetts has reviewed the manuscript and suggested numerous improvements to the language and the description of technical topics. Dr. Elizabeth Liebson of Tufts - New England Medical Center in Boston, Massachusetts has reviewed the chapter on biological vision.

There are a number of persons who have visited CVL for shorter or longer periods, persons with whom we have discussed various issues of common interest, persons who have in one way or other been influential in the course of our research. Among those we would like to mention: Prof. Edward Adelson, MIT, Cambridge, USA; Dr. Andrew Calway, Univ. of Wales, Cardiff, Great Britain; Prof. C. Forbes Dewey, MIT, Cambridge, USA; Prof Ramesh Jain, Univ. of Michigan, Ann Arbor, USA; Prof. Bernd Jähne, Univ. of California, San Diego, California; Prof. Murat Kunt, EPFL, Lausanne, Switzerland; Prof. Steven Tanimoto, Univ. of Washington, Seattle, USA; Prof. Torsten N. Wiesel, Rockefeller University, New York; Dr. Roland Wilson, Univ. of Warwick, Coventry, Great Britain. The authors would also like to acknowledge the beneficial influence from partners in the ESPRIT Basic Research Action projects 3038/7108: Vision as Process.

We are grateful for the permission to reproduce copyright figures as referenced in the text. In a few cases it has proved impossible to reach copyright holders. We regret this and hope that no offence has been given.

The major part of the research work described in this book has been supported by research grants from The Swedish National Board for Industrial and Technical Development (NUTEK).

Gösta H. Granlund

Hans Knutsson

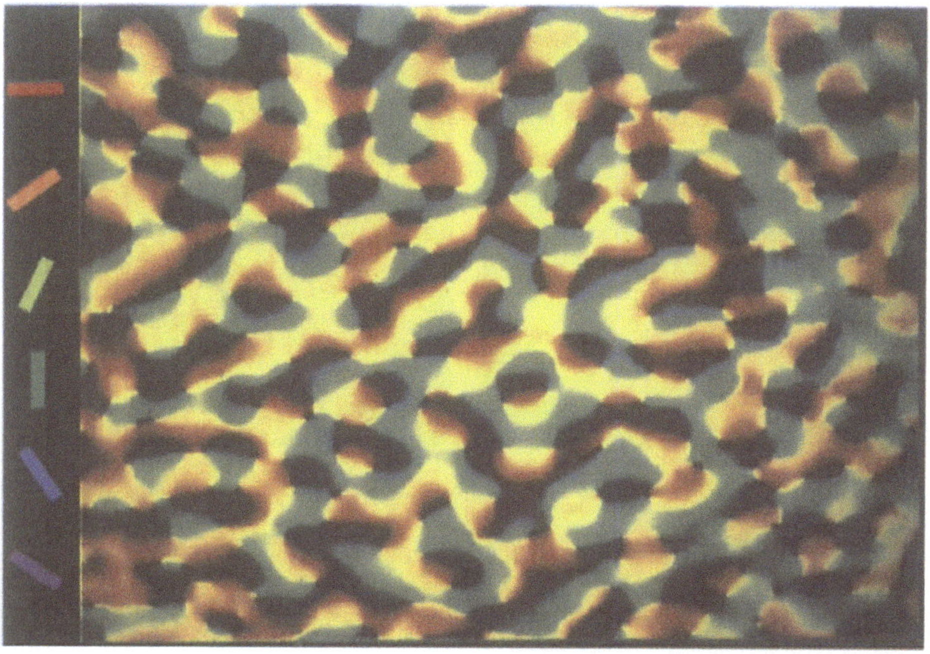

Color image 1 Color display of orientation preference map of the visual cortex.

Color image 2 Illustration of adaptive filtering. **Left image:** Original image. **Right image:** Result after adaptive filtering.

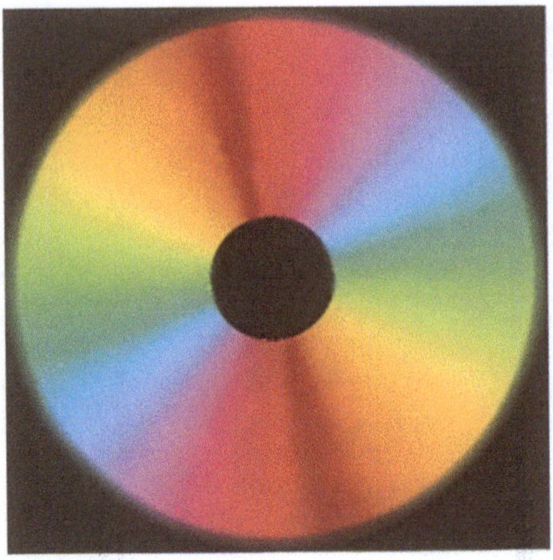

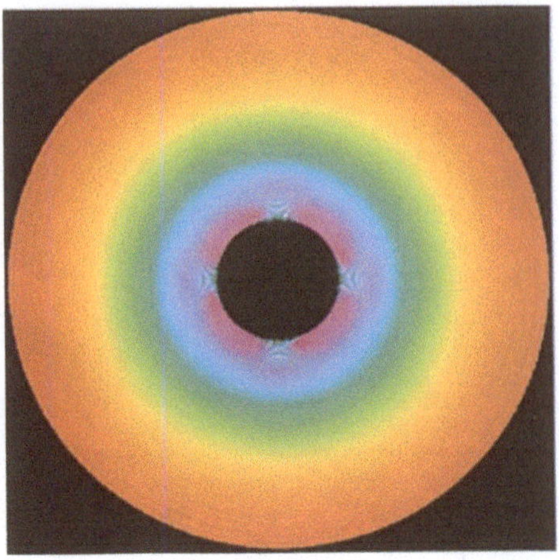

Color image 3 Low-level operations made upon the test image in Figure 3.9. **Top left:** Orientation extraction. **Top right:** Frequency extraction. **Bottom left:** Phase extraction.

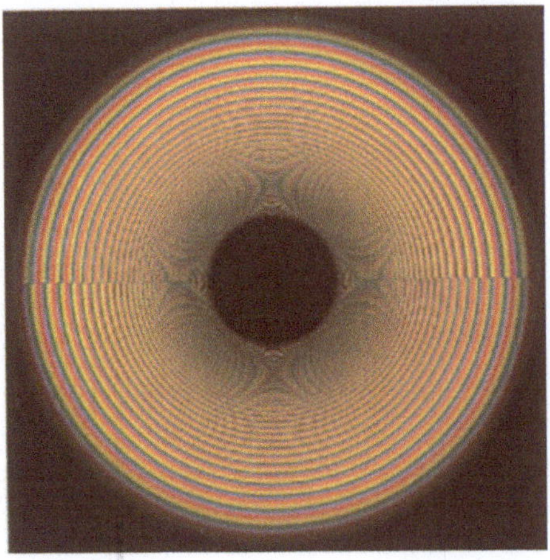

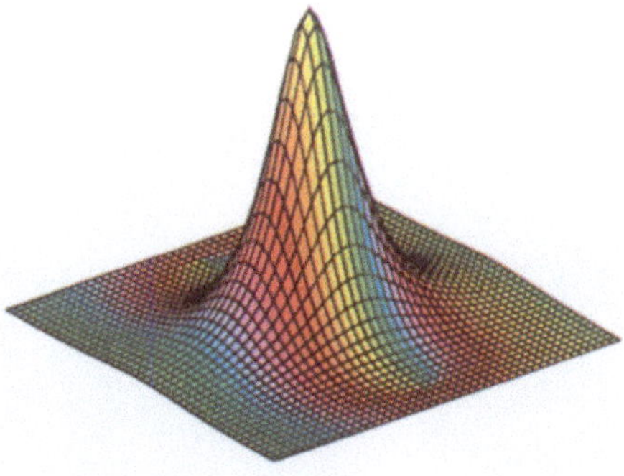

Color image 4 A two-dimensional quadrature filter in the spatial domain.

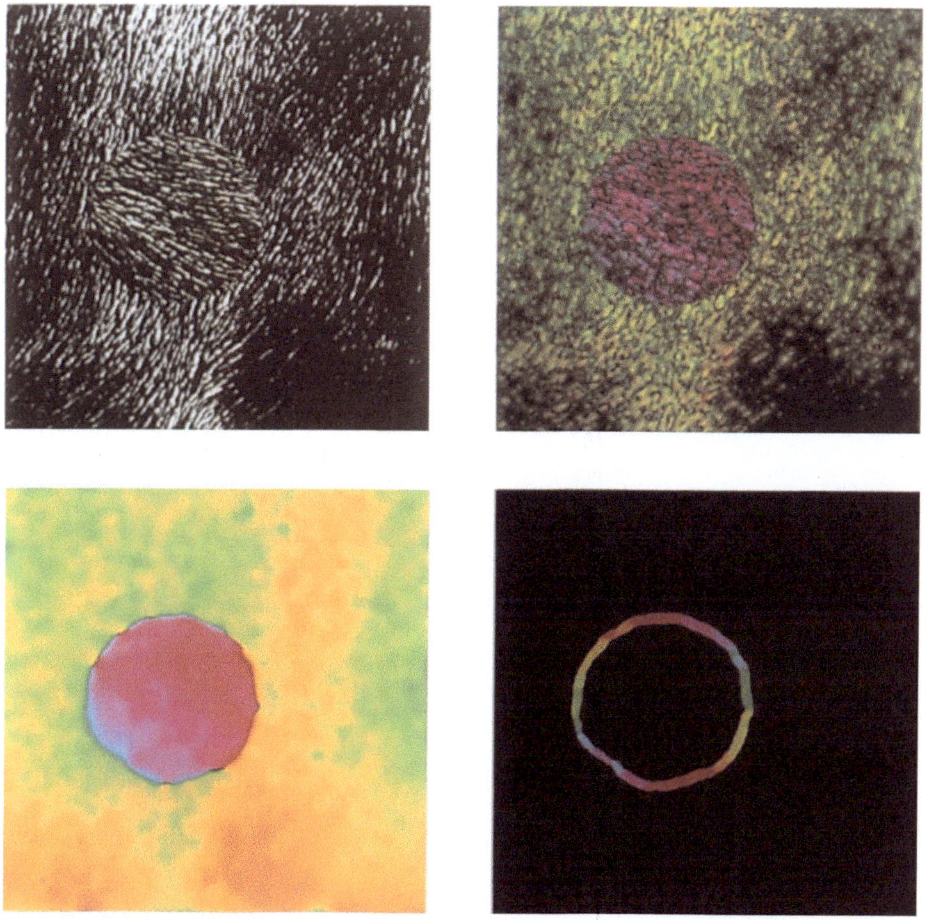

Color image 5 Example of two levels of transformation. **Top left:** image of a calf skin where a patch has been rotated a quarter of a turn. **Top right:** First level transformation representing the local orientation. **Bottom left:** Result of local averaging of the orientation description. **Bottom right:** Second level transformation representing borders between different orientations.

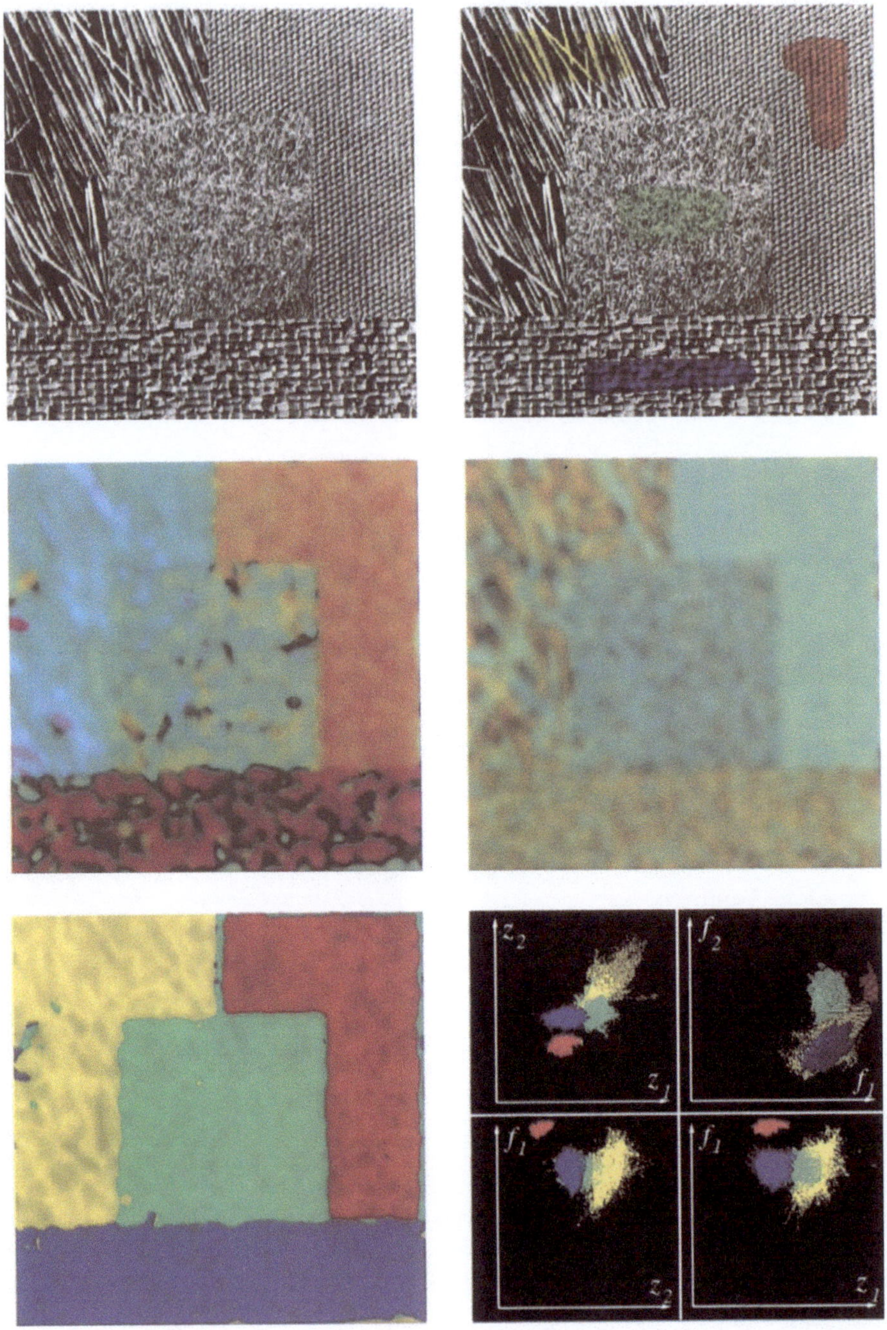

Color image 6 **Top left:** Original image containing four textures, from Brodatz. **Top right:** Training areas. **Middle left:** Local orientation estimates in vector representation. **Middle right:** Local frequency estimates in vector representation. **Bottom left:** Result from Maximum Likelihood classification: color depicts class membership. **Bottom right:** Histogram over training data in the feature space: color depicts class membership.

1

INTRODUCTION AND OVERVIEW

Computer vision has progressed considerably over the years. From methods only applicable to simple images, it has developed to deal with increasingly complex scenes, volumes and time sequences.

Different methodologies have been emphasized as crucial at different times. In the infancy of computer vision, it was believed that objects of interest could unequivocally be separated from the background using a few standard operations applied over the entire image. It turns out, however, that this simple methodology only works on images having a good separation between object and background. In the current state of the art, the prevalent view is that sufficiently efficient processing of complex scenes can only be implemented using an adaptive model structure. In the case of more difficult problems with noise, ambiguities and disturbances of different types, more sophisticated models or algorithms are required with provisions to adapt themselves to the image content. A further extension of this adaptivity is the current development of what is known as *Active Vision* [6, 131].

Classically, the methodology of image analysis contains many procedures to perform various tasks [37, 7, 56]. A common problem is that these procedures are often not suitable as components of a larger system, in which such procedures are combined and interacting. The reason is that information is represented in different ways for different types of features. It is difficult to have such descriptors combine their statements in a graceful way, and to have them control operations in a parametric way.

There is no indication that it will ever be possible to find some "simple trick" that miraculously solves most problems in computer vision. It turns out that the processing system must be able to implement a model structure, the complexity of which is directly related to the structural complexity of the problem under consideration in the external world.

The total spectrum of possible outcomes that a descriptive structure must accommodate is very large, however. It is obviously not possible to have different models or templates for each possible instance of an object or an event. A particular object may appear in many different orientations, sizes, projections, etc. It is necessary to encompass this variability in a more efficient manner than to have one model or template for each possible orientation of an object. It is necessary to find model representations that exhibit *invariance*, i.e. in which the descriptive statements do not depend upon what are considered irrelevant variations such as the orientation of an object.

What is irrelevant, or an *invariant mode*, will differ from one situation to another. In one case we want to recognize an object regardless of its orientation. In another case the orientation of the object is of major concern. In a systematic approach, we will try to design models to produce different modes separately, such that we can use the information if desired.

It turns out to be necessary to use what we may call different sub-algorithms or sub-models on different parts of an image. The selection of a particular sub-algorithm is typically based upon a tentative analysis of the image content. The techniques for such preliminary or *preattentive* vision have been extensively developed [6, 101, 135], although they are outside the scope of this book.

The reason for using different sub-algorithms is the simple fact that not all possible events can be expected in a particular context. As indicated earlier, this handling of sub-algorithms has to be implemented as a modification or a parameterization of more general algorithms. Without this, we would obtain a model computing structure which is totally unmanageable, which would exhibit a combinatorial explosion.

In order to allow the repeated use of existing models or model parts, we will try to design models that can be used for several purposes. These partial models will have some general properties of invariance generation and generality in representation. A substantial part of this book deals with this problem.

Our own work, as well as some of the work cited, has taken a great deal of inspiration from what is known about biological visual systems [58, 57, 92]. For this reason, Chapter 2 is devoted to an introduction to the main themes of biological vision. This is not to say that we always assume that the mechanisms presented in the chapters to follow are indeed models of phenomena used in biological visual systems. Too little is so far known to form any firm opinion. The ultimate criterion is simply performance from a technical point of view.

This introductory chapter is an overview of the main issues covered in the book. It focuses on signal processing aspects of hierarchical vision systems. It reviews some results, but mainly provides the motivation for the chapters to follow.

1.1 HIERARCHICAL COMPUTING STRUCTURES

An important topic of this book is the problem of how to assemble sufficiently complex models and the computational structures required to support them. In order for a system modeling a high structural complexity to be manageable and extendable, it is necessary that it exhibits modularity in various respects. This implies, for example, standardized information representations for interaction between operator modules. Otherwise the complexity will be overwhelming and the functional mechanisms completely obscure. One way to satisfy these requirements is to implement the model structure in a regular fashion. It is often useful to view this regular arrangement as a *hierarchy*, although we should not necessarily restrict the communication between levels to be only between adjacent layers of such a hierarchy. Hierarchical structures are in principle nothing new in information processing in general, or in computer vision in particular. A regular organization of algorithms has always been a desired goal for computer scientists. In order for such a structure to work effectively on spatial data, however, certain crucial requirements have to be fulfilled for information representation and for the structures of operations.

1.1.1 Locality, and the concept of frames

Most procedures used in image analysis and classification have the characteristic of *locality*. This is the assumption that we do not require the information from all pixels in an image to perform relevant analysis. One example is that we can recognize a book on a table without having all contextual evidence restricting the environment to an office, recognizing a particular language of the book, etc. The locality is valid in geometric space, as well as property space, contextual space, time space, etc. This is a comfortable relaxation from a requirement for complete references in our description, which would make recognition practically impossible.

Still, classification based upon local features may not be unique. In remote sensing, classification is often based upon the spectral signatures from different bands of an observation satellite. This means that we are likely to find the same color or combination of spectral signatures within objects which we would like to assign to different classes. This appears to be true for any type of local, sufficiently simple descriptor. The information content of a density value is limited, since we have a limited ability to discriminate between absolute levels of density, and since ambient factors such as illumination tend to decrease our ability of discrimination.

We may consequently, in order to establish the class membership of some object, need to make the observation over a sufficiently large region or *frame*. A convincing argument for this is to make a tiny hole in a sheet of paper, and put this paper on an unknown image. By moving the paper over the image we can observe it through the pinhole, element by element. When we have observed the last element of the entire image, we are still unlikely to have any idea about the content of the image. The situation is no easier for a computer vision algorithm.

It turns out that in order for us to recognize an object in an image, we may require access to information within a substantial part of the image. Fuzzy and complex images do in general require a larger frame for interpretation. This implies the interpretation of something which is an entire image, which can contain a large number of different objects or events.

We can conclude that although the postulate of locality allows us to make observations based upon limited information, it is often necessary to extend the region under consideration to a sufficiently large frame. The reason is that the isolated structure has to be be sufficiently complex to distinguish it from other alternatives.

By observing that we for efficient interpretation need access to a large frame or neighborhood, we have not really solved any problem. At best we have observed that we have a real problem. This is one of the important aspects we will address in the coming sections.

1.1.2 Scale and abstraction hierarchies

Among the first structured approaches to vision were those motivated by knowledge about biological visual systems. The perceptron approach of Rosenblatt [116, 102], has attracted new attention as neural network theory has become a hot research topic [55, 119, 82]. The work on layered networks has continued, where such networks would accept image data at their bottom level [129, 127, 50].

Burt introduced an approach to hierarchical image decomposition using the Laplacian or DOLP (Difference Of Low Pass) pyramid [27]. In this way an image is transformed into a set of descriptor elements. The image can then be reconstructed from its set of primitives.

Most of the work so far has dealt with hierarchies relating to size or scale, although they have indirectly given structural properties. Granlund introduced an explicit abstraction hierarchy [42], employing symmetry properties implemented by Gaussian wavelets in what today is commonly referred to as Gabor functions [40].

We can distinguish between two different types of hierarchies:

- *Scale hierarchies*
- *Abstraction hierarchies*

Scale hierarchies

The scale hierarchy implies that the image is represented as a set of image arrays having different resolutions and sizes. This is typically illustrated in the form of a *pyramid*, although the representation in the computing structure may have nothing to do with such a spatial organisation, as illustrated in Figure 1.1. The image at the lowest level may have a resolution of 512×512 pixels. The image at the next higher level may have a resolution of 256×256 pixels, etc. The lower resolution versions at higher levels of the pyramid are generated through averaging, typically performed by an optimal low pass filtering and undersampling.

The scale pyramid allows immediate access to the image in different resolutions. A large object can be quickly located in low resolution at a high level in the

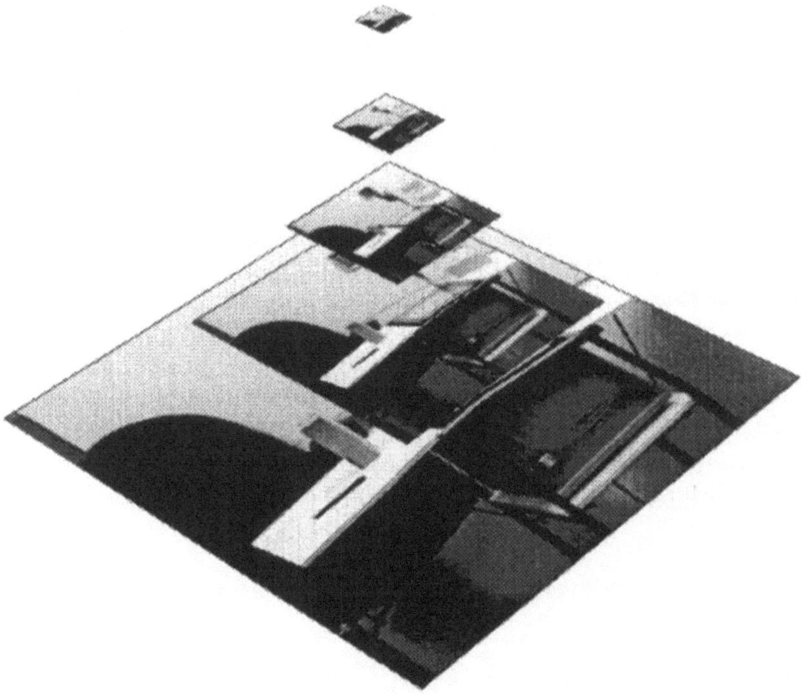

Figure 1.1 Representation of an image as a scale pyramid.

pyramid, after which a detail on this object is located at a lower layer of higher resolution in the pyramid. The averaging process produces a more compact representation of the image, but it also tends to smear and modify important structural attributes of the description.

The concept of scale or size as a dimension has been further extended in the so called *scale space* representation of images [148, 80, 90]. This deals with methods to describe, in a systematic way, changes in topology resulting from averaging.

Abstraction hierarchies

The abstraction hierarchy implies that the image can be considered as an expansion into *image primitives*, which can be viewed as conceptual building blocks forming the image. In this concept lies the assumption of a hierarchy, such that building blocks at a lower level form groups which constitute a single building block at a higher level. Building blocks at the two levels are viewed as having different *levels of abstraction.*

As an example we can look at Figure 1.2, which suggests a particular set of abstraction levels. At the lowest level we assume the image itself, describing a distribution of density and possibly color. At the second level we have a description of line and edge elements. At the third level we may have a description of curvature, or convexity and concavity. At the fourth level we may have objects. At the fifth level we may have relations between objects, and continue to higher levels as appropriate.

The example in Figure 1.2 only gives an intuitive idea of the notion of abstraction levels. This should not be taken as a statement of what the primitives in fact look like, or how they are represented. This will be the topic of most of the remaining parts of the book.

We furthermore assume a linkage between elements at different levels of abstraction. It is easy to accept the notion that line elements at an angle to each other form a convex structure element, representing curvature. Similarly, a number of convex structures will in combination infer an outline, which is a characteristic of an object.

We have seen that there are two different issues related to hierarchies, of which one has to do with level of abstraction, the other with size or scale. Although they are conceptually different, they are related. With increased level of abstraction generally follows an increase of the scale over which we relate phenomena. We need an array of very few pixels to define the existence and separation of two objects, while we need an array with a higher resolution to define the line or edge elements of an object. For that reason we often refer to the structure as an *abstraction pyramid.* We will deal with a typical such combination of abstraction levels and variation of scale in the following section.

It is in most cases of interest possible to describe the information at the output of a transformation with fewer samples than the input without any loss of information. This compression procedure is called *sub-sampling.* This means

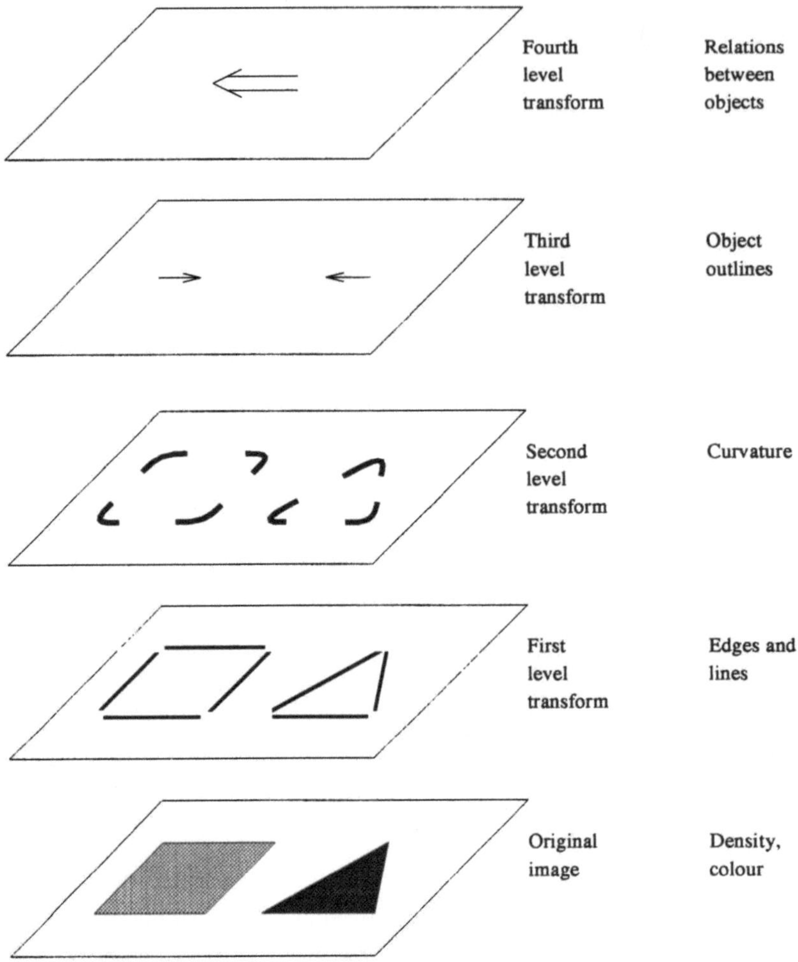

Figure 1.2 Conceptual representation of an image as an abstraction hierarchy.

that a 256 × 256 transform image at one level is, after the application of an operation, represented by 128 × 128 elements at the next level. The result of such a level-by-level decrease in the amount of data, generally by a factor of two in each dimension, is what is reflected in the pyramid representation.

1.1.3 Frames with a compact representation

We saw earlier that successful recognition of an object may require information within a rather large frame. Let us as a numerical example consider a frame of size, say, 128×128 pixels. This is equivalent to 64 kBytes for a color image, and is a large block of data to handle in order to classify each pixel in the image.

A hierarchical vision structure allows the representation of information at a particular point in the image as a description of properties within a number of frames of different resolutions and different abstraction levels. The consequence of this is a considerable reduction in the amount of data we must consider, when we describe the class of an object or an object part.

We can assume that the system of frames in Figure 1.3 is moved over the image we want to process. Our focus of attention point, or the element we want to classify, is indicated by the center element in this set of frames, located at some position in the original image. The center element or pixel may be represented at a resolution of one byte or eight bits if it is a gray scale image, and two bytes if it is a color image.

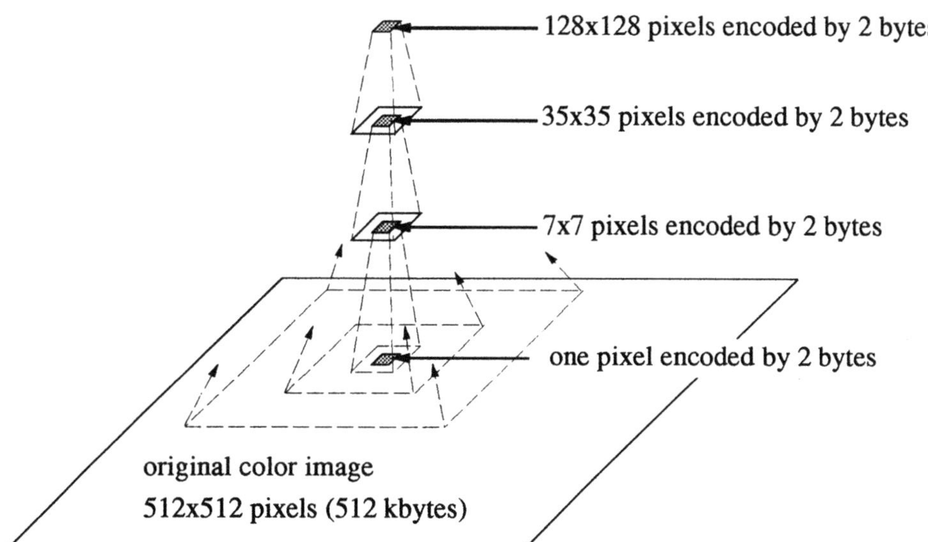

Figure 1.3 Set of frames with different sizes and resolutions.

From a neighborhood of a certain size, say 7×7 pixels, in the original image, we compute a descriptor and assign it to one element at the next higher level, which we call the *first level transform*. This first level transform may, for example, produce the local orientation description shown in Figure 1.2. We assume that this descriptor is also expressed with two bytes. What we have represented appears at a lower detail resolution than that of the 7×7 pixel window in the original image. This computation can be performed over the entire image to produce the entire first level transform of the image.

Next, we consider neighborhoods of the same size, 7×7 pixels, in the first level transform data. As before, a descriptor is computed, which we assign to one element at the next higher level, called the *second level transform*. This transform may, for example, produce a curvature description, as in Figure 1.2. We assume that this descriptor is also expressed with two bytes. What we have represented is given at a lower resolution of detail than that of the 7×7 pixel window in the first level transform image. This computation is performed over the entire first level transform image to produce the entire second level transform of the image.

Given the repeated compression, an element at the second level, representing a 7×7 neighborhood of the first level transform, will represent a region of approximately 35×35 elements in the original image. Because of the partial overlap required between neighborhoods this region will be smaller than the theoretically expected 49×49 element region.

Similarly, a region of 7×7 pixels in the second level transform data will produce one element of the *third level transform*. This will in turn represent a region of approximately 128×128 elements in the original image, and similarly for higher level transforms.

We can refer to the discussion of abstraction levels and assume that the different level transforms are intended to produce primitives that correspond to the suggested abstractions (Figure 1.2). The first level transform may then produce primitives such as lines and edges, the second curvature, the third object descriptions, etc. Note that this particular sequence of primitives is by no means absolute, and should serve only as an example. Various alternatives are possible as are combinations of these.

In this way we derive a set of descriptors with different resolutions over different size neighborhoods. Given a case of four levels, including the original image, this last discussion and Figure 1.3 suggest that the label of the center element of attention could be derived from eight bytes rather than the entire content of the outermost frame of size 128×128 elements, containing 64 Kbytes. It should be emphasized that a particular position of the frame structure will *not* classify all elements within the 128 × 128 frame using the eight byte set, but classify the center element of this frame structure. Still this procedure will utilize the information within the 128 × 128 neighborhood for the classification of the element. By moving the frame around the entire iamge, we can consequently classify all pixels in the image. The sizes quoted for regions, neighborhoods, etc, should not be taken as definitive values, but as typical examples of values.

1.1.4 The frame structure as a hierarchy

The primitives just described in general terms can be used for classification of objects and parts of objects. We will give an intuitive account of how this might be accomplished, but will defer a more detailed treatment to Chapter 12.

Let us assume that we have an original image with a square in blue color. We want to identify a point on the vertical edge of the square, next to a corner. The recognition frame structure is moved over the image in a scanning fashion. Sooner or later, the frame will be located in the crucial position, in accordance with Figure 1.4.

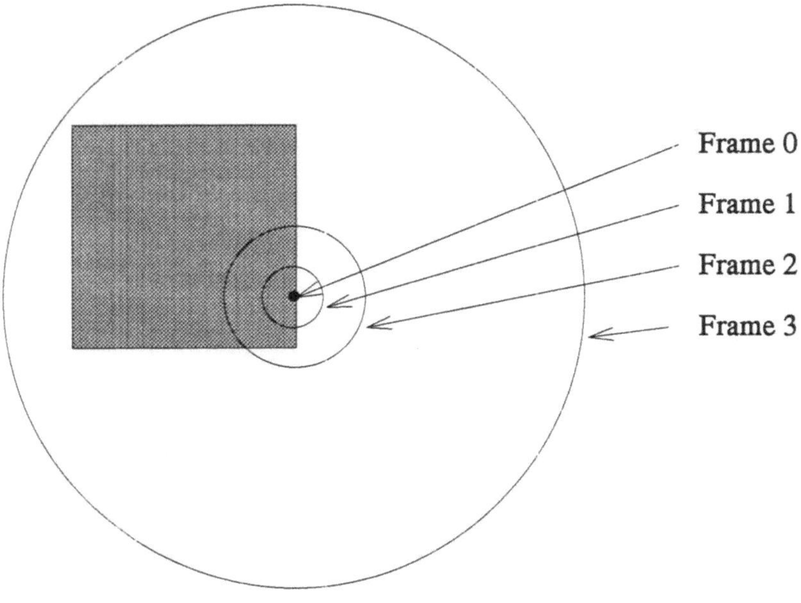

Figure 1.4 Characterization of a particular point of an object.

The outputs from the different frames will give a statement describing the point under consideration as:

```
Frame 0:    blue_pixel        AND
Frame 1:    vertical_edge     AND
Frame 2:    close_to_corner   AND
Frame 3:    object
```

Frame 0 containing one pixel of the original image, provides the information about the color of that pixel at the center of the frame structure. The first level transform, computed within Frame 1, provides a statement of the fact that we have an edge within the window, and the orientation of this edge. The second level transform, computed within the even wider Frame 2, gives a statement of curvature in the neighborhood of the point. The third level transform, valid within the outermost Frame 3, indicates that we have an aggregation of curvature contributions inferring that we have the outline of an object.

This is a fairly detailed statement, which can be used to discriminate between different parts of the object having different combinations of properties. It should be emphasized again that this discrimination is based upon only 8 bytes, two from each frame, but because of the different abstraction levels involved and different degrees of globality, this gives a very good characterization [8].

The window frames in illustrations in this book will be interchangeably square or round, with no particular significance intended. Generally the windows are square in their definition as an array, but the weights usually define an average shape that is round, as will be seen in later chapters.

1.2 LOW LEVEL REPRESENTATION AND OPERATIONS

The preceding has been a discussion in fairly intuitive terms to give an idea of the structures involved. So far we have said nothing specific about *what* we will represent in an image, or *how* we will represent it, except that the representation should describe some relevant primitive properties of the image. It must be based upon some measurements or tests upon the image. The measurement can be viewed as a matching between some suitable templates and local neighborhoods, or *windows* in the original image. Depending on the degree of matching, we will assign some measure to the descriptive transform; see Figure 1.5. If the matching is performed with a single template under simple conditions, the measure derived will be a scalar. In general we may want to match with several templates to obtain a better definition of what happens within the window, which leads to a measure in the form of a *vector.* This is what is indicated in Figure 1.5, and is generally the case in our future discussions.

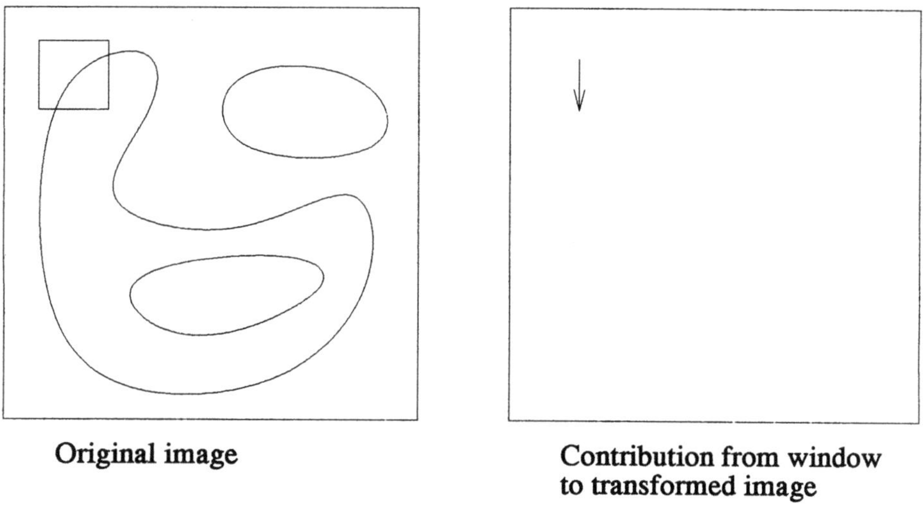

Figure 1.5 For every neighborhood (window) of the original image, a measure is defined in the image transform, representing the content within that neighborhood.

The matching is usually performed as a *convolution* between a template and the image. For a particular window, at a particular position of the image having pixel values x_h inside it, we can write the convolution as a product sum

$$q = \sum_k f_k h_k \tag{1.1}$$

where h_k are weight coefficients defining the template or *kernel.* Further details on the process of convolution will be given in Chapter 3 and Chapter 4.

As stated earlier, the measurements should describe the occurrence of important image primitives. Chapter 2 will show that lines and edges are important low level primitives for biological vision systems. These can be derived by convolution, using *line* and *edge* kernels. A convolution kernel, designed to describe near-vertical edges and consisting of 10×10 elements, is illustrated in Figure 1.6. Its application to the image in Figure 1.7 produces the transform result in Figure 1.8.

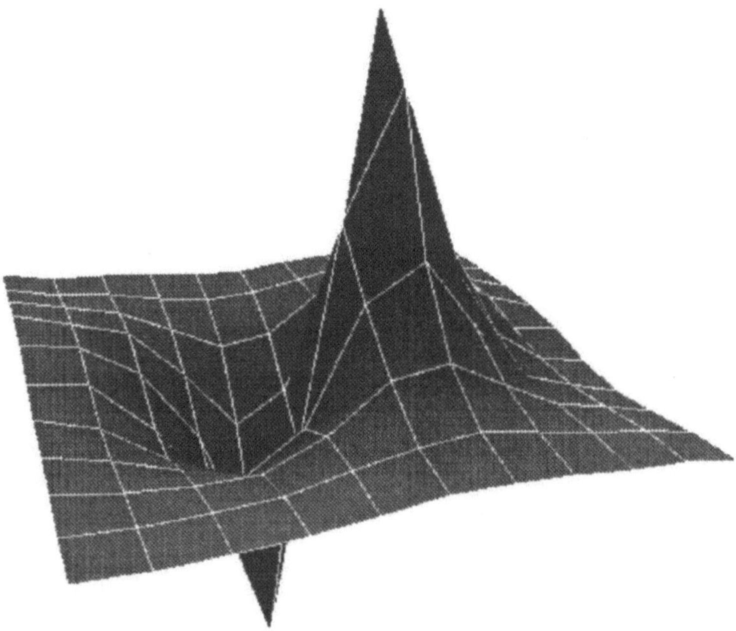

Figure 1.6 Convolution edge kernel with near-vertical orientation.

In this case the description derived is scalar, and the output or the degree of match is indicated by increased or decreased density, depending on what type of density transition we have, from left to right, in the original image.

The measurement just mentioned describes only the amount of vertical edges in an image. In addition, there may in a given window be horizontal lines, as

Figure 1.7 Test image for the application of one convolution kernel.

well as lines and edges in any other orientation. In order to describe all these possibilities it turns out to be necessary to test for all these cases, by convolving with a set of line and edge kernels in a number of orientations. The number of orientations is typically four, as indicated in Figure 1.9.

For practical reasons, there are no kernels representing exactly vertical or horizontal structures. By organizing the convolution kernels in the directions $22.5°, 67.5°, 112.5°$ and $157.5°$ all 4 kernel sets can be obtained from one single optimized set due to the mirror symmetry around the lines $y = x$ and $y = -x$. The convolution kernels are implemented on a 15×15 grid.

For every position of the window we have an output in the format of an 8-dimensional vector $\mathbf{q}$. The indices l and e refer to line and edge respectively.

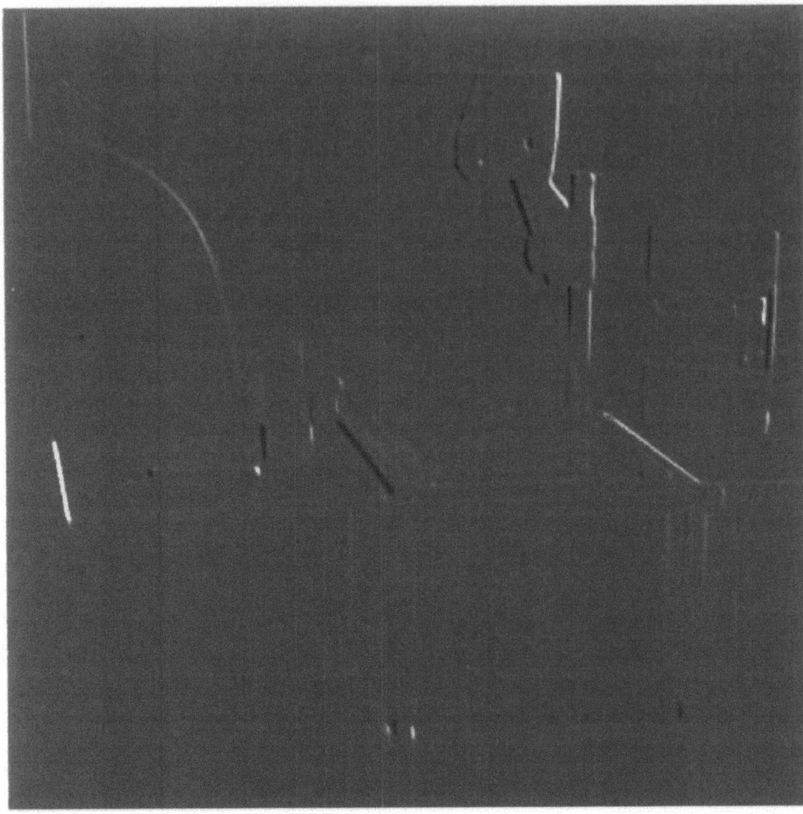

Figure 1.8 Convolution using the filter kernel in preceding figure.

$$\mathbf{q} = \begin{pmatrix} q_{1l} \\ q_{1e} \\ \cdot \\ \cdot \\ \cdot \\ q_{4e} \end{pmatrix} \tag{1.3}$$

It turns out that there is a great deal of correlation between components; this, as well as other restrictions, can simplify the representation from an 8-dimensional vector to a simpler entity.

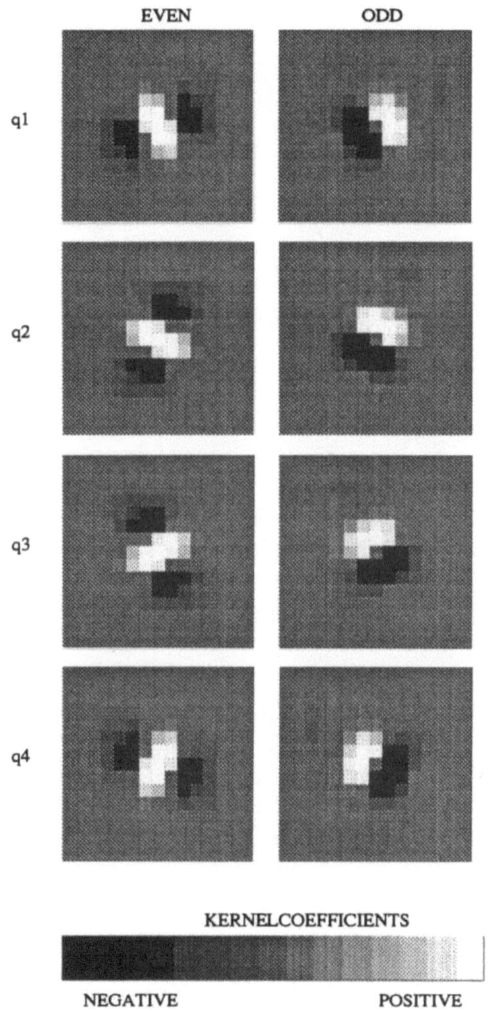

Figure 1.9 Two-dimensional convolution kernels for line (even) and edge (odd) elements in four different orientations.

There are in principle several ways in which such a combination of components can be made. The way in which it is done is crucial, as it will determine the ability to represent relevant mappings in space. We will for the ensuing discussion use the *compatibility representation* developed at Linköping University [42, 43, 48]. The particular feature of this representation is that information in the hierarchical structure is represented in the same way at every level. This means that the representation symbols look similar at every level, while the meaning of such a symbol varies depending on the level of the hierarchy where it is located.

Representation of phase

For a starter, line and edge components for a particular orientation, k, can be combined into a magnitude and phase representation

$$\begin{cases} q_k = \sqrt{q_{kl}^2 + q_{ke}^2} \\ \theta_k = \arctan\left(q_{ke}/q_{kl}\right) \end{cases} \tag{1.4}$$

We can consideer q_k as an appropriate measure for the combination of line and edge, as the corresponding kernels are orthogonal and the individual components can consequently be added geometrically. We say that the components exhibit the property of *quadrature*, a topic which will be treated in detail in later chapters.

We can illustrate the relation between line and edge components as magnitude and phase according to Figure 1.10.

The phase makes it possible to distinguish, for example, between a bright line against a dark background and its complement, or to identify a particular combination of line and edge. If we take into account only the magnitudes, the output from a neighborhood is reduced to 4 components, one for each orientation.

$$\mathbf{q} = \begin{pmatrix} q_1 \\ q_2 \\ q_3 \\ q_4 \end{pmatrix} \tag{1.5}$$

Representation of orientation

It is possible to reduce the data representation further by employing the following assumption of *local one-dimensionality* :

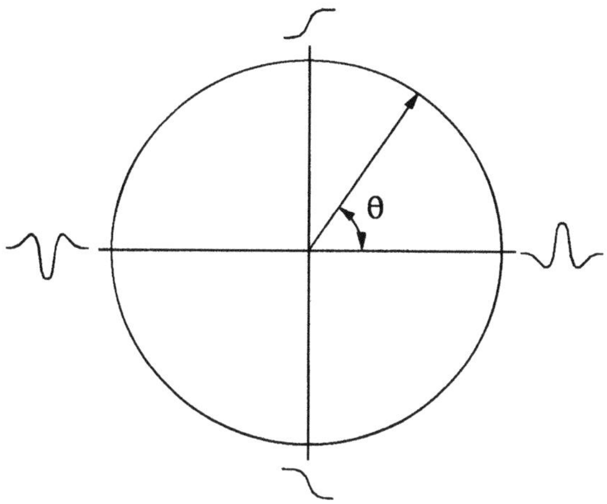

Figure 1.10 Relation between line and edge components expressed as a phase.

The content within a window, measured at a sufficiently small bandwidth, will as a rule have a single dominant component

This implies that we will generally find only one of the components q_k, or an adjacent pair, q_k, q_{k+1}, having substantial values. In addition, the orientation variation of the kernels spans only 180° as apparent in Figure 1.9, while we want to represent the output with vectors spanning the full 360° range.

This allows us to formulate a more compact descriptor $\mathbf{z}$ as a function of all features q_k for $k = 1, \ldots, 4$:

$$\mathbf{z} = \begin{pmatrix} z_1 \\ z_2 \end{pmatrix} = \begin{pmatrix} q_1 - q_3 \\ q_2 - q_4 \end{pmatrix}. \tag{1.5}$$

We can visualize the output vector $\mathbf{z}$ as a sum of vectorial components, each one weighted by the component for orientation k, q_k, as indicated in Figure 1.11.

The correspondence between orientation of line or edge segments and the vector representation is illustrated in Figure 1.12. We can see how the vector representation rotates with twice the speed of the orientation of the corresponding line or edge segment. For this reason we call it *double-angle* representation, a property which will be dealt with extensively in the following chapters.

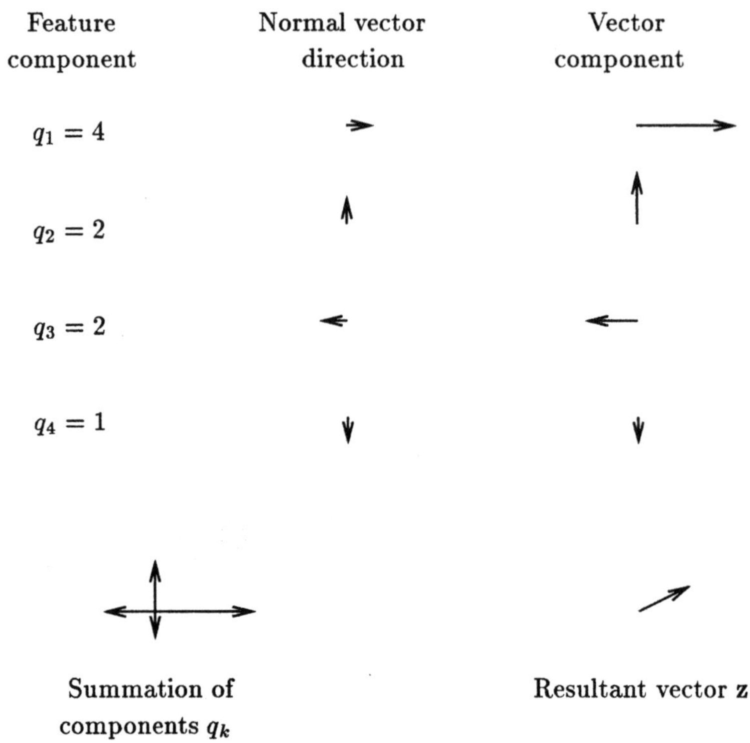

Figure 1.11 Generation of resultant vector output **z** from feature components q_k.

1.3 DESCRIPTION IN TERMS OF SYMMETRY

The next important question is: In what general terms should image information be described? We have earlier assumed that lines and edges are important descriptors at low levels, but we would like to have some more general guidance in our search. There are various possible choices, but the alternative explored in this book describes information in terms of *symmetry* relations [54, 13, 11, 12]. There are several reasons why this is useful. Symmetry implies continuity, and most processes which we are interested in are characterized by various aspects of continuity as opposed to randomness. This includes the type of continuity represented in edges, lines, regions, motion, as well as rotation. It

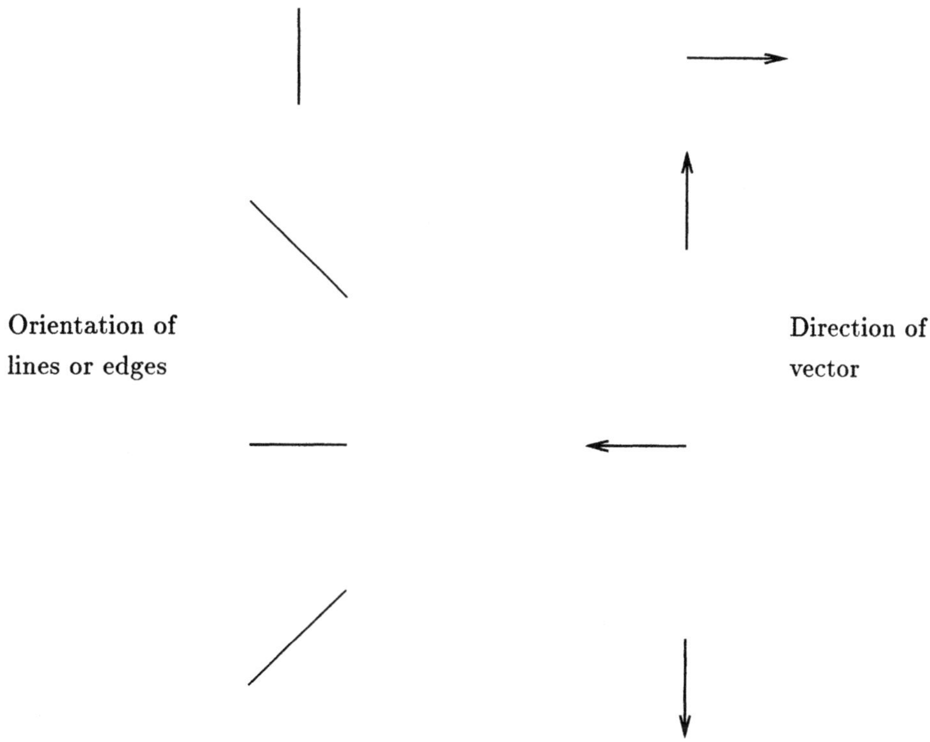

Figure 1.12 Notation used for orientation of lines and edges, and the corresponding direction of the descriptive vector.

is often possible to represent a given instance of a pattern as a transformation of a prototype, which makes that class of patterns exhibit what is called *group* properties[86].

Another important reason is that description in terms of symmetry allows a compact representation. The recognition or existence of symmetry implies a model, according to which a data compression can be made. Symmetry operations can be viewed as producing reciprocal reference pointers between hierarchically ordered data structures.

Symmetry descriptions have an inherent robustness. Operations in computer vision often imply differentiation. These would generally be sensitive to noise. Symmetry properties, however, imply an anisotropy that allows operations to differentiate in one dimension while they integrate in another. The symmetry operations proposed imply continuous functions of the input data, a property that appears necessary in a hierarchical system so as to allow a feedback adaptation of models.

There appear to be two important classes of symmetry:

- *One-dimensional or linear symmetry*
- *Central or rotational symmetry*

One-dimensional or linear symmetry

One-dimensional or linear symmetry of a pattern can, as a first intuitive approximation, be seen as a property that allows us to describe the pattern using only half of it together with its mirror image [12]. In reality, the class of similarity is wider than just mirror symmetry. We look first at how the description is done for patterns in one dimension.

It turns out that such patterns can be described in terms of bisymmetrical and antisymmetrical relations. This implies that the patterns can be decomposed into functions of the types illustrated in Figure 1.13. The bisymmetrical relation is what is most similar to the case of mirror symmetry mentioned earlier.

The degree of bisymmetry and antisymmetry of a given function in one dimension is estimated by matching or convolving it with the functions in Figure 1.13. These functions can also be seen as cosine and sine functions multiplied by a smooth window function. At low levels these functions assume the function of conventional line and edge detectors, as discussed earlier. At higher levels they indicate the presence of abstract symmetry relations of patterns.

Most patterns or signals have a combination of bisymmetry and antisymmetry. As the components which are similar to cosine and sine are orthogonal, they can be combined geometrically along orthogonal axes as was indicated in Figure 1.10. The relation between the bisymmetric and the antisymmetric components is now described as a phase angle. This phase angle describes whether we have a positive or negative line, or an edge and its direction of transition [48].

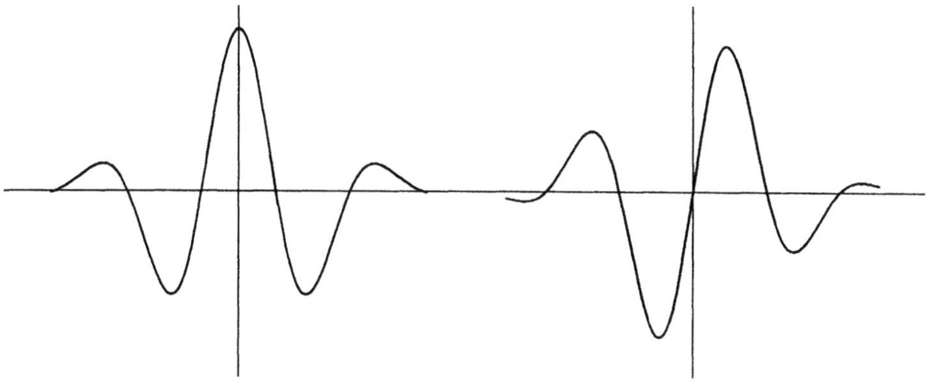

Figure 1.13 One-dimensional symmetry functions. Left: a bisymmetric kernel equivalent to a weighted cosine acts as a line detector. Right: An antisymmetric kernel equivalent to a weighted sine acts as an edge detector.

Patterns in an image are two-dimensional, which makes it necessary to use symmetry descriptors which are correspondingly two-dimensional. In Figure 1.9 we saw an illustration of a set of such symmetry descriptors or convolution kernels in four different orientations.

In the left side of the kernel column we have bisymmetric functions with symmetry axes in different orientations. These will, for example, detect lines in different orientations. In the same way, the antisymmetric functions on the right side will detect edges in different orientations.

There may be some cause for confusion in that the preceding discussion uses the term *one-dimensional* for different purposes. The signal under consideration may be defined in one dimension, such as in Figure 1.13, or in two dimensions, such as in Figure 1.9. Still, the symmetry aspect of interest is in both cases one-dimensional, as the cross-section of a two-dimensional kernel such as in Figure 1.9 is equivalent to what is indicated in Figure 1.13. In Chapter 4 we will refer to such functions as *simple functions.*

Central or rotational symmetry

Another class of symmetry which has been found to be important is that of *central* or *rotational* symmetry [9]. This is useful for the description of stationary objects, as well as for description of motion such as rotation and expansion. In this context we shall only suggest that this implies basis functions of the types indicated in Figure 1.14. We will deal with these descriptors in further detail in Chapter 11.

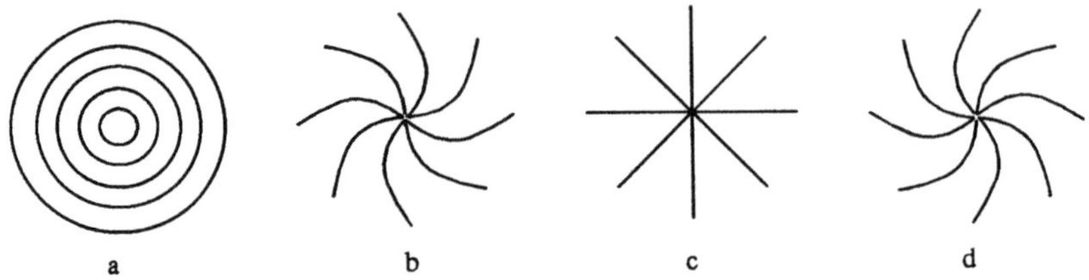

Figure 1.14 Examples of basis functions corresponding to rotational symmetry.

1.4 CASCADING OF OPERATIONS

One important property of the hierarchical computing structure is that operations are performed in a cascade. Figure 1.15 illustrates how this can work out for a particular case.

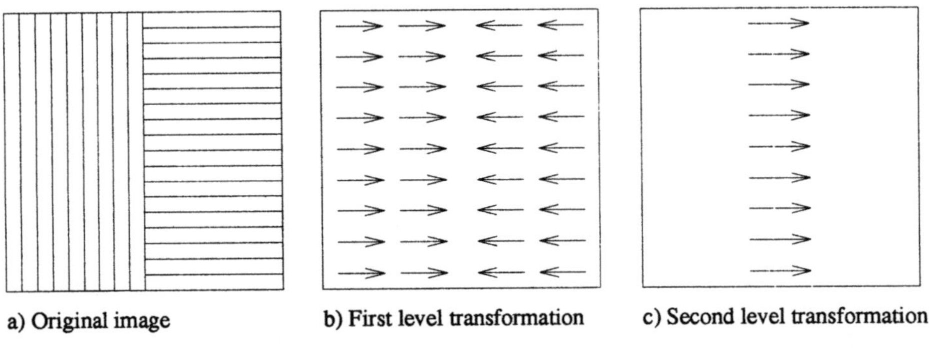

Figure 1.15 Results of two transformations.

In this case, the original image, Figure 1.15a), is a stylized pattern with two segments containing lines in two different orientations. The result from the first-level transformation in b) gives the orientation of feature vectors according to the earlier discussion. The same type of operation can work upon a vector field as well as a scalar field. This too is illustrated in Figure 1.15, where the transformation image in b) acts as input producing the result given in c).

The differentiation property of the operator results in a zero output within the regions where the vector field is constant. We obtain a non-zero output only in regions where we have a transition of the vector field. This is the case for the boundary between two regions.

This illustrates one important property of the hierarchy, to simplify the structure of the image by grouping primitives together to give them a collective representation. In the example given in Figure 1.15, orientation is the property characterizing the regions. However, as all types of information are represented in the same way using vectors, the same procedure can be applied to features such as color, spatial frequency, texture, etc. This is one important aspect of the generality obtained using this representation.

1.5 COMPATIBILITY REPRESENTATION

A problem in relation to hierarchical structures has always been to divide the tasks properly among the levels and to communicate between different levels. If the division into different levels is to be meaningful, the system can not be doing the same things at all levels. In spite of this, information must be usable at a different level from where it was generated. This implies that symbols have to somehow carry their own code book for interpretation. This generally would require complex translational interfaces, which produce a cumbersome structure.

If we look at the situation in traditional image processing, different features are traditionally represented in different ways. Edges are represented in one way and lines are represented in a different way. Texture is represented in a third way and regions in a fourth, and so on. This creates significant problems when we want different types of information to interact over levels, to be combined from different sources, or to control adaptive models.

In the representation introduced in this chapter and further developed throughout the book, information in the hierarchy is represented in its basic form by two-dimensional vectors or complex numbers:

- *The angle or argument of such a vector points out the dominant class membership of an event*
- *The length or magnitude of a vector indicates the certainty or confidence of the class membership identified*

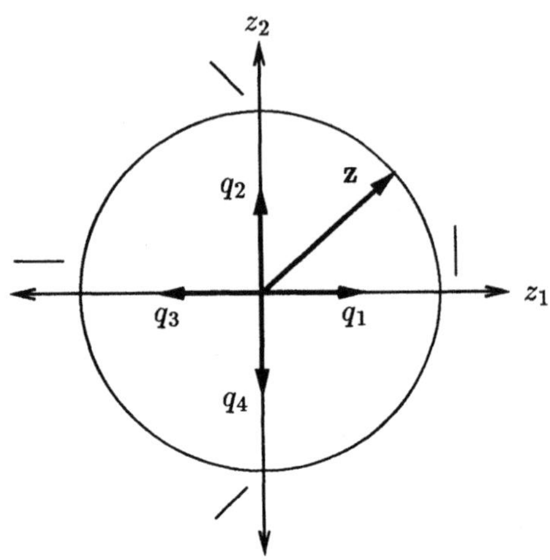

Figure 1.16 Illustration of direction of output vector **z** as a function of dominant orientation of structure.

To further clarify the notation used in this case we refer to Figure 1.16, which represents, in a different way, the correspondence between orientation of a line or an edge and the direction of the descriptive vector.

We can observe that lines and edges are arranged in terms of similarity of orientation. Orientations that are maximally different, e.g. at right angles, are positioned at opposite sides of the *compatibility* circle. Orientations that are relatively similar are put next to each other on the compatibility circle.

What we have observed as a special case concerning operations at low levels illustrates a general principle that is used at every level of the hierarchy:

Events are represented in terms of relative compatibility of occurrence, which is in terms of how they occur together. Events that are relatively incompatible are positioned at opposite sides of the compatibility circle, while events that are relatively compatible are put next to each other on the circle. This gives a representation of events in terms of how they occur together or in terms of how they associate. The result is a truly associative *information representation.*

We will later see the advantages of this type of information representation for communication between levels in the hierarchy.

1.6 DESCRIPTION OF SIZE AND SCALE

To most phenomena we can associate a measure or at least a sense of size or scale. In practice we usually encounter this property for discrete entities, size of an object, width of a bar, size of grain, etc. Measurement methods on the other hand generally prefer that the property in question is constant over a sufficiently large region, a sufficiently large *field*.

A property that can often be used for estimation of size and scale is *spatial frequency*. In Fig. 1.17 we have three patterns with different spatial frequencies indicated.

Figure 1.17 Three patterns with different spatial frequencies.

1.6.1 Frequency properties of operations

In order to describe these three patterns optimally we need matching kernels having corresponding spatial frequency characteristics. For a constant relative bandwidth we require the same number of periods, say 2 - 3, within the window. This implies that the kernel describing low spatial frequencies will be large, while the kernel describing high spatial frequencies will be small; see Figure 1.18. The bases for these mechanisms will be considered in Chapter 4 on Fourier methods.

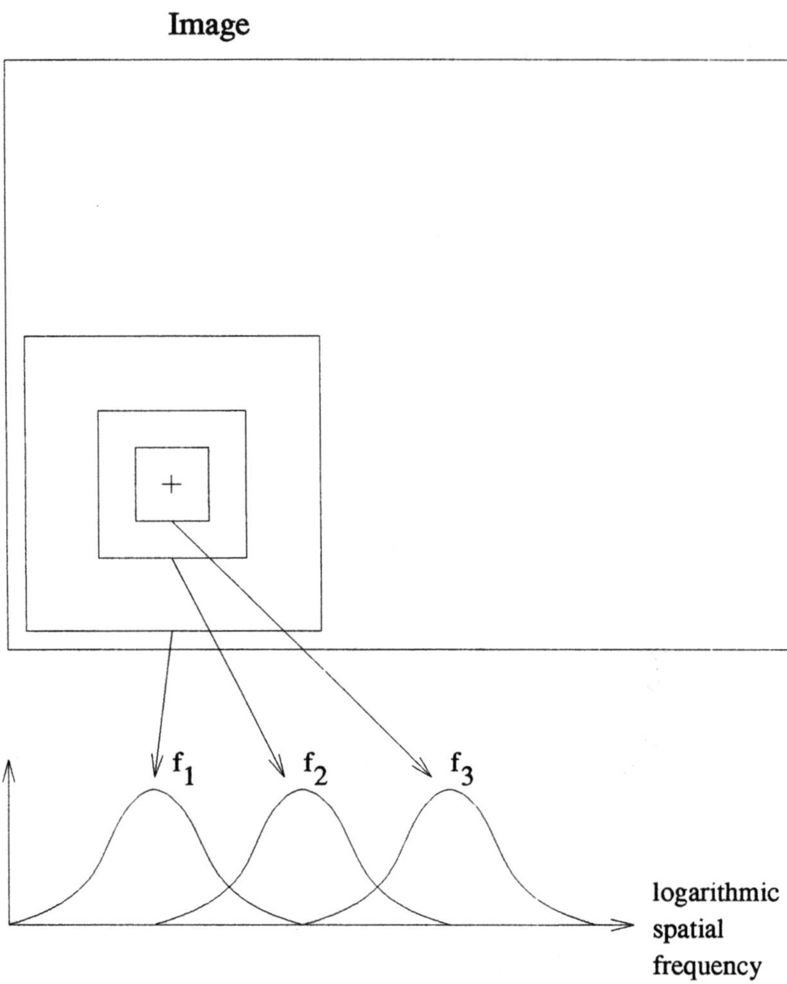

Figure 1.18 Operators with windows of three different sizes, and the corresponding band pass filter responses.

1.6.2 Vectorial description of size

Explicit description of size or scale has traditionally not been a well developed topic in computer vision. The advent of scale space methods, [148, 80, 90], has brought a great deal of attention to the issue. There are two important reasons why considerations of size or scale are important.

An operator of a particular size and design has a pass band such that it best represents objects or events of a particular size. Objects smaller or larger than that are less well represented. In most images we have objects or events of different sizes. We generally do not want this to affect our ability to describe them; neither do we want a dependence upon the quantization of the image.

Another reason why size is important is that it is a descriptive feature. A small object is something different from a large object, and it is often important to base discrimination upon this.

Each operation in the system is implemented using kernels in different sizes. This provides a particular feature within different pass bands as illustrated in Figure 1.18. The filter responses can be designed to overlap in a regular way, to produce a continuous description of the size of the particular phenomenon.

It would be advantageous if size or scale is represented in the same way as other features. It turns out that this property can indeed also be represented as a two-dimensional vector or a complex number [49]. The outputs from the different filters are combined geometrically along three different vector directions as indicated in Figure 1.19. As the spatial frequency increases from low frequencies to high frequencies, the outputs from the different filters will vary in magnitude and the combined frequency descriptor vector will rotate.

We formulate a compact descriptor $\mathbf{z}_f$ as a function of magnitudes for different frequencies q_{f_k} for $k = 1, 2, 3$:

$$\mathbf{z}_f = \sum_{k=1}^{3} q_{f_k} e^{i(k-2)\frac{2\pi}{3}}. \tag{1.6}$$

The angle or argument of the vector indicates the dominant local spatial frequency, and the length or magnitude of the vector indicates the certainty of the frequency identified; see Figure 1.20.

Because it combines the filter responses from three different pass bands the frequency descriptor vector rotates as the spatial frequency increases from low frequencies to high frequencies. An alternative and computationally more efficient implementation takes the components from differently subsampled versions of the image [49]. This allows a wider spatial frequency coverage as well, as more than three frequency components can be used.

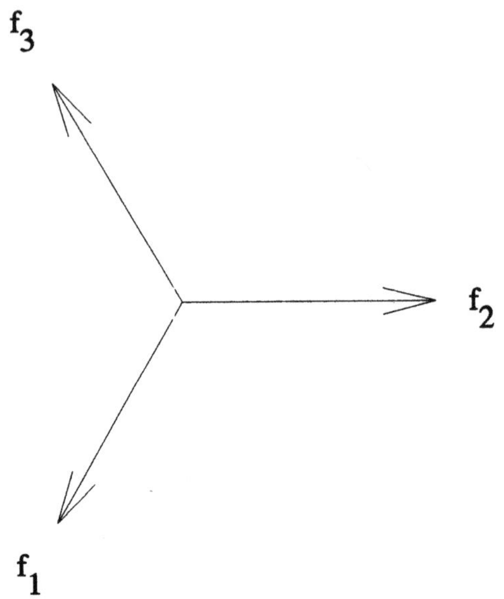

Figure 1.19 Vectorial combination of filter responses to produce a single description vector.

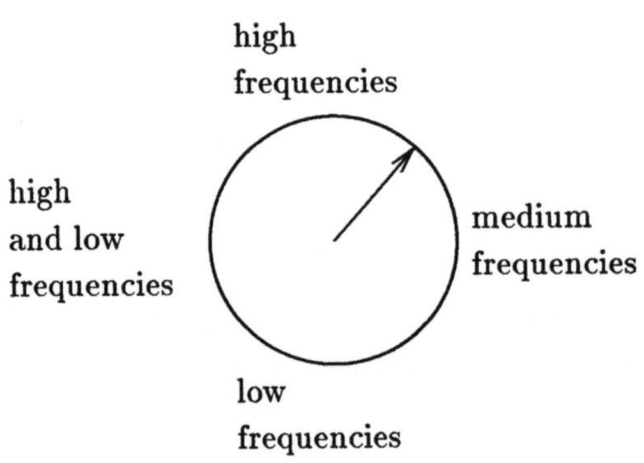

Figure 1.20 Arrangement of a vector representation for spatial frequency.

An important property of this descriptor is that it is self-normalizing. If the contrast of the pattern observed decreases by one half, the output from each filter will also decrease to one half of its earlier value. In the representation of

Figure 1.20 this will result in the vector decreasing its magnitude while retaining the same argument. A change of contrast consequently does not change the measured value of the dominant frequency, but only the statement of certainty. This measured size property is applicable not only to linear patterns, but quite generally to the measurement of size of objects or object parts. We will deal further with this description in Chapter 9.

1.6.3 Optimization of Operators and the Uncertainty Principle

We know from Fourier theory, Chapter 4, that every operation which we perform in the spatial domain has its corresponding consequences in the frequency domain. As spatial frequency is only one possible property that we may want to compute, we can extend this statement to saying that:

- *To every operator in the spatial domain, there is a corresponding operator in the feature domain. A change of a parameter of an operator in one domain leads to a reciprocal change in the corresponding parameter of the operator in the other domain.*

The qualitative situation is illustrated in Figure 1.21. An operator of a certain size performs a measurement of a particular property within that size region. If we measure frequency, the accuracy of the measurement will be inversely proportional to the space over which we measure. A small operator will give a large uncertainty, while a large operator will give a small uncertainty of the property measured [143]. The uncertainty aspect is treated further in Section 4.3.2.

While these properties are most familiar in the measurement of frequency, they are as well valid in the measurement of any feature useful in image processing. The principle consequently relates to any operation customarily performed. One particular example of interest is the widely used 3×3 operation for filtering. It turns out that this is a too small neighborhood to allow good performance. It produces a feature or frequency spectrum that is so wide that we obtain an overlap of parts of the spectrum, an effect known as *aliasing*. Computations show that for typical choices of band-width, it is necessary that an operation have a minimum size of 7×7 or 9×9 pixels to avoid these adverse effects.

It is preferable that operations used are optimized to take into account properties in the spatial domain and the frequency or feature domain simultaneously

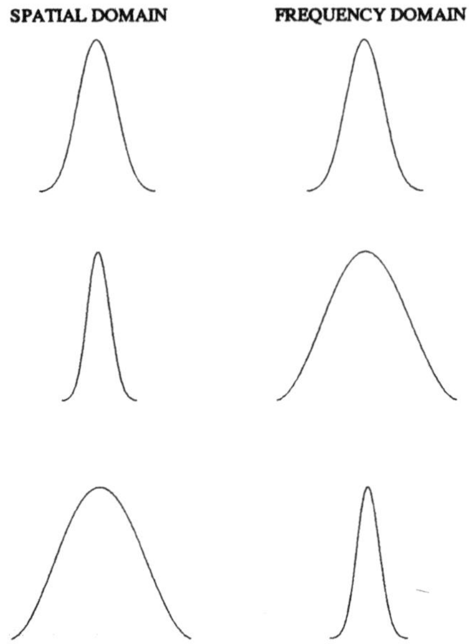

Figure 1.21 Correspondence between window sizes in the spatial domain and pass band in the frequency domain.

[65]. We observed earlier that an operator that is small in the spatial domain leads to a great uncertainty in the feature property. Conversely a small uncertainty in feature leads to a large size operator. A large size operator, however, leads to a great uncertainty in our knowledge of the spatial position of the event.

Operators can be designed to have maximal compactness in the spatial domain and in the feature domain simultaneously. This is equivalent to their uncertainty product being as small as possible. These issues will be treated in Chapter 5 for optimization of filter kernels.

1.6.4 Separation into form and size

We have seen earlier how we can measure properties which describe form or structure, as well as properties which describe size or scale. We can view this as giving a phenomenon a description which separates form and size. See Figure 1.22.

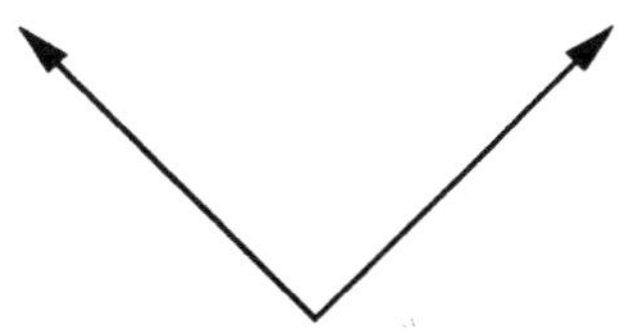

Figure 1.22 Separation of description in terms of size and form.

It is often very important to be able to separate these properties. In certain cases we want to recognize objects with respect to their form or structure, regardless of their size or their distance from the observer. In other cases size is an important feature. This is an example of the invariance properties required for different types of descriptors.

1.7 MODEL BASED PROCESSING

What we have discussed so far are the properties of processing as we go upward in the processing pyramid. This is often referred to as *bottom-up* processing. This has involved issues of how information is represented and of how operations are performed. In this discussion it has only been possible to illuminate certain aspects, while it is necessary to defer most details to the following chapters.

What we shall deal with next has even more profound effects on derivable results than the mechanisms discussed so far. We will also give some of the reasons for the information representation selected as well as for other peculiarities of the structure. In the same way that context information affects our interpretation of a more local event, context information determines the possible alternatives of events within a local region, and consequently the operations we want to perform. The issue is how information at a higher level can be used to control the processing at a lower level. This is often referred to as *top-down* processing. A hierarchical structure allowing this is illustrated in Figure 1.23.

For the general case we have to abandon the hope that processing can proceed only in one direction. Instead we have to assume a structure similar to Figure 1.23, where processing proceeds in a number of stages, and the processing in one stage is dependent upon the derived results at higher levels. This leads

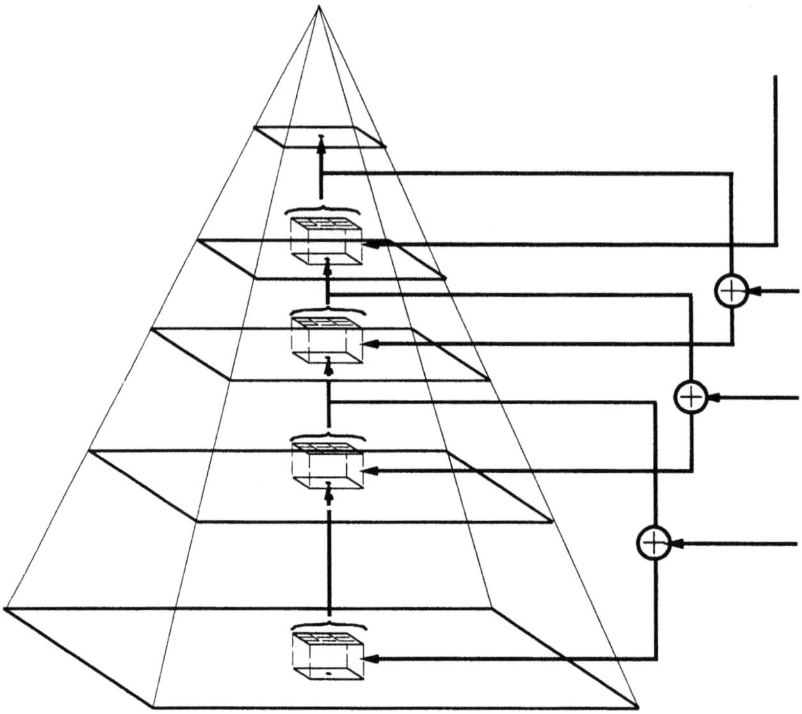

Figure 1.23 A hierarchical structure with bottom-up and top-down flow of information.

to a model based analysis, where models are assembled from combinations of primitives from several levels. The new and difficult step is that these models do not remain constant, but that they have to adapt to the data.

1.7.1 Adaptive filtering

To be more specific, let us take a closer look at the example of Figure 1.24. This structure performs context-dependent filtering [76] upon a noisy fingerprint, controlling the pyramid structure from the top down.

A controlling transform is computed from the original image. In the simplest case the controlling transform may be an ordinary first order transform giving the dominant orientation of structures in the image. In other cases it may also contain information about the dominant spatial frequency as well as informa-

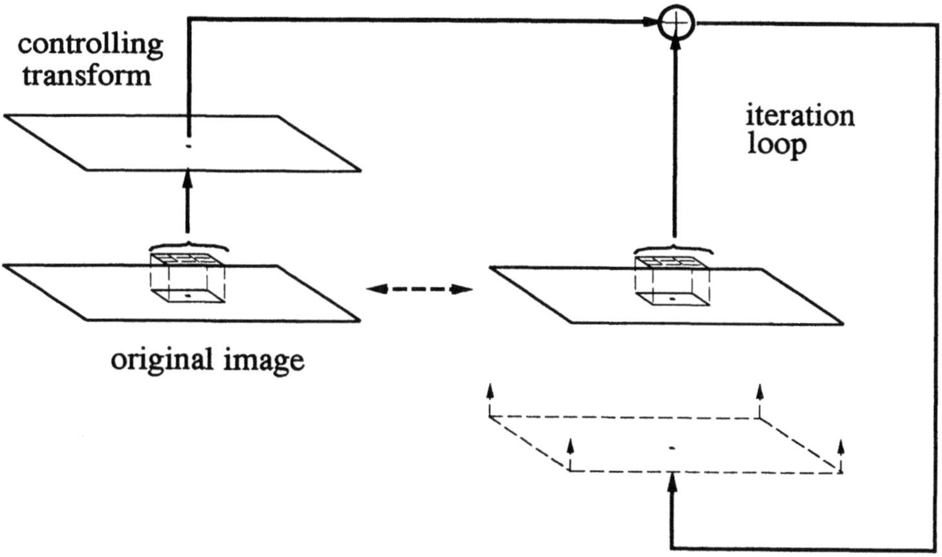

Figure 1.24 Top-down controlled structure for content dependent image filtering.

tion about the phase structure of the image. The original image to be filtered is now brought into an iteration loop. The image is convolved with the set of filter functions identified earlier.

Now the information from the controlling transform can be used to impose restrictions on what types of variations are allowed in the description. In this case the permitted orientation will be restricted to what has been determined as the dominant orientation within a particular region of the image. Consequently, only structures compatible with this restriction will be described. In Figure 1.25 is an example of such a non-isotropic filtering of a fingerprint after one and two iterations. The effects on noise suppression are apparent. We will deal with these mechanisms in further detail in Chapter 10.

1.7.2 Rule based processing

If we look at a particular line element we can test whether the model hypothesis of a line going along this element is supported by adjacent elements. We can see that this is true in certain regions, which is why such elements are enhanced

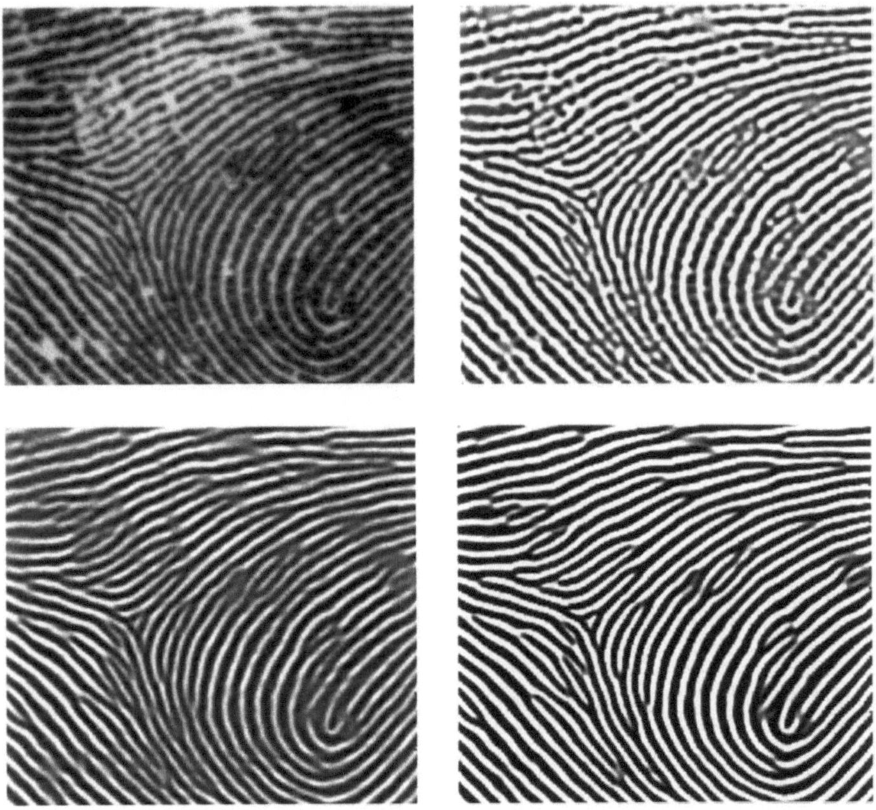

Figure 1.25 Non-isotropic filtering of a fingerprint. Top left: original image. Top right: the original image filtered using an isotropic filter with corresponding frequency characteristics. Bottom left: the original fingerprint after one iteration of non-isotropic filtering. Bottom right: After two iterations.

in the first and second iterations illustrated in Figure 1.25. In other regions of the image, the model hypothesis of a line is not supported, as elements have various different orientations. In such regions the certainty or confidence is decreased. After two iterations, only the regions having elements consistent with the behavior of a line are left.

There are two ways to view the model based processing just presented. One way is as an adaptive filtering, where filters are controlled in terms of orientation, spatial frequency, phase, etc. by the image content [76]. That was the interpretation implied in the preceding subsection.

The other way to view it is as a *consistency* or *relaxation* operation [117, 51]. In this case the *class membership* is updated given a hypothesized model and relative *compatibility* of class membership of surrounding elements.

The implementation and the effect of a particular compatibility function are given in the example that follows (see Figure 1.26). Let us assume that the consistency with respect to the hypothesis of a horizontal line is to be determined for a line element. Let us denote this dominant orientation with a label w_1. Line elements lying along this orientation are examined. If these have the same label w_1, they are considered compatible with respect to the hypothesis posed. This should give a positive contribution to the certainty of the class membership of both events. The compatibility function r_{ij} illustrates the contribution to the certainty from one adjacent line element, at a distance Δ_{ij}.

If one of the events has a different label, w_3, corresponding to a perpendicular orientation, we have a relationship that is incompatible with the model where they lie along a horizontal line. This should give a negative contribution to the certainty of class memberships of both events (Figure 1.26b). An orientation between these two is considered as relatively independent, and gives no effect (Figure 1.26c).

Relaxation operations are ordinarily very cumbersome in computational terms, and require a great number of iterations to converge. A very attractive feature of this implementation of consistency operations is that the associative information representation described makes it possible to implement such operation with ordinary convolutions using the standard symmetry kernels. The normalization property of this implementation provides convergence after 2 to 4 iterations in most cases.

This procedure is useful not only for the restoration of lines in noise, but for any type of feature. As we have seen earlier, images are decomposed into hierarchies of symmetry descriptions using the same type of information representation at every level. This allows the consistency operation described to be used between any adjacent levels of the system, even with different compatibility models. In this way very complex models can be implemented.

Such adaptive image content dependent models are very important for image enhancement, where they allow sharpening of contours at the same time as noise is removed. They are even more important for image analysis and classification, although the detailed effects are more difficult to demonstrate visually.

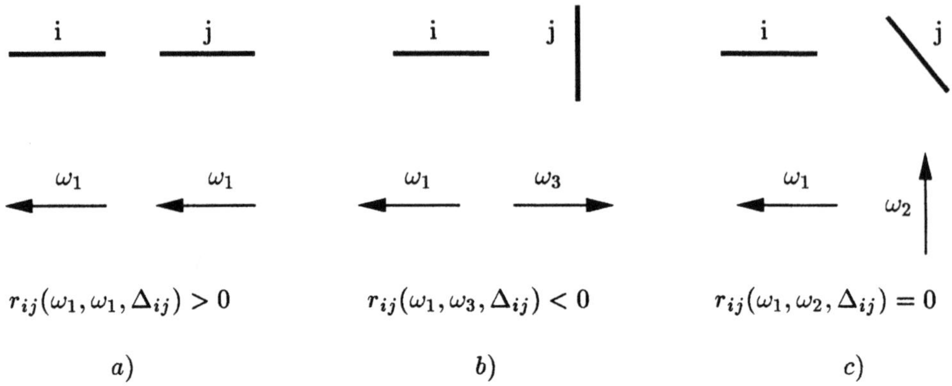

Figure 1.26 Illustration of compatibility function. a) Compatible events b) Incompatible events c) Relatively independent events

1.8 REPRESENTATION OF VOLUMES AND TIME SEQUENCES

The discussion has so far, for simplicity of argument, dealt only with two-dimensional images. The principles discussed can be extended to multi-dimensional data such as volumes and time sequences. There are, however, some complications, due not only to the increased volume of data, but also from a more fundamental point of view. We will only briefly discuss some aspects and give references to other documentation.

The mapping requirements of operations in multiple dimensions are more severe than for two dimensions [66]. When the original space is a hemisphere, an exact equivalent of the complication encountered in two dimensions occurs: planes that differ by a small angle can end up being represented by vectors that are very far apart, i.e. close to opposite sides of the rims of the hemispheres. This is definitely unacceptable.

It turns out that information can for this purpose be represented by matrices of a particular form called *tensors* [67]. The tensor representation can be used for filtering in volumes and in time sequences, implementing spatio-temporal filters [140]. It can also be used for computation of higher level features such as curvature [10] or acceleration. The tensor mapping can be controlled by transforms to implement enhancement of volume data or time sequences [72]. The details of this will be presented in the following chapters.

1.9 CLASSIFICATION AND RESPONSE GENERATION

The last stage of a computer vision system generally produces some action in the external world. It may control the motors of an industrial robot. It may point out defective units on the conveyor belt of a quality control system. Or it may point out suspected cancerous regions in an X-ray image.

All these situations require combining all the descriptive properties derived in the following chapters, into an output control signal of suitable form. This issue is very complex; for a successful completion it requires using additional invariance mechanisms, taking geometry into account, including object modelling, etc. This is a vast field, which will not be covered in any depth in this book. Still, it is of interest to bring the structure to some form of logical completion in the combination of feature properties for response generation.

In summary, the task of a computer vision system is to produce a response. Whether this response is immediate or delayed, whether it acts on the input image to identify a particular object, produces a modified version of the original image, or causes some action in the external environment is conceptually less important.

In the text adjacent to Figure 1.4 we had a first discussion of the mechanisms related to the identification of an object part with certain properties. We continue this discussion in Chapter 12, after having presented more details on the feature generation mechanisms.

2

BIOLOGICAL VISION

2.1 MOTIVATION

The most, some would say the only, successful vision systems are the biological ones, ranging in complexity from those found in lower animals to the very flexible and highly integrated system for human vision. Although it has not, in general, proved feasible to employ biological functions for technical systems, the situation may be quite different in the field of information processing. The computational structures which have been developed in our current computer technology are primarily designed to handle information in strings. This is very different from spatial and cognitive information, and it has been found that the current organization of computers does not support an efficient representation and computation for spatial data. For this reason it is of great interest to look at the organization of biological visual systems. The study of such systems may suggest more efficient information representations, computation structures and suitable primitives for representation of spatial information.

The purpose of this chapter is to serve as an introduction to the physiology and the psychophysics of vision, with an emphasis on aspects that are pertinent to the design and understanding of technical computer vision systems. This treatment, even if reasonably accurate according to recent research, is of course not complete, and interested readers should consult the biology and psychology literature, such as [57], [112], [15], [33], [94], [151], [89] and [26].

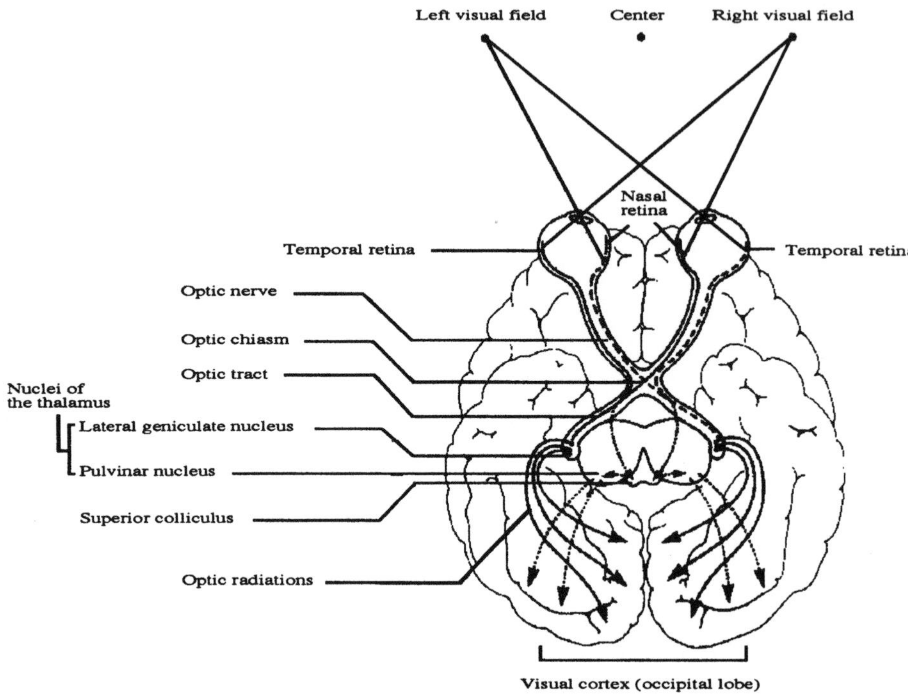

Figure 2.1 Overview of the visual pathway from the eye to the visual cortex. (After [31].)

2.2 OVERVIEW OF THE VISUAL SYSTEM

The visual system, shown in Figure 2.1, starts with the eyes, where an optical system forms an image on the rear wall containing a set of light-sensitive detectors, the *retina.* The retina contains about 125 million light sensors plus some processing capabilities. The processing in the retina achieves among other things a compression of information allowing it to be transmitted along the *optic nerve.* The optic nerve contains only about 1 million fibers and is approximately the diameter of your little finger.

The retina of each eye can be divided into two parts, based on the parts' registration of the left and the right halves of the visual field around the central vertical line of division. The optic nerves from the left and the right eyes meet deep in the brain in what is called the *optic chiasm.* There the optic nerves combine and divide to form the *optic tracts* which contain the axons with information from the left or the right visual fields from both eyes combined.

As shown in Figure 2.1, fibers from the right visual field go to the left optic tract, and vice versa. The optic tracts lead to the *lateral geniculate nuclei*, (LGN). For the moment we will see the LGN as intermediary connecting stations on the way to the *primary visual cortex* in the occipital (neck) lobe.

In cold-blooded vertebrates, such as frogs, the optic nerve ends up in a midbrain structure called the *optic tectum*. In mammals there is a homologous structure called the *superior colliculus* that receives a portion of the signals from the optic tracts. This structure is associated with the generation of eye movements.

The major part of the processing, as well as the major focus of our interest, is the processing in the visual cortex.

2.3 PROPERTIES OF NEURONS

The processing mechanisms of the visual system are constructed of primitive elements called *neurons*. There are a great variety of neurons in the brain, and there may be no such thing as a typical neuron. However, there are properties that are characteristic of most neurons, and we will consider these characteristics in this section.

The neuron or nerve cell in Figure 2.2 has a morphologically defined region called the cell body, which contains a nucleus as well as the reproductive and metabolic machinery required for the usual housekeeping functions of a cell. A unique property of nerve cells is that the body gives rise to a cylinder-shaped, signal conducting nerve fiber, called the *axon*. The axon can have a length from less than a millimeter to more than a meter. An axon usually splits into several branches near the end point. The cell body also gives rise to a different type of extension called the *dendrite*. The dendrites generally have a length in the order of a millimeter.

The typical information transmission process within a neuron is that the cell's dendrites receive information from other nerve cells and transfer this information to the cell body; here the signals are summed and then transmitted along the axon where they can be transferred to other nerve cells. The transmission between neurons takes place at regions called *synapses*, where the axons come very close to dendrites of other nerve cells. We say that information is transmitted from one nerve cell, the *presynaptic* cell, to the next, the *postsynaptic*

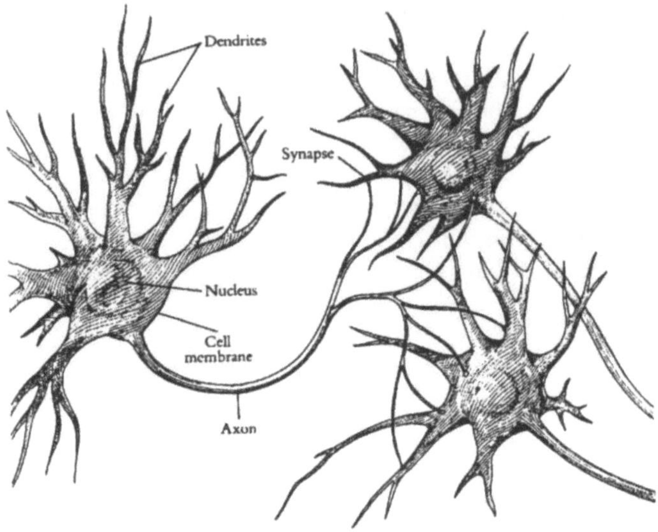

Figure 2.2 Illustration of the principal parts of a nerve cell. (From [57].)

cell. The transmission of information at the synapse from the axon of one nerve cell to the dendrite of another nerve cell occurs through *chemical transmission.*

Although we often visualize a typical neuron like that illustrated in Figure 2.2, we should caution that there are many different types of nerve cells. It is estimated that there are more than 100 different types of neurons in the brain, but the exact number is not known. One problem is that no two nerve cells are identical, and it becomes very difficult to say whether any particular difference distinguishes between classes or individuals. Figure 2.3 illustrates a few different types.

Neurons are often organized in clusters containing the same cell type. The reason for this is certainly computational efficiency. The brain consequently contains thousands of clustered cell structures which may take the form of irregular clusters or of layered plates. One such example is the cerebral cortex which is such a plate of cells with a thickness of a few millimeters.

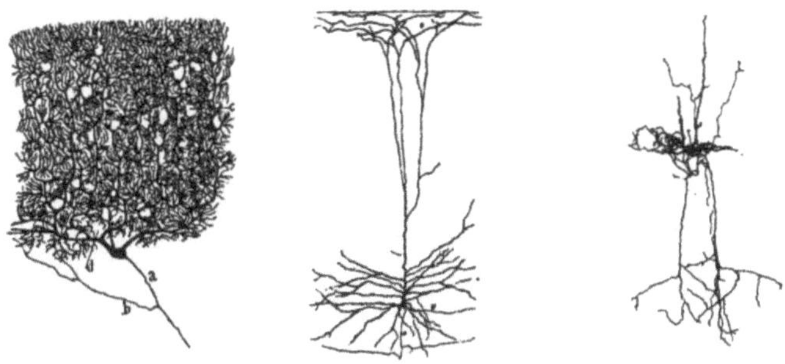

Figure 2.3 Examples of a few different nerve cell structures. From left to right: a Purkinje cell of the cerebellar cortex, a pyramidal cell and a stellate cell, both from the cerebral cortex. (From [114].)

2.3.1 Impulse transmission

In computers and electronics we are used to the fact that signals are electrical, i.e. they are carried by electrons. Although there are electrically measurable effects from the transmission mechanisms in neurons, the mechanisms themselves are chemical.

The most immediate function of the neuron is to receive signals over its dendrites and to sum or *integrate* this information. The result is then delivered to other cells, usually in the form of discrete events called *nerve impulses.* A nerve impulse is assumed to be identical to any other nerve impulse as regards pulse shape, amplitude and duration. The property that varies in order to convey useful information is the instantaneous rate of impulses at a given time. For different types of nerve cells in different states the impulse rate may vary from one every few seconds up to about 1 000 impulses per second.

First we consider the typical transmission mechanism within a particular cell. In some way, which we will look at later, an impulse is generated and travels down from the cell body along the axon to its terminals. The axon can here be viewed as a tube with a thin membrane separating its inside from its outside. In the rest state, protein structures called *active ion pumps* in the membrane produce a potential difference over the membrane to the equivalent of approximately 70 millivolts negative on the inside of the axon. The ion pumps accomplish this by exchanging three sodium ions from inside the membrane for every two potassium ions they transport in through the membrane, leading to an interior deficiency of sodium ions and an abundance of potassium ions. An impulse is nothing but a traveling wave of local reversal of this potential difference. When an impulse approaches a certain part of the axon, the membrane potential difference decreases, causing a number of voltage sensitive *ion channels* - protein structures – to suddenly open within a short length of the nerve fiber. This leads to an influx of positive sodium ions resulting in a reversal of the potential difference, bringing it to about 40 millivolts, positive inside. The sodium channels then close. Meanwhile potassium channels have also opened and remain so as long as the potential reversal persists, which leads to an outflux of potassium ions and a rapid restoration of the resting state with a negative potential inside. This process takes only about 1 millisecond. The process that leads to the opening of the voltage sensitive channels and the subsequent potential reversal is referred to as *depolarization.*

The event described can be measured using electrical instruments, although the process itself is chemical. Thin electrodes can be inserted into the nerve fiber to measure the potential differences. After the sodium channels have closed, they are unable to reopen for a few milliseconds, the *refractory time.* This assures that the impulse cannot propagate back towards the cell body, and it determines the highest rate at which a cell can fire impulses. High activation of nerve fibers typically means rates of a few hundred impulses per second, but there are cells that can reach about a thousand impulses per second. It is important, again, to observe that in the transmission of an impulse, no matter is moving along the axon, but a disturbance wave of the chemical activity and potential is traveling over the membrane.

The speed of impulse transmission is typically some meters per second, but may be a great deal slower. Even this faster speed may, however, be too low for many situations of transmission over larger distances. For this purpose the nervous system has developed a special mechanism for increasing the transmission speed. There are special cells that wrap themselves around the axons like a sheet of insulation. This substance, which is rich in fat, is called *myelin.* The sheet is interrupted every few millimeters at *nodes of Ranvier*, which allow the ion currents to enter and leave the axon. The result is that the nerve impulse will jump from one node to the next. This allows for a great increase in conduction velocity. Most of the large cables of fibers in the brain are myelinated.

2.3.2 Synaptic transmission and impulse generation

We saw in the last subsection how an impulse, once initiated, propagates along an axon. We now look at how such an impulse can be triggered and what happens when an impulse reaches the end of an axon.

From the discussion in the preceding subsection we can assume that an impulse will sooner or later reach the end of an axon, and we may ask ourselves what happens there. The end of an axon constitutes the first part of the synapse, and it contains very sophisticated mechanisms. The arrival of the impulse at this endpoint leads to the emission from the endpoint of packets of substances called *neural transmitters.* These neural transmitter substances cross the small cleft between the axon and the dendrites. On the dendrite membrane surface they attach to special protein structures called *receptors.* The transmitters activate the receptors that in turn cause ion channels to open, which depolarizes or oppositely, stabilizes the post-synaptic cell.

Some tens of neural transmitter chemicals have thus far been identified, and the number may well prove to be between 50 and 100. Acetylcholine and noradrenaline are typical examples of neural transmitters. The effect of a neural transmitter molecule diffusing over the synaptic terminal gap may be either to lower or to raise the membrane potential. In the first case we call a synapse *excitatory.* If, as in the second case, the neurotransmitter causes the membrane to stabilize, the synapses are referred to as *inhibitory.* Generally the type of neurotransmitter released determines whether a given synapse is excitatory or inhibitory. However, some transmitters such as acetylcholine may in some synapses be excitatory and in others inhibitory, which certainly complicates the situation for the model builder. As far as is known, a particular synapse will remain either excitatory or inhibitory.

The impulses are usually initiated at the point where the axon leaves the cell body, the *axon hillock.* This is where the different contributions transported along the dendrites are summed. A very important property of nerve impulses is their "all-or-nothing" quality. This provides a noise immunity similar to that in digital systems. A particular nerve cell may, via its dendritic tree, be in touch with numerous other nerve cells over tens, hundreds, or thousands of synaptic terminals. At any moment a given cell will be told to depolarize at some of its terminals, and be told not to do so at other terminals. At any moment the level of the membrane potential is the effect of all excitatory and inhibitory inputs combined. If the membrane potential is reduced sufficiently, the depolarization will be sufficient to generate an impulse. Usually this will be sustained in the form of a train of impulses. A higher degree of membrane potential depolarization will cause a greater number of impulses to be initiated every second.

In many cases the impulse coming from a single axon terminal has an insignificant effect on the next cell, producing a depolarization lasting only a few milliseconds before dying out. When, on the other hand, impulses come from many nerve cells, the receiving nerve cell tends to sum or integrate the effects. In this way the excitatory and inhibitory influences are summed together. We can, however, not assume a simple additive effect between different influences. Different types of neurotransmitters have different effects by themselves and in combination. The influence on a particular cell depends on where in the dendritic tree the synaptic transmission occurs.

A very important property of synapses is that their strength can be *modified*, i.e. the amount of transmitter substance released by a presynaptic impulse may be altered, and/or the postsynaptic efficacy of a given amount of transmitter in raising or lowering the membrane potential may change. This plasticity is essential for learning and memory. A classical example is the study by Kandel and his colleagues [64] of the habituation of a withdrawal reflex of *Aplysia california*, a marine snail. After repeated stimulation of the gill, the snail eventually ceases to withdraw it. It was found that this short-term learning depends on a decrease in transmitter release from sensory neurons across synapses to motor neurons. The reflex returns after a few hours. Another important finding is that of long-lasting strengthening of synaptic couplings (referred to as *long-term potentiation*, LTP) between simultaneously activated cells of the hippocampus, a part of the brain essential for the consolidation of memories and their transfer to long-term memory [19]. This event might be a physiological basis for associative learning.

The axons we have seen thus far carry all-or-nothing impulses. There are, however, exceptions to this rule. A depolarization will always spread along a fiber, declining with time and distance; this is how potential differences are spread in dendrites. There are indeed nerve cells that act with depolarization only, without generating any impulses. In such neurons the depolarization at the synaptic terminals causes a release of neurotransmitters. These nerve cells are found where the propagation distance for the signal is very short. They also appear to use the effect that the magnitude of the depolarization varies with the strength of the stimulus, a property usually referred to as *graded* signals. A larger signal leads to a large depolarization at the terminal, and a larger release of neurotransmitters. We may say that graded potentials represent an example of continuous signals, while the impulse conduction is a pulsed, or, rather, discrete-event representation. In the latter case we have a pulse-magnitude restoration mechanism which makes the system insensitive to transmission over larger distances, similar to the situation in digital systems.

2.4 THE RETINA

The first element in the chain for biological vision processing is the eye with its light sensitive retina. Although there are a number of interesting features about the construction of the eye and its optical system, we will be very brief in its description.

The retina covers the rear wall of the eye and contains the light sensitive receptors upon which an image is projected. There are two types of receptors, *cones* and *rods*. The cones are used for daylight color vision, and there are about 5 million of them. The cones are concentrated in a particular region near the optical axis of the eye called the *fovea*. The rods, of which there are about 120 million, are used for night vision and dominate in the near periphery of the visual field. Figure 2.4 shows how the density of different types of receptors varies over the visual field. In one part of the retina, where the optic nerve exits from the eye, there are no receptors at all resulting in a blind spot. Because of advanced processing at high levels, we are not aware of this blind spot except in particular test situations. The receptor densities vary considerably over the retina. The fovea is the most important section of the human retina. In the center of the fovea there is a small depression which is about 1/3 mm in diameter. The reason for this depression is that the light passes a much thinner cellular layer before it reaches the photoreceptors. The cones in the fovea have a different shape than in the periphery. They are much longer and thinner, down to only 1 μm in diameter. This section of the retina contains only cones. For that reason we have no night vision in the central field of vision.

The receptors are made light sensitive through different pigments belonging to the *rhodopsin* family. The rod pigment most effectively absorbs photons at a wave length of 496 nm (blue-green), but is substantially less sensitive to long wavelengths, which is why red light may be used without compromising night vision. In the normal human retina there are three distinct cone pigments with different characteristic peak wavelengths to which they are most sensitive. The absorbance spectra of the receptors are shown in Figure 2.10.

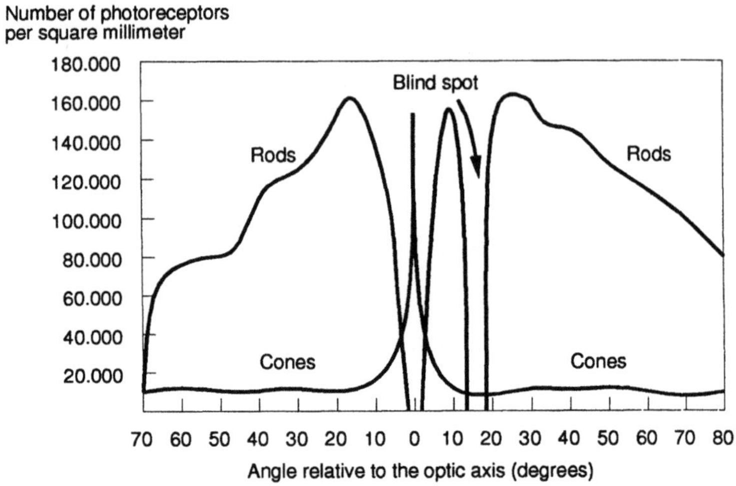

Figure 2.4 Density of different photoreceptors over the retina. (Adapted from data of [108].)

The rod and cone receptors are examples of cells with a graded or analogue response. Receptors are spontaneously active when not stimulated by light. When stimulated, they become ***hyperpolarized***, which means that their activity is reduced. Hyperpolarization is thus the opposite of depolarization. In daylight, rods are constantly hyperpolarized and thus inactive.

2.4.1 The neural structure of the retina

In addition to light sensitive receptors, the retina contains a fairly complex processing structure. An immediately apparent effect of this is that the information from the approximately 125 million photoreceptors is compressed for transmission along about one million fibers in the optic nerve. In this section we will examine the structure of the processing part of the retina. For this discussion we refer to Figure 2.5.

The retina consists structurally of three major layers of neurons and has about the same thickness as a sheet of paper. The first layer under consideration contains the photoreceptors as discussed in the preceding section. It may come as a surprise that this layer is not the one which light first enters, but it is the layer closest to the back wall of the eye.

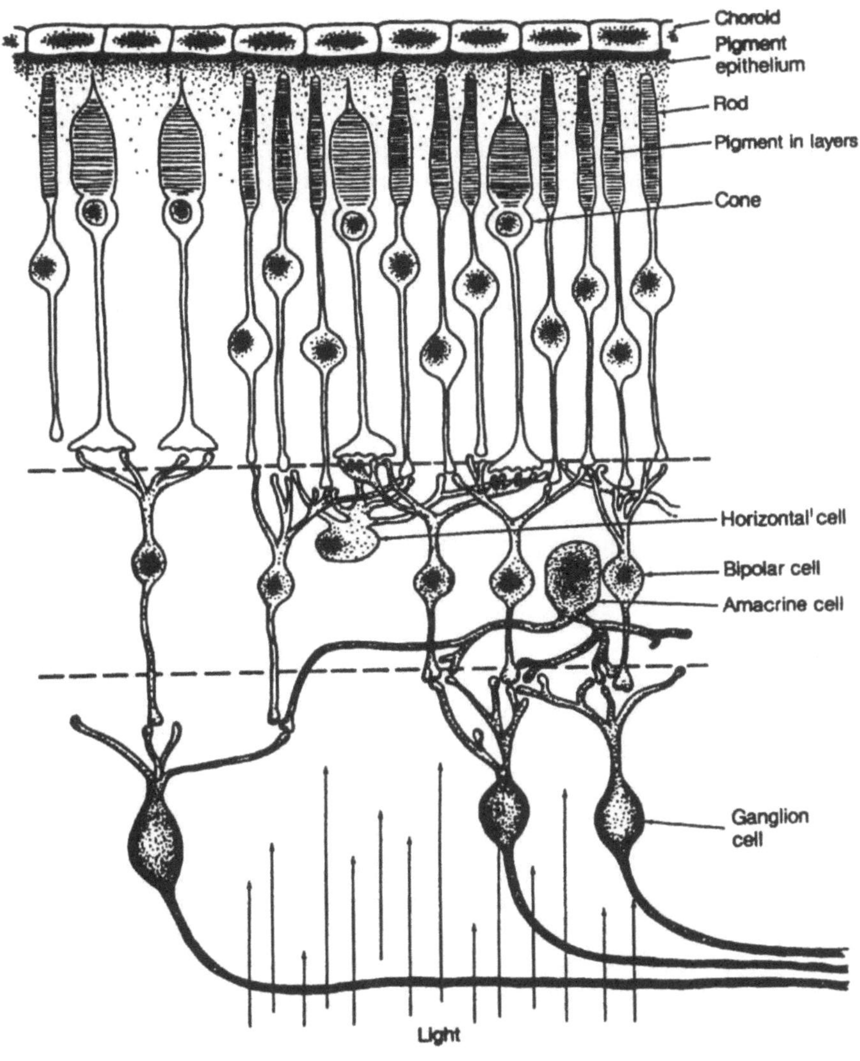

Figure 2.5 Schematic diagram of the human retina. (From [31].)

The second layer of the retina is characterized by *bipolar cells*. These neurons have two, long, extended processes. On one end, the bipolar cells make synapses with the photoreceptors. On the other end, they make synapses with large retinal *ganglion cells*, which constitute the third layer of the retina. The axons of the ganglion cells form the fibers of the optic nerve.

In addition to the bipolar cells, which connect between the photoreceptors and the ganglion cells, there are another two types of cells. These two types of cells have lateral connections. Closest to the layer of receptors are the *horizontal cells*. They have short dendrites and a long horizontal process over some distance across the retina. Cells of the other type are called *amacrine cells*. These are large, laterally interconnecting cells, located between the ganglion and bipolar cells. There is a great variety of amacrine cells differing in size as well as chemical properties. More than 30 different types have been isolated.

Just like the receptors, bipolar, horizontal and amacrine cells have graded responses, whereas the ganglion cells fire pulses.

2.4.2 Processing in the retina

We now look further at the processing occurring at the level of the retina and at the type of output signals which are transmitted along the optic nerve. In this discussion we begin with a description of the output from the retina, which consists of signals traveling along the retinal ganglion cell axons. A suitable way to describe this is by the light stimuli which provoke a response from a ganglion cell.

Let us first define an important concept. The *receptive field* of a neuron consists of the receptor area, in this case a part of the retina, whose signals influence the output of the neuron. Equivalently, we may define it as the part of the visual field to which the neuron reacts. Sometimes the description of a receptive field includes the *type* of stimuli that affect a neuron, and not only the location in space. We will adopt this extension. The first measurements of receptive fields started around 1950. A typical measurement implies that a simple pattern is projected onto the retina, and the output from ganglion cells is observed. First of all it turns out that any steady and diffused background light will not change the frequency of impulses fired by the ganglion cells. They will emit pulses at a rate of a few pulses to about 20 impulses per second for any illumination including complete darkness. Hence, even without stimulation, there is a base level of firing.

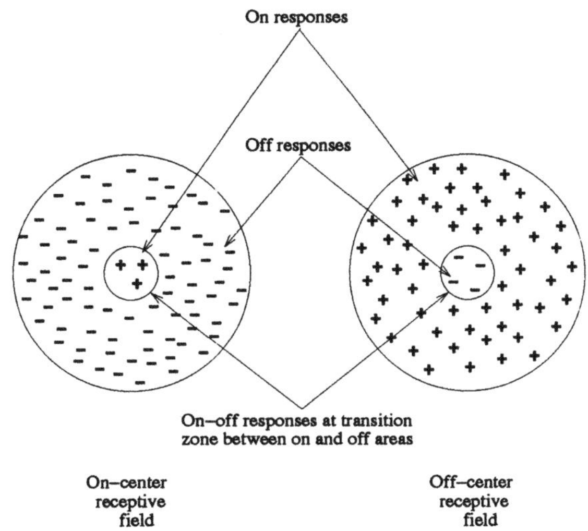

Figure 2.6 Circular center-surround receptive fields.

It was then found that by projecting a small spot of light onto the retina, it is possible to identify the group or single ganglion axon that is influenced. This influence can be either an increase or a suppression of the firing rate. The ganglion cells are consequently of two types, which are called *on-center cells* and *off-center cells.*

An on-center cell responds with an increased rate when a small spot is turned on anywhere within a well defined area. Such a response is called an *on-response.* Moving the spot of light a small distance away from this active region, results in the light's suppressing the spontaneous firing of the cell. This process implies a suppression of the impulse rate during stimulation by light, and is called an *off-response.* A more careful exploration shows that such an on-center cell contains a circular *on* region surrounded by a much larger *off* region, Figure 2.6. An *off-center* cell has the opposite behavior. Its receptive field consists of a small center from which off-responses will be obtained, and a surrounding region which produces on-responses. It appears that these two types of cells are intermixed and about equally common.

If we make the light spot increasingly larger, the response improves until the center of the receptive field is covered, then starts to decline as more and more of the opposing surround is covered. The visual system has very high resolution for details projected on the fovea. The smallest field centers measured in a monkey take up about two minutes of arc or about 0.01 mm on the retina. As mentioned earlier, the cones in the fovea have a diameter and a center-to-center spacing of about 2.5 μm. A circle of that size on the retina corresponds to a coin viewed from a distance of about 150 m.

Although the spacing between the smallest cones in the fovea is 30-35 sec arc, the visual system is able to discriminate between lines at a separation as small as 5 sec arc. This is termed *hyperacuity*, and depends upon cooperative mechanisms involving several orientation sensitive filters [136].

The resolution decreases as we go out toward the periphery of the retina. Here the receptive field centers are composed from thousands of receptors and they can have a diameter of one degree or more. All of this creates a variable resolution perception, which is maximally high at the center of the optical axis and decreases toward the periphery of the visual field.

We may now ask ourselves how these center surround receptive fields are constructed, using the neurons that we know are available. The hypothesized structure is depicted in Figure 2.7.

It turns out that the bipolar cells themselves have receptive fields of center surround type, which means that the generative machinery has to precede these cells. The receptive fields then appear to be generated by connections among receptors, bipolar cells, and horizontal cells.

It has been demonstrated that the bipolar cell always sends a single dendrite in the direction of the receptors. This dendrite may either synapse with one or a few receptors. If it synapses onto more than one receptor, these receptors will still occupy a very small local region of the retina. It is believed that these few receptors synapsing directly onto the bipolar cell form the center of a receptive field. The area they span also matches the size of a typical field center. The horizontal cells, on the other hand, synapse onto receptors within a considerably larger area, as indicated in Figure 2.7. The horizontal cell then sends an axon which synapses onto a bipolar cell. As is seen in the figure, a horizontal cell takes its input from both field center and surround. It is assumed that the horizontal cell output has an effect on the bipolar cell that is opposite to that of the bipolar cell's direct input from the receptors.

As mentioned above, bipolar cells, like receptors and horizontal cells, are of the graded or continuous type and, consequently, do not fire impulses. Still, they do have responses in terms of depolarization, and cause an increased transmitter release from the cell's terminals. When inhibited they will be hyperpolarized

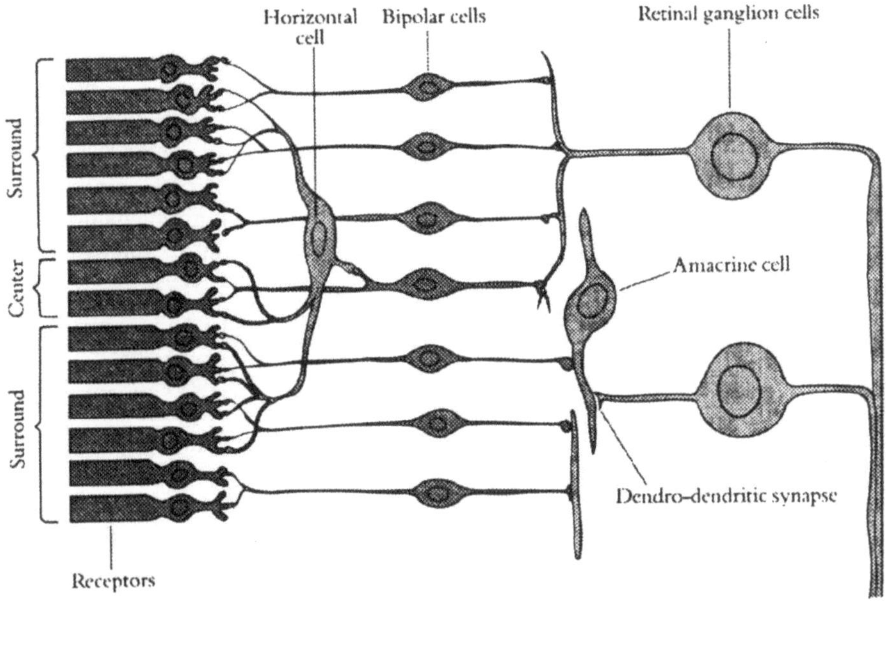

Figure 2.7 A hypothetical circuit diagram for the generation of center-surround receptive fields. (From [57].)

and release less transmitter substance. Bipolar cells are of two types, on-center and off-center, just like the ganglion cells. It is not known in detail how the receptive field is organized, but current evidence favors the view that the two different types of bipolar cells have different receptor molecules, determining the type of synapse.

The organization and structure of horizontal cells is still more complicated as these cells appear in several subtypes and can differ very much from one species to another. As an example, there is no ordinary, easily recognizable axon process. The only consistent interpretation given all varieties available is that the processes from horizontal cells, like those from amacrine cells, can simultaneously serve the functions of both axons and dendrites. We can see in Figure 2.7 that there are processes between the horizontal cells as well as the center receptors. It is not known whether they are to be viewed as input dendrites or output axons. For some species it is known, however, that horizontal cells do feed information back to receptors. Despite all the uncertainty about exact functional mechanisms, we can conclude that horizontal cells do connect between receptor cells over an area that includes the surround, and are the only cells to do so.

As described earlier, the center-surround receptive fields are already defined at the level of the bipolar cells. We ask ourselves what happens between the bipolar cells and the ganglion cells besides the change from an analogue representation in the bipolar cells to a pulse representation in the ganglion cells. What do amacrine cells do?

The function of the amacrine cells is not well understood. They have a surprising variety of shapes, and they use a number of different neurotransmitters. It can be seen that they link bipolar cells and retinal ganglion cells to form various types of indirect routes between these. It is known that amacrine cells form the major path from rod-driven bipolar cells to ganglion cells; there is virtually no direct path. That the amacrine cells lack axons implies that their functioning may be symmetrical, using their processes for input as well as output. It is believed that the amacrine cells participate in some of the sensitivity control systems of the retina. They are also thought to generate some of the differences between ganglion cell responses to be described below.

2.4.3 Distinctions between different classes of retinal ganglion cells

We have in the earlier discussion assumed that all ganglion cell responses are essentially of the same type. It turns out that there are actually several types of responses which differ in a number of important characteristics. Three different classes of cells found in the cat retina have been given the names of X, Y and W cells. The major characteristic distinguishing these cells is the speed with which neural impulses travel along their axons. The Y cells are the fastest, with a conduction speed of about 40 meters per second. X cells have a conduction speed of about half that of the Y cells. The W cells, whose existence and characteristics are debated among researchers, have a very low conduction speed of about 10 meters per second. W cells are considered to have more complex receptive field responses than those discussed earlier, with some responding to stimuli moving in a particular direction. The response rate of the cells assumed to belong to the W category is very sluggish, and most researchers believe that the basic analysis of the visual scene is mainly done by the X and Y cells.

X cells are physically smaller than Y cells, and have smaller center surround receptive fields. There are a great many more X than Y cells, and they differ in their distribution over the retina. Virtually no Y cells are found in the central fovea, and the number of these cells increases as we move towards the periphery of the retina. The cells also have a different response pattern. X cells respond in a sustained manner when stimulated, with their neural activity continuing as long as the stimulus remains. Y cells give a more transient response. They give only a brief burst of pulses when the stimulus starts or when it ends.

Figure 2.8 shows another difference between X and Y cells. In part A we see a receptive field of a ganglion cell, where half of the receptive field is illuminated with light and the other half is dark. We can now switch, or quickly rotate, the stimulus to assume the patterns shown as B or C. An X cell will show no noticeable response, while a Y cell will produce a strong response.

It is believed that the large receptive fields of Y cells are generated by amacrine cell inputs, whereas the X cells may take their input more directly from bipolar cells. Transient behavior is also characteristic of amacrine cell responses, further indicating their possible connection to Y cells.

The distinction into X, Y and W cells applies to the cat, seemingly the most studied animal in biological vision. The retina of the monkey, like that of man, is somewhat different from that of the cat, since monkeys have color vision.

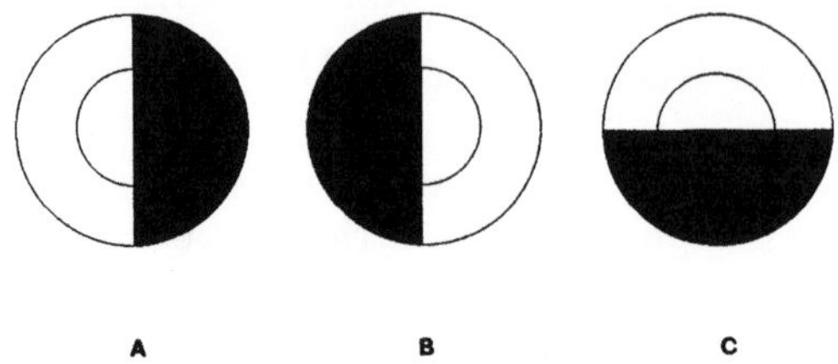

Figure 2.8 Effects on X-cells and Y-cells of different illumination changes.

There is, however, a similar distinction between ganglion cells found in the monkey retina: this has two types roughly corresponding to the X and Y cells of the cat except that the X cells of the monkey are color-sensitive.

Recent research indicates that the X cells and Y cells form the basis for two different subsystems of the visual system. One subsystem (Y) is used for detection and analysis of movements. The other one (X) is specialized for detailed, high resolution analysis of stationary patterns and for color vision.

2.5 COLOR VISION

The science of color is quite old. Newton showed in *Opticks* in 1730 [106] that white light consists of all colors, and Young proposed in 1801 that the retina has three types of sensors [91]. Helmholtz and Maxwell studied color mixtures in the 1850s and demonstrated that almost every color, as we perceive them, can be mixed out of only three different monochromatic sources, provided the sources have wavelengths fairly well separated. At first this appears to be a very surprising effect. Does this imply that we "create" new wavelengths merely by mixing the sources additively? This is certainly not the case. However, the physical stimuli, i.e. the three sources, create the same visual sensation as the actual wavelength. There have been many attempts with additive mixtures, using two or more than three sources. The results obtained show that it is impossible to find two sources that can give the visual sensation of all colors. When more than three sources are used, all color sensations can be produced but nothing is gained. As mentioned above, there is a good biological reason why: we only have three different types of cones in the retina, sensitive to different wavelength intervals. Our vision system uses three different filters in order to get a rough idea about the energy distribution along the wavelength axis. This implies that the visual perception system probably works with a 3D-like representation. The most well-known example is the classical Young-Helmholtz RGB-model, where R, G, B stand for Red, Green and Blue. If we want to analyze colors in the same way the human eye does, there is probably no point in working in more than three dimensions.

2.5.1 Color receptors

We have already mentioned in Section 2.4 a few basic facts about the light receptors. The combined spectral sensitivity curve for the three daylight receptors is indicated by the *photopic* sensitivity curve shown in Figure 2.9 which also shows a very similar curve, the *scotopic* (night vision) sensitivity function, which indicates the sensitivity function of the rods. The daylight sensitivity function has a maximum at approximately 550 nm whereas the rod sensitivity function is centered at 500 nm. This difference in wavelength centers gives rise to the *Purkinje shift.* Two colors that during daylight look equally bright, say a 520 nm blue-green, and a 600 nm orange, will, when the luminance is lowered so that rods are activated, look distinctly different in brightness, the blue-green appearing much brighter.

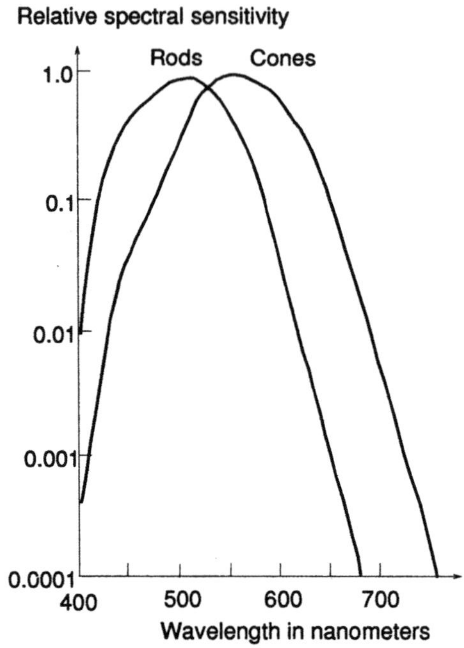

Figure 2.9 The photopic and scotopic light sensitivity functions of the eye.

The basis for color vision is the existence of cones with three different types of photo pigments. These photo pigments have absorbance spectra as indicated in Figure 2.10. Note, incidentally, that the sensitivity peaks actually are not at red, green and blue, but rather at yellow-green, blue-green and violet. The RGB names are used for convenience and historic reasons. It turns out that only one out of ten cones is of the blue type, whereas the red and green cones occur in approximately equal numbers.

2.5.2 Spectrally opponent spaces

In the late nineteenth century Evald Hering proposed the existence of three opponent processes in the human vision system: One for red/green sensation, one for yellow/blue, and a third, qualitatively different from the first two, for black/white. Hering made these assumptions on the basis that we can see a yellowish-green or reddish-yellow color, but we can not see or even imagine a color that is reddish-green or bluish-yellow. He found it very strange that some colors could appear at the same time (we can imagine that orange is a mixture

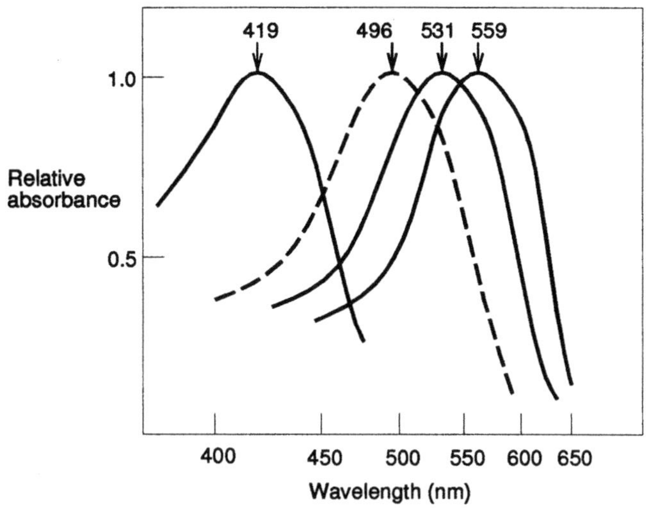

Figure 2.10 The absorbance spectra of the four photo pigments of the normal human retina. The dashed curve is for the rods.

between red and yellow) and some combinations could not. He concluded that blue in some sense must be an opposite color to yellow and red opposite to green, see Figure 2.11.

Hering's opponent-color theory competed with the Young-Helmholtz trichromatic theory for a long time, but eventually it turned out that both are largely correct as models of the color representation in man, but at different stages of processing —the Young-Helmholtz RGB-model for the receptors, and Hering's model for the bipolar and ganglion cells of the next stages.

The receptive fields of color coding X ganglion cells are of the center-surround type described in subsection 2.4.2. However, in the case of color vision the small center is fed by a *single type* of cone. Though it was initially believed, in conformance with Hering's ideas, that the opposing surround and the center were fed from different types of cone, recent results suggest that the surround receives input from all three cone types [132]. Using the notation *center/surround* we consequently obtain the following opponent processes: R/RGB, G/RGB and B/RGB.

In primates, including man, more than 90 % of all axons leaving the eye carry color information. Thus only a small percentage of the ganglion cells receive a nonspecific RGB/RGB input, ideal for spatial luminance discontinuity detection independent of spectral content. But note that since the spectral sensitivity curve of each cone is quite wide, small spots of light still excite a color coding on-center cell, (almost) irrespective of its spectral content. Consequently color-coding axons carry both color and luminance information.

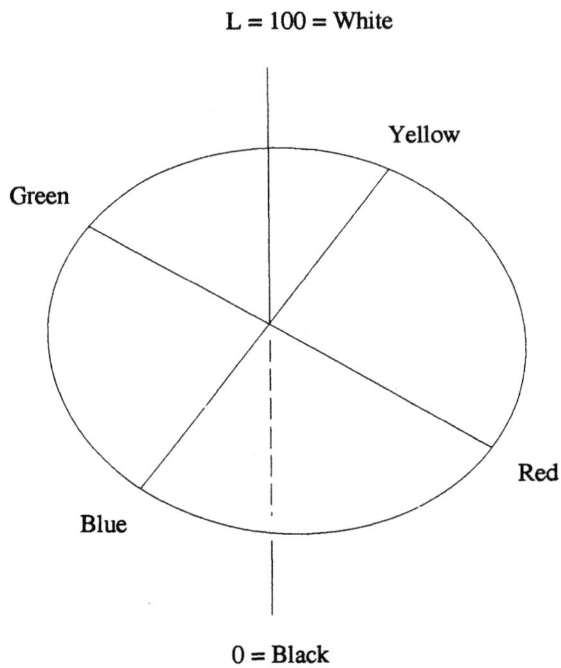

Figure 2.11 Opponent color space.

2.5.3 Color constancy

The spectral composition of light from a light bulb differs distinctly from that of sunlight. This also means that the spectral content of the light reflected from a surface will depend on whether the surface is placed indoors or in sunlight. The fact that we hardly register this difference is referred to as *color constancy*. Edwin Land [84] has demonstrated the remarkable properties of this effect.

In a typical experiment a patchwork of differently colored surfaces is illuminated by light from three different sources, e.g. red, green and blue light, whose relative intensities can be adjusted. Suppose now that a surface A in one case reflects light according to a certain spectral intensity distribution. It is then possible, by using a photometer, to adjust the intensities of the light sources so that the spectral composition of light reflected from another surface B will become identical to that of surface A in the first case. It is an experimental fact that B will *not* appear to have the same color as A had before the adjustment. The perceived colors of the surfaces typically have not changed at all, and will not change no matter which relative intensities of the light sources are chosen, as long as all three sources contribute. On the other hand, if we have just a single uniformly colored surface and the rest is black, the color of this surface will vary with the relative intensities of the light sources. This and other experiments indicate that the brain not only uses color opponency locally, but in fact compares signals globally.

2.5.4 Modeling of color perception

A great deal of work has been spent on the problem of how to describe and measure color for representation in technical systems. La Commission Internationale de l'Eclairage (CIE) is an organization concerned with standardization of measurements of light and color. In 1931 the CIE adopted a color specification method which for the first time gave the concept of color a scientific basis. The CIE presented a pool of data that shows how to mix three sources of light to achieve a given visual sensation. It should be noted that the results apply to what the CIE calls the Standard Observer, which of course means that there may be differences in how individuals perceive color mixtures. An often used graphic representation is to plot the proportions, or *chromaticity coordinates*, as functions of wavelength (see Figure 2.12).

The rgb *tristimulus* values indicate the proportions of the responses from the cones in the retina, from a *monochromatic* source of light with a wavelength λ. Similarily, if we mix r, g and b with the proportions α, β and γ, we get the visual sensation of λ.

There are colors that do not correspond to a single physical wavelength. You do not, for instance, find purple (red plus blue) or brown (red or orange plus black, i.e. brighter surround) in the rainbow spectrum. When we deal with such colors and also include hue and saturation, another representation is needed. A well-known way of displaying color coordinates is to use the RGB color space. Since this space is three-dimensional it is difficult to reproduce it on a two-dimensional medium. To avoid this problem one usually normalizes the color space.

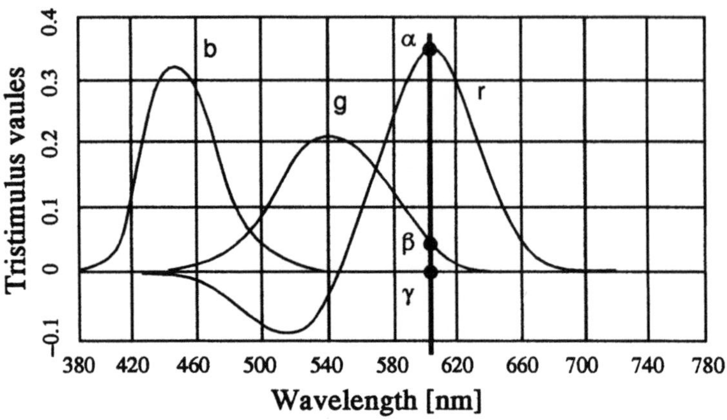

Figure 2.12 Chromaticity coordinates, tristimulus values r,g and b of spectral stimuli of equal radiance for the Standard Observer.

In Figure 2.13, R, G and B are the absolute values of the filter responses and r, g and b are the relative values. When the normalization is done, only the information about hue and saturation remains. The information about the corresponding luminance is of course lost.

$r + g + b = 1 \qquad \Rightarrow \qquad$ if r and g are given, b is also known.

One immediately notices a problem: some wavelengths show negative contributions. Since energy is always positive, there are wavelengths that cannot be implemented with the three primaries r, g and b.

CIE suggests three new non-physical primaries X,Y and Z. These primaries exist only in a mathematical sense. The variable Y is chosen so that it corresponds to the luminance sensitivity of the human eye. The relations between the RGB space and the XYZ are:

$$X = 2.77R + 1.75G + 1.13B$$
$$Y = 1.00R + 4.59G + 0.06B$$
$$Z = 0.00R + 0.06G + 5.59B$$

The XYZ color space has the nice property that all colors can be represented with positive coordinates, as shown in Figure 2.14.

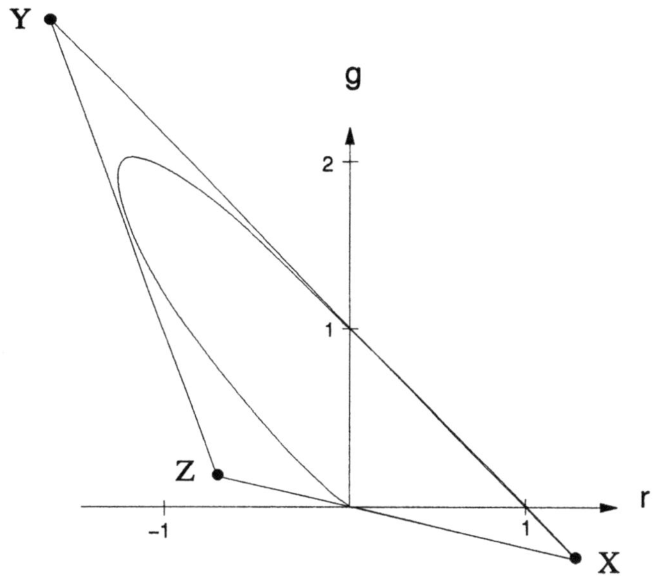

$r = \frac{R}{R+G+B}$ $g = \frac{G}{R+G+B}$ $b = \frac{B}{R+G+B}$

Figure 2.13 Normalized RGB color space, chromaticity diagram.

Using the same sort of normalization as above, one obtains the CIE1931 chromaticity diagram of Figure 2.15. The colors that can be produced with RGB are marked in the diagram. A television system can only reproduce chromaticities falling within the triangle formed by connecting the chromaticities of the three display-tube phosphors. The light is fully specified with x, y and Y, since

$$z = 1 - x - y \qquad \frac{Y}{y} = X + Y + Z$$

A number of improved color spaces have been suggested to overcome the CIE1931 diagram's perceptual nonuniformity, i.e. the lack of correspondence between numerical metric and perceived color differences. These modifications are, however, outside the scope of this presentation.

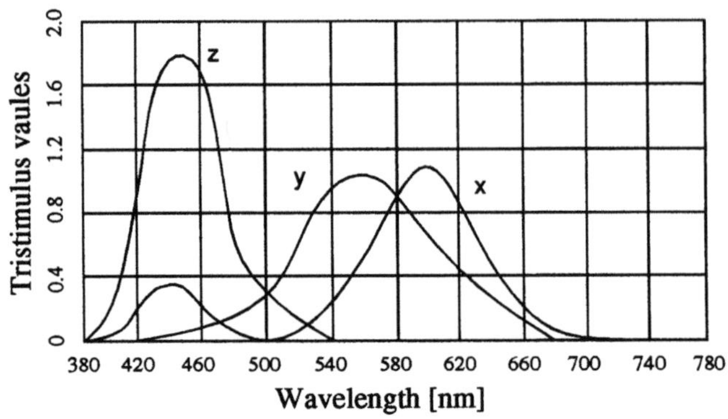

Figure 2.14 Tristimulus values x,y and z.

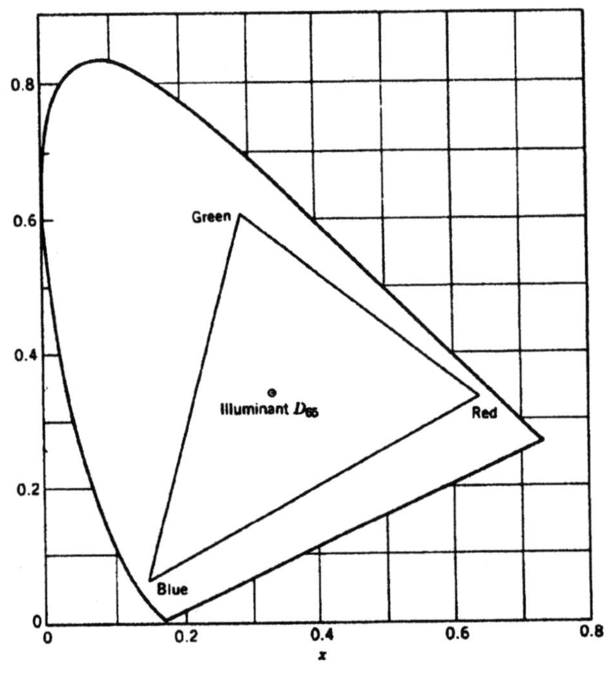

$$x = \frac{X}{X+Y+Z} \qquad y = \frac{Y}{X+Y+Z} \qquad z = \frac{Z}{X+Y+Z}$$

Figure 2.15 CIE1931 chromaticity diagram. The triangle indicates colors reproducible with television systems.

2.6 THE VISUAL PATHWAYS AND THE LATERAL GENICULATE NUCLEUS

The axons of the retinal ganglion cells form a bundle that exits from the blind spot of the eye to constitute the optic nerve. From there lead two different paths which carry somewhat different information.

The primary visual pathway is called the *geniculostriate system.* This arises from the optic chiasm, where the two optic nerves come together. As mentioned earlier, the bundle of fibers are propagated in such a way that the right visual fields from the left and the right eye are combined into one bundle forming the left half of the *optic tract* and correspondingly for the left visual field. Consequently the two visual fields will be analyzed in different hemispheres of the brain. For most primates the major termination for the optic tracts is a part of the *thalamus* called the *lateral geniculate nucleus* (Figure 2.1).

The lateral geniculate nucleus (LGN) is divided into six layers stacked on each other. Each one of these layers appears to receive input from only one eye, and there is also a degree of interaction between the layers. Each layer contains a topographic map of the visual field, implying that adjacency between points in the visual field will lead to adjacency of their maps in the layer. The function of the LGN is largely unknown. In general, the receptive fields of the lateral geniculate cells are similar to those of the retina, which means that there are center surround cells of off- or on-type. It has been found that the X and the Y cell axons synapse with cells in different layers of the LGN, the X axons going to the four upper *parvocellular* layers, and the Y axons going to the two lower *magnocellular* layers with the names indicating a difference in cell size for the two groups. This means that the X signals, appropriate for carrying color and form information, are kept separate from the Y signals, carrying information about motion in the visual field. As we will see later, this separation is maintained in the cortex, indicating a division into separate pathways for form, color and motion detection. The form and color pathway is often referred to as the *parvo* system and the motion pathway as the *magno* system. One often regards the LGN as a simple relay station, although the immense number of feedback axons returning from the primary visual cortex seems to contradict this.

When the approximately one million axons of the lateral geniculate neurons travel away from the cell body, they distribute into a fan of fibers called *optic radiations.* The fibers eventually reach cells in the cortex in the posterior portion of the brain, known as the *occipital lobe.* The particular area in the

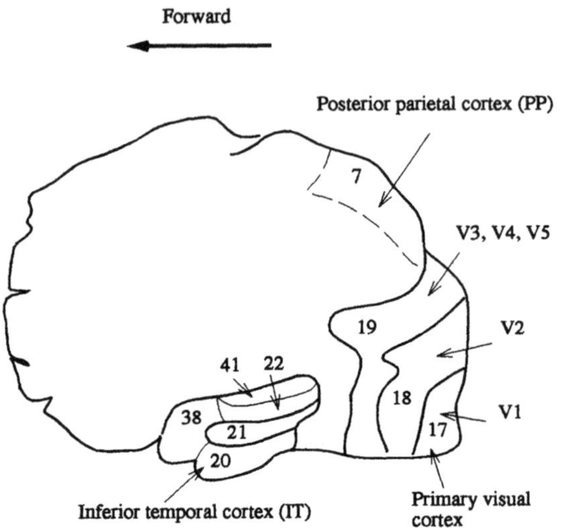

Figure 2.16 The conventions for labeling of the parts of the visual cortex. (From [31].)

occipital lobe that receives the majority of the axons has a banded appearance in cross-section, which qualifies it for the name *striate cortex*, also referred to as the *primary visual cortex*. Cortical visual areas have been labeled in various ways over the years, and the two main conventions are indicated in Figure 2.16. The system devised by Brodmann designates the major areas with numbers 17, 18 and 19, which was based upon the appearance of cells in the area. Later researchers have begun to use different labels for these areas, namely V1 and V2 for area 17 and 18 respectively, and divided area 19 into subareas V3, V4 and V5. Areas V1-V5 are presumed to constitute the first visual maps of the cortex.

There is also a second pathway to the visual cortex, starting with the *retinotectal* path. This consists of a number of fibers from the optic tract which branch off towards the brain stem. This particular region of the brain stem contains in an evolutionary sense a much older, more primitive, visual center called the *tectum*. This part is responsible for most of the visual processing in cold-blooded animals. The part receiving most of the incoming fibers is a cluster on the back surface of the brainstem known as the *superior colliculus*. It has been found that only Y-and W-type inputs come to this region via the retinotectal path. It does, however, receive signals from a wide range of other locations, such as the primary visual cortex and the association cortex, parts of the brain involved in high-level computations. Given the types of input cells (Y and W), it appears that the superior colliculus is involved with dynamic aspects of perception. Indeed, it is known that signals from the superior colliculus generate coordinated eye and body orientation towards a visual target. Experiments show that the superior colliculus triggers *saccadic eye movements*, i.e. the short, jerky jumps made when focusing on an object of interest. The neural structure of the superior colliculus is, at least to some extent, understood [124]. Simply put, it can be described as consisting of two interconnected neural sheets stacked on each other. The upper sheet, which receives the retinotectal signals, has a topographic map of the visual field. Artificial stimulation of cells in the lower sheet, using inserted microelectrodes, generates eye movements towards the part of the visual field mapped by neurons immediately above them in the upper sheet. Many axons from the lower sheet go to centers further down the brainstem directly involved in the execution of eye movement commands.

Other axons from the superior colliculi reach the thalamus via the *tectopulvinar* pathway (Figure 2.33), but unlike the axons in the primary visual pathway, they do not go to the LGN. Instead they synapse with cells in the *pulvinar nucleus* and the *lateral posterior nucleus*. From there signals reach the visual cortex, but not the primary visual cortex. Very little is known about the function of this pathway. It may be involved in attention control [149] and compensation for eye movements [105], i.e. the fact that the environment does not appear to move during eye movements.

We now turn our attention to the other visual pathway that performs the major part of the processing and will be at our focus of attention for the next few sections.

2.7 THE PRIMARY VISUAL CORTEX

2.7.1 The neocortex

The visual cortex is a collective name for a number of distinct interconnected regions of the *cerebral cortex*, which consists of a 2–3 mm thick folded sheet of neurons covering the outer surface of the brain. The sheet is often referred to as *gray matter*. Below this sheet there is *white matter*, which consists of myelinated axons connecting different cortical regions with each other, or connecting a cortical region with lower centers of the brain, situated deeper inside the skull.

The major and evolutionary most recent part of the cerebral cortex, called the *neocortex*, is quite regular in its architecture and consists of 6 layers, with layer I at its surface and layer VI at the junction with white matter (Figure 2.22). Layer I, alternatively called the *plexiform layer*, has very few neurons, but is packed with axons, dendrites and synapses.

The input-output relations of cortical areas also follow a regular pattern. In the cortex there is a clear pattern of feed-forward processing, from primary sensory areas, via secondary sensory areas, to motor command areas. However, there are also very strong connections in the opposite direction – feedback – and important couplings with lower brain centers – side-loops. It turns out that feed-forward axons from one area to the next tend to synapse with cells in layer IV of the target area, and to a lesser extent, layer VI. They originate in layers II and III of the sending area. Feedback axons, on the other hand, go to layers I and VI, and originate in layers V and VI. Inputs from lower brain structures typically go to the upper layers, while outputs to lower centers originate in the deep layers.

What distinguishes different parts of the neocortex is largely the sources of their inputs, the destinations of their outputs, and the relative thicknesses of the different layers. The primary visual cortex appears to be one of the most complex cortical areas of all, illustrated by the fact that its density of neurons is more than twice that of other areas.

The primary visual cortex, V1 or area 17, is assumed to contain over 100 million neurons in each hemisphere. It is quite apparent that it is a nearly impossible task to find out how such a complex structure works. What is known about these cells is largely based on investigative research with microelectrodes on animals or the experiences of accident patients with local lesions in the cortex.

The pioneering work was done by David Hubel and Torsten Wiesel who received the Nobel Prize in 1981 for their research. They developed a procedure to map receptive fields of cortical cells in animals. This consisted in inserting recording electrodes into cortical cells in V1, and measuring the cell's responses to stimuli projected onto a screen in front of the animal.

2.7.2 Receptive fields in the visual cortex

When cortical cells were measured using different types of visual stimuli projected in front of them, the familiar center surround characteristics of ganglion

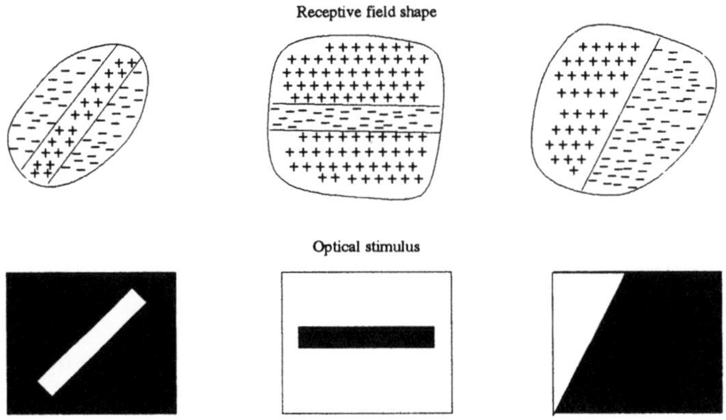

Figure 2.17 Receptive fields of simple cortical cells.

cells and lateral geniculate cells were occasionally still found. However, the majority of the measured receptive fields had different characteristics.

Receptive fields of one class of such cells are shown in Figure 2.17. This type of cell is usually labeled a *simple cell.* These cells have no spontaneous activity and produce no response to diffuse illumination. It turns out that the response is critically dependent on the presence of an appropriately oriented luminance-discontinuity (an edge or a thin bar) in the receptive field. We say that such cells have *orientation specificity* in addition to their usual *position specificity.*

Another class of cells has been labeled *complex cells.* They appear to be tuned to more complex stimulus patterns. They tend to have larger receptive fields than simple cells, although the size varies. Complex cells respond optimally to a particular orientation of the stimulus. They also respond preferentially to motion within the receptive field, and the direction of this movement is important as well. The exact location of the stimulus is, on the other hand, less important. This appears to be the first example of the generation of *invariance* to produce increasingly generalized responses over wider areas.

To continue the characterization of classes of cells, there are certain cells that respond not only to the orientation and to the direction of movement of the stimulus but also to more particular features such as length, width, angles, etc. Figure 2.18 shows some examples of responses of such cells. Hubel and Wiesel originally characterized these as *hypercomplex* cells, but it is no longer clear whether they constitute a separate class, or, rather, represent a spectrum of more or less complicated receptive fields. It should be emphasized that cells with very intricate receptive fields are atypical for the primary visual cortex, but become more abundant in secondary visual areas.

2.7.3 Response characteristics of feature detectors

We saw earlier that the response of a simple cell is critically dependent upon the orientation of the stimulus with respect to the orientation of the receptive field axis. This orientation of the stimulus is called the cell's *preferred orientation.* As the stimulus line or edge rotates from the preferred orientation, it will activate increasingly more of the inhibitory parts of the receptive field structure, causing a reduction in the rate of firing. When the stimulus is at a right angle to the preferred orientation, excitatory and inhibitory regions are likely to cancel each other causing no output from the cell. We can observe a function which relates the rate of firing of the cell to the angle of tilt of the stimulus with respect to the preferred orientation, called the *orientation-tuning function.* For different cells we will have different functions, and it is convenient to define an *orientation selectivity* of a particular cell. This can arbitrarily be defined as the half-width of the tuning curve at the point where the response is half the maximum amplitude (Figure 2.19). In Figure 2.20 we can see the distribution of tuning selectivity of simple cells and of complex cells of the monkey's visual cortex.

Simple and complex cells have been investigated with respect to the linearity of spatial summation. It has been found that simple cells behave in a linear fashion. This can be demonstrated by measuring the cell's response when the stimulus consists of a series of optimally oriented light and dark bars (a sinusoidal grating) moving across the receptive field. The linearity is indicated by the fact that the responses from simple cells vary in synchrony with the stimulus over a wide range of frequencies. Complex cells, on the other hand, appear to be nonlinear as their responses are not governed by the spatial frequency of patterns passed across their receptive fields. In the monkey, it has been found that X ganglion cells, which have linear receptive fields, feed into simple cortical cells, and that Y cells having nonlinear receptive fields feed predominantly into complex cells. Another characteristic of a cortical cell may be obtained by measuring the contrast sensitivity of a cell as a function of the position of an optimally oriented line. This function can be viewed as a *line-spread function,*

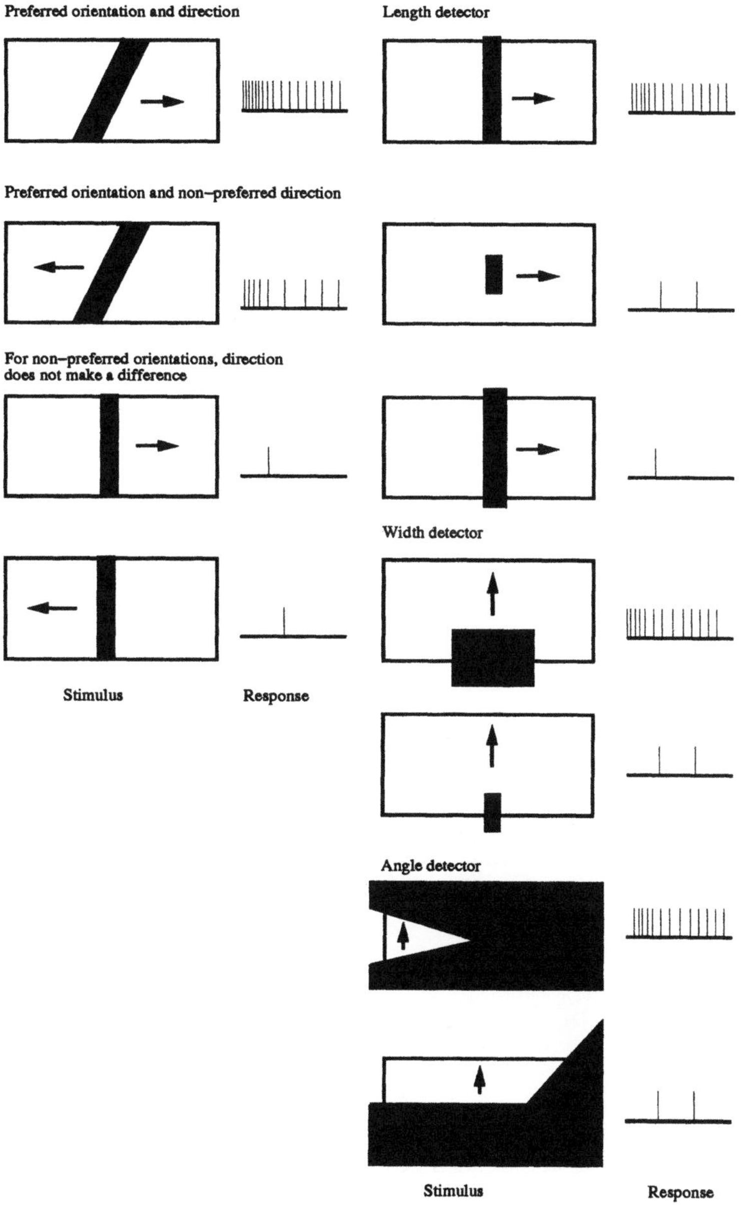

Figure 2.18 Examples of response situations for complex and hypercomplex cells. (After [58] and [59].)

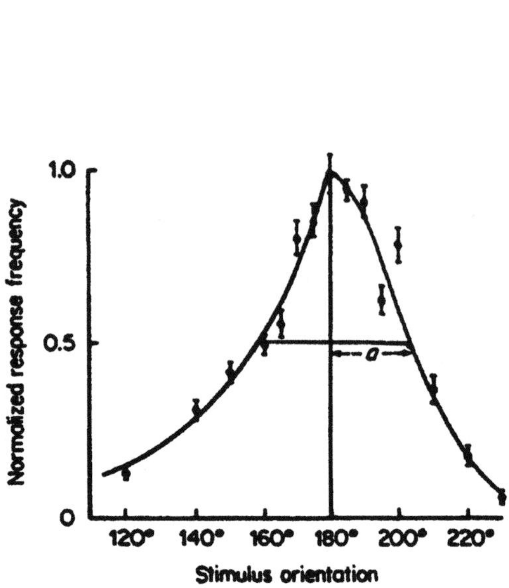

Figure 2.19 Orientation tuning curve of a simple cell of a cat. (From [52].)

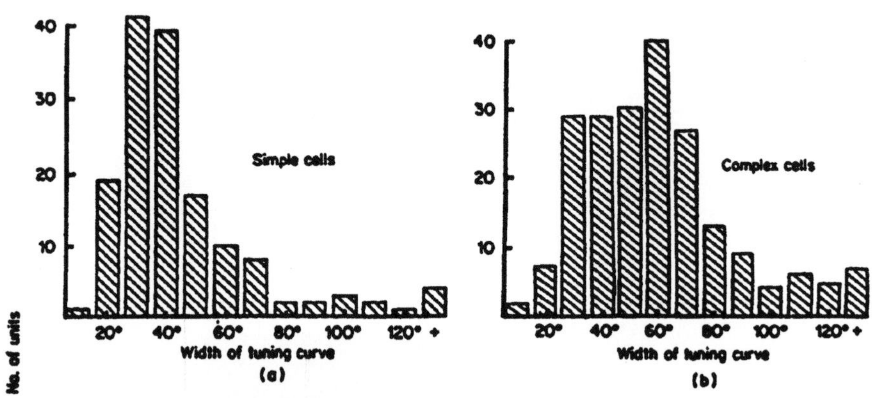

Figure 2.20 Distribution of tuning selectivity of (a) simple cells and (b) complex cells from the monkey's visual cortex. (From [121].)

and is in this case also called the *sensitivity profile* of the cell. We can see measurement results of two such cells in Figure 2.21.

Cell (a) has an excitatory center part and inhibitory flanks. Interestingly enough, it has more peripheral facilitatory flanks which are referred to as *disinhibitory flanks*, as they give a small positive contribution when stimulated.

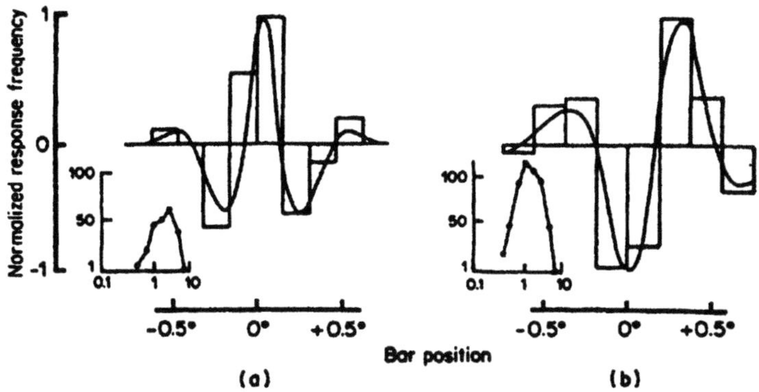

Figure 2.21 Histograms representing sensitivity profiles of (a) a bisymmetrical and (b) an antisymmetrical simple cell in the cortex of a cat. (From [104].)

In computer vision we recognize this as a line detector . Cell (b) looks like an edge detector in that it responds to light on one side and dark on the other side of the receptive field. With regard to their mathematical form, we will call the line detectors *symmetrical* or *cosine cells*, while edge detectors are called *antisymmetrical* or *sine cells*. It should be emphasized that the characterization into edge and line detectors should not be taken literally: there is a whole spectrum of detectors with different optimal phases, just as there exist detectors for a wide range of orientations, not just vertical and horizontal. Another characteristic of visual cortex cells is their *response variability*. This is the standard deviation of the cell's response for a given stimulus condition over several tests on the same cell. It turns out that the response variability of complex cells in the cat is on the average half that of simple cells. This is another example of *invariance*, or independence with respect to certain effects.

2.8 COLUMNAR ORGANIZATION AND THE LAYERS OF THE CORTEX

We have seen earlier how the visual field maps onto the retina. Various types of processing are performed in the retina after which the information is transported via the lateral geniculate nuclei to the visual cortex. As the information travels from one part of the visual system to the other, the spatial relationships displayed in the various locations are more or less maintained. We recognize this as a sequence of spatially *topographic* maps, one after the other. This spatial ordering appears to continue for at least a few more stages beyond the visual cortex. Its origin can be traced to the particular connection structure of the cortex, which is characterized by short (4–5 mm) lateral couplings within each area. This means that a neuron can communicate with its neighbors within a short range, but not with neurons further away. We see that a topographic map of the visual field onto a cortical area is ideal for *spatially local operations* on the visual data, e.g. detection of edge and line segments. Similarly, when a visual area does not display such a topographic map, this is an indication that its computations are not devoted to spatially local operations.

2.8.1 Connections between the layers of the primary visual cortex

The main interlaminar axon couplings in a cross-section of V1 are shown in Figure 2.22. As mentioned, V1 is unique in its complexity and this is especially evident in the layering. We notice that layer IV is unusually thick and divided into a number of sublayers. Another unusual feature is that sublayer IVb is an output layer. This layer receives its major input from sublayer IVCa, which in turn is fed by axons from the magnocellular (Y) layers of the LGN. From the discussion in Section 2.6 on separate pathways for form-plus-color and motion, we may guess that the output from layer IVb is related to motion detection, while the output from the upper layers may be related to form and color detection. This is indeed the case, and we will say more about this when discussing the continuation of the visual pathways.

The complexity of the receptive fields of cells varies between the layers, and increases with the distance from the inputs. Center surround receptive fields are found in the lower part of layer IVCb, simple cells in the upper part of IVCb, in the lower part of IVCa and in layer VI. Standard complex cells appear in the upper part of IVCa, as do direction selective cells. Direction selective cells are abundant in layer IVB but are absent in layers II and III, where instead length-sensitive, *end-stopped cells* are common (10–20 %). Layer V is characterized by having complex cells with receptive fields that are much larger than those in the upper layers. Layer VI has cells with extremely elongated, narrow receptive fields.

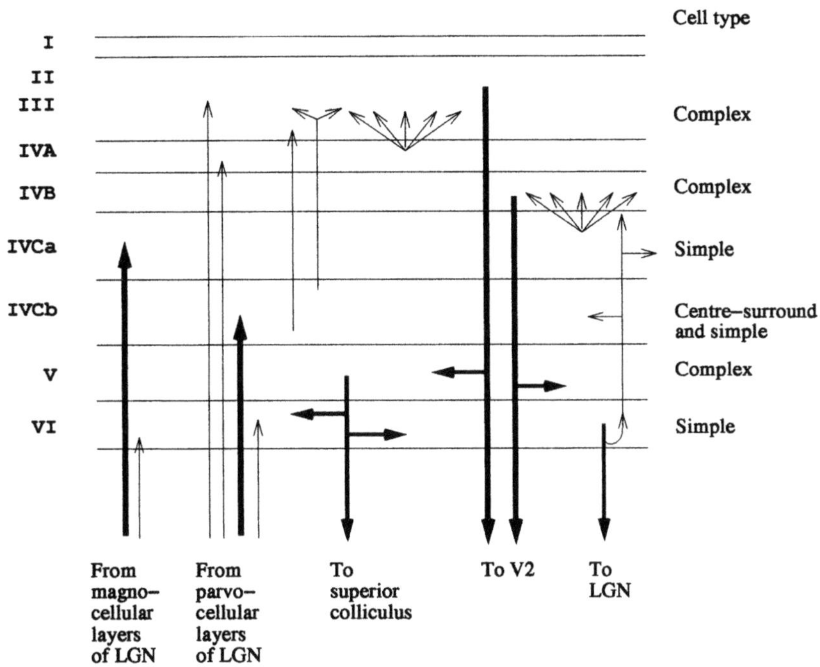

Figure 2.22 Main pattern of interlaminar axon couplings in V1. (After [57].)

2.8.2 Orientation columns

In their study of responses of cells in the visual cortex to different stimuli, Hubel and Wiesel found certain clear, unambiguous patterns in the arrangement of cells with particular responses. It turned out that as the measurement electrode moved through the gray matter, at right angles to the surface, cells encountered had the same orientation of their receptive field axis.

Secondly, it was found that if the electrode penetrated at an angle, the axis of the receptive field orientation would slowly change as the tip of the electrode moved through the cortex. From a large series of experiments in cats and monkeys it was found that:

Neurons with similar receptive field axis orientation are located on top of each other in discrete columns; moving into adjacent columns results in a continuous change in orientation of the receptive field axis.

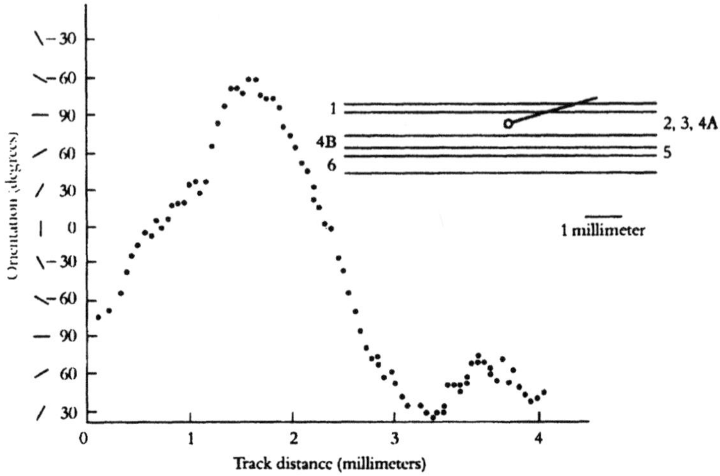

Figure 2.23 Representation of orientation versus track distance for an electrode penetration at an angle. (After [60].)

This situation is displayed in a stylized fashion in Figure 2.23. A very elegant experiment was performed by Gary Blasdel [18] who applied voltage sensitive dye to the striate cortex of a monkey and stimulated the visual field with different orientations, one after the other. The dye causes responding cells to change color and this may be registered with a video camera. The different contributions corresponding to stimulation of different orientations were then combined and represented in different colors to give the orientation color map indicated in Color image 1. We can see a fairly regular change of orientation locally, such that within a local neighborhood a complete set of orientations is available. We can also see an irregularity on the global scale, in that this pattern is completely random.

2.8.3 Ocular dominance columns

Another system of column organization is that for *eye preference*. In the striate cortex of the monkey and the cat, the inputs from the two eyes are at first kept separate, such as they are in the lateral geniculate bodies. The majority of geniculate fibers end in layer IVC. In this particular region, neurons receive input from one eye only. Outside layer IV, simple and complex cells appear to be driven by both eyes, but any particular cell may have a preference for input

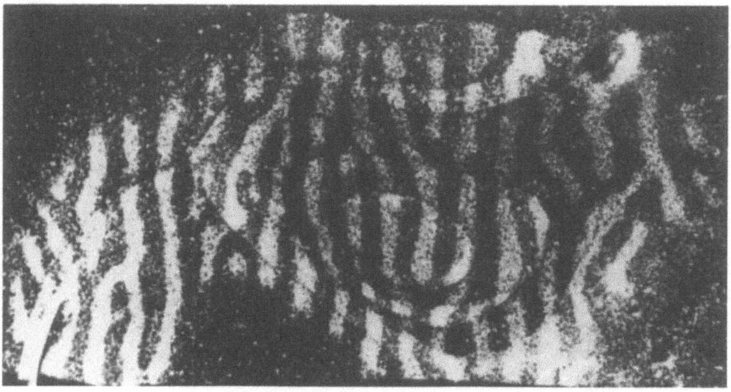

Figure 2.24 Ocular dominance map of the visual cortex. (From [62].)

from one eye, and there is a continuous range of dominance among the cells for one eye or the other. These groupings are called *ocular dominance columns.*

The first discovery and description of ocular dominance columns was made through very tedious microelectrode measurements and was later demonstrated by anatomical methods in the same way as for orientation columns. The pattern is demonstrated for only one of the eyes, while the metabolic activity of different parts of the cortex is registered using radioactive markers. The display of such activity is indicated in Figure 2.24. This pattern is reconstructed in a more stylized way in Figure 2.25. The black regions or stripes in that figure represent, say, the left eye, while the white stripes represent the right eye. There is, again, a fairly regular transition locally between left and right eyes, while there is a near randomness in the global arrangement of this pattern.

We can see a distinct difference between this dominance pattern and the earlier one for orientation columns. The pattern for ocular dominance is basically one-dimensional and consists of linear structures, because there is only a variation along a linear dimension between two values: left and right. The orientation columns, on the other hand, have a more patch-like pattern. The crucial difference is that we here represent orientation, which is a variable that is continuous modulo π. The orientation field cannot be represented continuously with a scalar, but it can be represented by a two-dimensional vector.

The relation between ocular dominance columns and orientation columns is illustrated in Figure 2.26. We have here illustrated orientation columns and

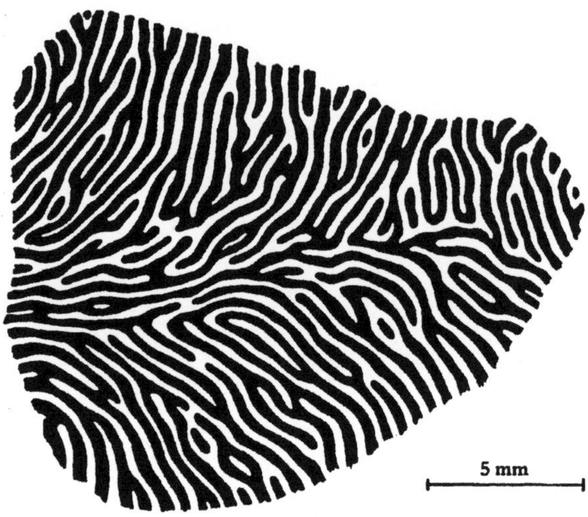

Figure 2.25 Stylized version of ocular dominance map of the visual cortex. (From [88].)

ocular dominance columns having orthogonal orientation to each other to emphasize the fact that they are two essentially independent ordering structures. The columns for eye dominance are found to be about 0.25 - 0.5 mm wide, while the columns for orientation dominance are about 0.15 - 0.35 mm wide.

2.9 POSSIBLE IMPLEMENTATIONS OF FILTERS

We have seen in earlier sections how the different maps of the visual field extend from simple features, such as single receptors in the retina, to more complex features such as orientation detectors in the form of simple cells and complex cells. We now ask ourselves how these processing structures or filters are implemented.

The first conceptual stage going from single receptors to the center surround representation was already discussed in the section on the retina. These center surround representations of the visual field are then carried through the optic nerve over the lateral geniculate nucleus to the visual cortex. A simple cell

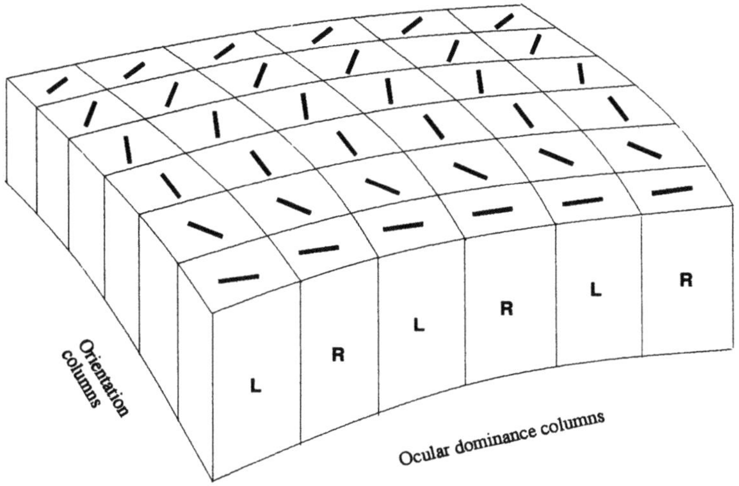

Figure 2.26 Relation between ocular dominance and orientation columns.

receptive field might be implemented as indicated in Figure 2.27. A number of center surround cells with their receptive fields along a line are connected to a cell performing a summation. The receptive field of the simple cell will consist of a long, narrow excitatory region consisting of the individual excitatory regions of the center surround cells. This will be surrounded by an elongated inhibitory region, with components from the inhibitory regions of the individual center cells. In addition to the construction of more complex properties, we can also see an efficiency of computation. A particular center surround cell can, obviously, be an element in several detectors in different positions and orientations.

Complex cells represent the next step in the analysis. They turn out to be the most common cells in the striate cortex, constituting about three quarters of the population. We saw earlier that complex cells have an orientation preference but are less sensitive to the exact position of the stimulus. We here assume that a number of simple cells having the same orientation preference make excitatory synapses with a complex cell. An edge of the right orientation falling anywhere within the region will evoke a response from one or several of the simple cells, leading to activation of the complex cell. In this case, there appears to be an adaptation at the synapses, such that only a moving stimulus will activate the complex cell.

Many complex cells in layers IVCa and IVB respond better to one direction of movement than to the opposite direction. It is not completely known how such directionally selective cells are implemented, or exactly at what level the asymmetric response to direction occurs. Regardless of the exact level of introduction, or the details of implementation, we can assume that *delay* elements consisting of inhibitory interneurons are introduced in a fashion indicated in Figure 2.28.

A cell on the bottom line in Figure 2.28 receives input from a simple cell immediately above it, and a delayed inhibition from another cell whose receptive field is adjacent to that of the first one. Now, if the stimulus moves to the left, the bottom line cell is excited by one of its inputs just as the inhibition arrives from the adjacent cell which has just been activated. The two effects will now cancel, and the cell will not fire. For a movement from left to right, the inhibition will arrive too late to have any effect. A complex direction-selective cell is obtained if the axons of several cells with similar inhibitory directions converge on a single cell. No such structure has been observed in the cortex, but some such arrangement is a necessity. It is known that neurons with very thin axons have a slow conduction speed and these could act as the appropriate delay mechanism.

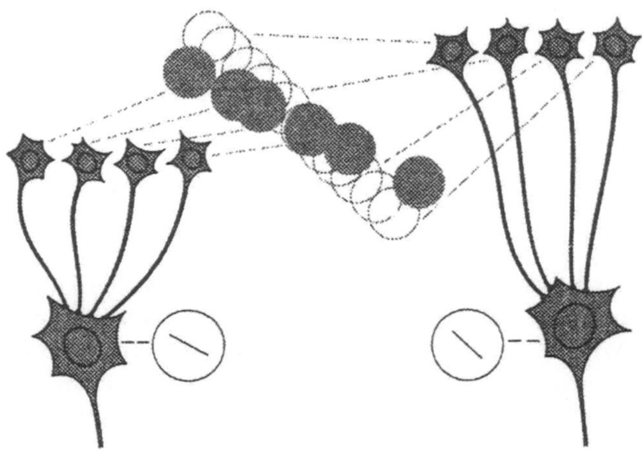

Figure 2.27 Assumed implementation of simple-cell receptive field from center surround receptive fields. The figure shows receptive fields of two adjacent simple cells. (From [57].)

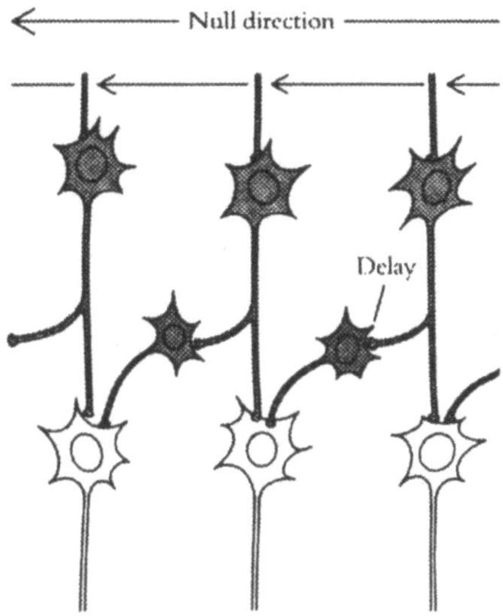

Figure 2.28 Assumed implementation of direction-sensitive cell. (From [57].)

Similar mechanisms can be assumed for higher order features such as complex length-sensitive cells, as in Figure 2.18.

Models similar to those presented above were originally proposed by Hubel and Wiesel [58]. Recent results indicate that these models cannot reproduce the sharp orientation tuning found in experiments. It appears [133] that the response of a cell is modified by inhibitory signals from (1) neighboring cells in other orientation columns that see the same spatial area (cross-orientation inhibition), and (2) neighboring cells in the same orientation column that see other spatial locations.

2.10 CHANNEL ORGANIZATION OF FEATURES

Earlier we saw how the output from a particular simple cell varies depending on the position and orientation of a stimulus. We attributed a particular selectivity curve or tuning curve to the dependence of output upon variation of different stimulus parameters. This appears to be a general characteristic for

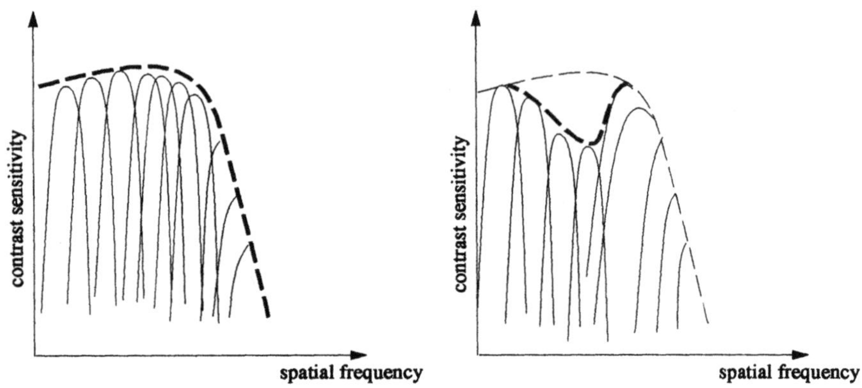

Figure 2.29 Illustration of how the MTF (dashed) could be composed of a number of separate channels. In (b) we can see the effect of adaptation on the sensitivity of channels. (From [16].)

all parameters that have been isolated in the cortex. In this section we discuss some such parameters and their implications.

2.10.1 Receptive field size and frequency response

Measurements indicate that the receptive fields of simple cells and complex cells have not only different orientations but also different sizes; this means they will optimally detect events and objects of different characteristic retinal sizes. The small receptive field is better adjusted to small phenomena, while the large receptive field describes larger objects and phenomena. Directly related to the size of a particular receptive field is its bandpass function in the Fourier domain. The mathematical aspects of this topic will be treated in Chapter 4. Here we briefly describe some of its effects.

The combination of all orientation detectors produces the overall *modulation transfer function*, MTF, of the visual system. Individual simple and complex cells combine into frequency selective channels to form the MTF. This is illustrated in Figure 2.29.

That this is the case is suggested by perceptual adaptation experiments. By *adaptation* we simply mean that cells that have been strongly activated become

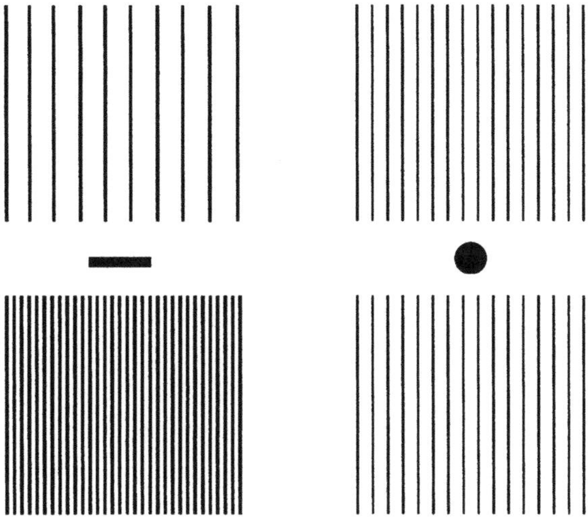

Figure 2.30 Demonstration of spatial frequency adaptation. (From [17].)

fatigued and less responsive. Figure 2.30 is a demonstration of spatial frequency adaptation. If we first look at the two square wave gratings on the right, we notice that the top and bottom ones are of the same spatial frequency and are in fact identical. To adapt to the two gratings on the left, fixate on the black bar between the two and move your fixation point back and forth along the bar for about a minute. Then look at the dot between the identical gratings on the right. The grating at the top should appear to have a higher frequency than the one below. Adaptation to the low frequency pattern on the upper left has caused the appearance of the medium frequency on the upper right to shift toward higher frequencies.

To understand this effect, let us revisit Figure 2.29. When a stimulus is applied at some spatial frequency, the corresponding channel is activated, and, as the stimulus is maintained, this becomes less sensitive because of adaptation. The adjacent channels, on the other hand, will be less affected. If we now apply a stimulus with a spatial frequency which is lower than the adapting frequency, the channels will produce an output which is relatively higher, the further away they are from the frequency of the adapted stimulus. This will result in the interpretation that the stimulus is even lower in frequency. The same reasoning can be applied to stimuli with a spatial frequency higher than the one of the pattern used for adaptation. It is not known whether these channels are well organized and fixed in their center frequency and bandwidth, or if this

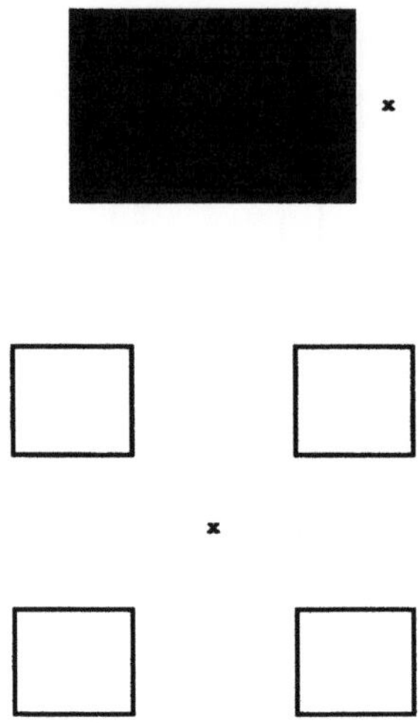

Figure 2.31 A figural after-effect of size. (From [81].)

is an arrangement which has developed locally among detectors with randomly different frequency characteristics. Still, the effect of the channel arrangement and the mechanism that relates adjacent channels to produce a percept, is quite apparent.

2.10.2 Size

Another closely related phenomenon is called *figural after-effect* of size. This is illustrated by Figure 2.31. We start by fixating on the cross next to the black rectangle at the top for a minute or two. After that, we look at the cross between the open squares at the bottom. We should then observe that the two squares on the left side of the figure appear further apart than the two on the right.

2.10.3 Color and orientation

Channels may also be color-selective and be used to associate color features with structural features. If we look at an orange and black vertical grating and a blue-green and black horizontal grating for a minute or two, we adapt to some channel structure relating the two. If we then look at a black and white

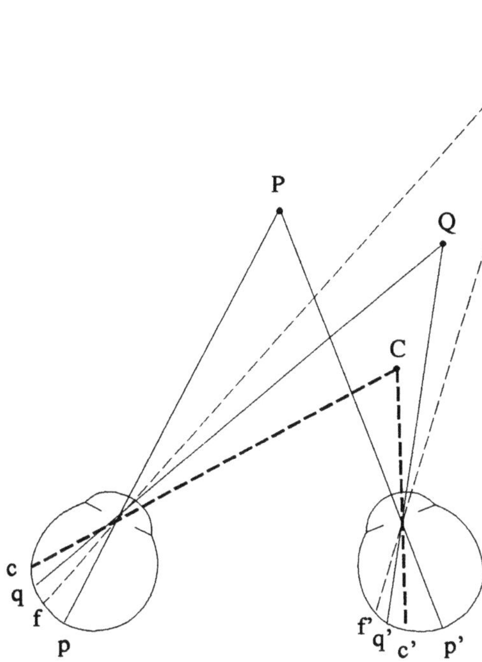

Figure 2.32 Illustration of disparity.

vertical grating, this will appear bluish-green while a black and white *horizontal* grating will appear slightly orange. This is the McCollough effect [100]. It can be extended to even more complicated patterns such as a combination of color and curvature.

2.10.4 Disparity and stereoscopic depth perception

For the following discussion we refer to Figure 2.32.

Fixating the gaze on a point P is equivalent to adjusting the eyes so that the images of the point fall on the foveas p and p'. The images of another point Q, at the same distance from observer, will fall on positions q and q' outside the foveas. Note that the positions are at equal distance from their corresponding fovea.

Now consider a point C closer to the observer than the point of fixation. Its two retinal images c and c' are at unequal distance from their corresponding foveas. Similarly, a point F farther from the observer produces retinal images that are also at unequal distances from their corresponding foveas, but the deviations are in opposite directions from those of a close point. These deviations are referred to as positive and negative binocular *disparities* and we conclude that it is possible to tell whether a point is closer to or farther from the point of fixation by analyzing the corresponding disparity.

Hubel and Wiesel found cortical cells driven by both eyes that were sensitive to disparity, "far cells" responding to lines or edges at distances beyond the point of fixation, and "near cells" to distances closer to the observer. It is thought that such cells are involved in stereoscopic depth perception, *stereopsis*. It is not yet clear whether it is the magno or the parvo system that processes stereopsis [122].

Recent results indicate that there is a whole spectrum of cells sensitive to different degrees of disparity. Lehky and Sejnowski [85] have in a computational study estimated that there must be at least 20 independent channels that carry disparity information.

2.10.5 Channels and coding

From the discussion above we see a pattern that looks like this: a cortical cell is typically sensitive to a specific class of stimuli, such as orientation, and has a characteristic sensitivity function – a peak centered on a particular class member – within this class. We may call this a *channel representation*, or, alternatively, an *interval coding* or *value-unit representation* [5], and it appears to be quite common in the visual cortex. The brain seems to use what is called *population coding*, which means that information is represented by the collective activation of a group or population of cells. As we saw in Section 2.7, orientation selective cells typically have quite wide tuning curves. This means that a stimulus with a distinct orientation will activate many cells, not just the one with optimal receptive field orientation. This coding pattern repeats itself again and again in different parts of the brain and has several important consequences.

First, population coding is *robust*. It does not really matter if a few cells die, since this will only slightly change the activity. Second, it is *efficient*, in that it means that the brain does not have to allocate a separate channel for each possible instance of a class of stimuli. Color vision and disparity are excellent examples.

In color vision we found that a particular color is coded as the combined activity of 3 different receptors, RGB, forming a 3-dimensional vector. Similarily we find that disparity is coded as a vector with some 20 components. What is relevant here is that the sensitivity curves of the different component detectors *overlap.* Each channel has a limited resolution, but by combining the different channels, e.g. in color sensitive center surround cells, we attain *hyperacuity,* a precision that is greater than that of a single channel. If just one color channel were activated at a time, it would have to have a very narrow tuning curve to match human accuracy, but this would require many more channels to cover the whole range of possible colors.

As a further example of population coding, consider the representation of visual space used in the superior colliculus. Positions and directions in space are population coded by n cells that each has a preferred direction $\mathbf{u}_i$ and the actual vector $\mathbf{u}$ being represented is determined by the activations α_i of the cells, so that $\mathbf{u} = \alpha_1\mathbf{u}_1 + \alpha_2\mathbf{u}_2 + \ldots + \alpha_n\mathbf{u}_n$. Our previous simplified description of the function of the superior colliculus omitted the fact that the lower layer receives population coded information about the present *direction of gaze,* as well as signals from the upper layers about the position of an interesting object at a certain position. The gaze direction information is vital since, for a given retinal stimulus position, different muscle distributions must be activated depending on the original direction of gaze. The superior colliculus can combine several populations and convert the distributed description of a position in space into a localized activity in its lower layer. The mechanisms that accomplish this are quite sophisticated [149].

2.11 THE VISUAL PATHWAYS BEYOND THE PRIMARY VISUAL CORTEX

The preceding sections have examined certain fragments of functions of different parts of the visual systems. We may now wonder how all of this is put together into a functional structure that allows for control of muscles to activate responses. It is depressing to acknowledge that we do not really know. However, there are a number of interesting clues that may lead the way to the understanding of biological vision systems and, in addition, suggest an ap-

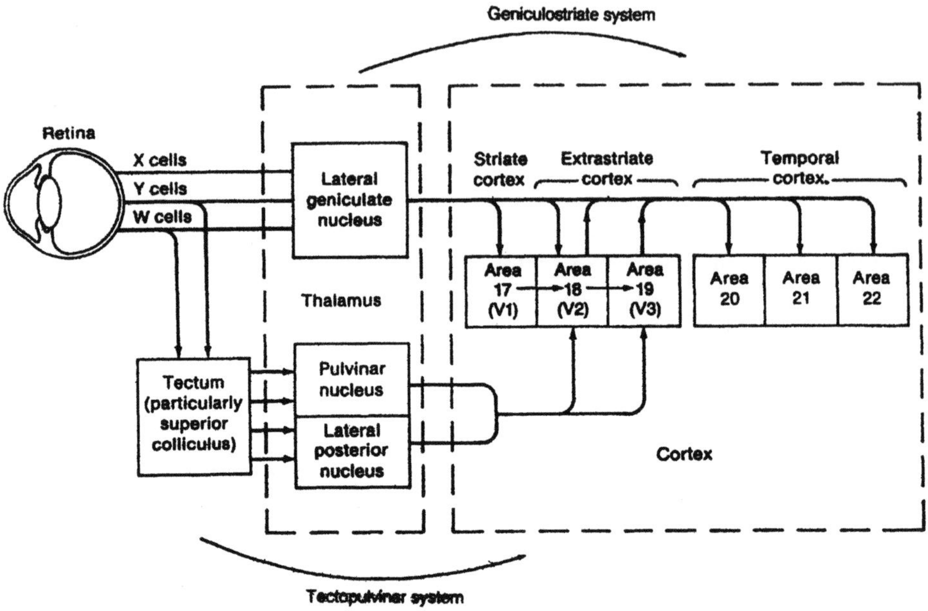

Figure 2.33 An overview of the pathways of the visual system. (From [31].)

proach to the design of efficient technical systems. As discussed earlier, these clues are sufficiently fascinating to warrant a discourse.

In the cortex there are multiple representations of the visual field in the form of different maps. In primates, more than 20 such maps have been found. More than 10 of these are located in the temporal cortex and at least three are in the occipital region, while others are scattered elsewhere in the brain. Some of the lowest-level maps can be seen in Figure 2.33. These maps are supplied by X, Y, and W axons as indicated in the figure.

Why are there so many different maps of the visual field? It is believed that each map is designed to extract some subset of properties from the visual image. It has been found that certain maps deal with lower level primitives, while others can be assumed to handle higher level primitives. Certain maps are relatively specialized for color, others for orientation or movement. It has also been found that the destruction of any one of these cortical maps may produce pronounced disruptions of particular aspects of visual processing, rather than a complete inability to see.

A further focusing on the mechanisms and the interactions between different maps can be made with reference to Figure 2.34. This figure shows five different maps, V1, V2, V3, V4 and V5 (see Figure 2.16). The last four areas are sometimes referred to as *secondary visual cortex.* There appear to be four different perceptual pathways or description systems. One pathway, which deals with motion, is located in area V5. One deals with the interpretation of color and is located in V4. There are two systems which deal with form. One of these uses color and is as well located in area V4. The other one, which is often denoted as dealing with "dynamic form" – visual perception of complex movements such as dance – is located in V3. The information to these areas is supplied by areas V1 and V2, and we will take a further look at some of the characteristics of these areas.

Some axons from the parvocellular layers of the LGN synapse with cells in layers II and III of V1. These cells tend to be grouped in local clusters which in certain staining procedures appear as more heavily stained *blobs*, separated from each other by *interblob* regions. The blob-cells are not sensitive to orientation, but appear to be wavelength selective. The blob regions are highly active when the visual input consists of stationary, colored patterns. Axons from layers IVCb and IVA synapse with cells in the interblob regions of layers 2 and 3. These cells are orientation sensitive complex cells.

The axons of Y-cells, from the two magnocellular layers of the LGN, synapse with cells in sublayer IVCa. The axons of these cells in turn synapse with cells in the neighboring IVB layer. We know from earlier sections that the Y-cells give transient responses to stimuli, and that the responses are essentially independent of the color of the stimulus. Some cells in layer IVB project to areas V3 and V5. In particular the cells that connect to V5 are clustered into small patches which are surrounded by cells connected to other visual areas, a system reminiscent of the blob-interblob system in the upper layers.

Area V2 is characterized morphologically by the way it stains with respect to the same metabolic enzyme that reveals the blobs in V1. In V2 there is a very regular pattern of alternate thick and thin stripes, separated from each other by more lightly staining interstripes. It has been found that cells selective for color are located in the thin stripe regions, while cells selective for directional motion are found in the thick stripe regions. There are in addition cells sensitive to form, which are distributed in both the thick stripes and the interstripe regions.

The major observation to make from this is that there is a remarkable segregation of functions between different subsystems. Still, these are somehow integrated to form a perceptually manageable and fairly complete representa-

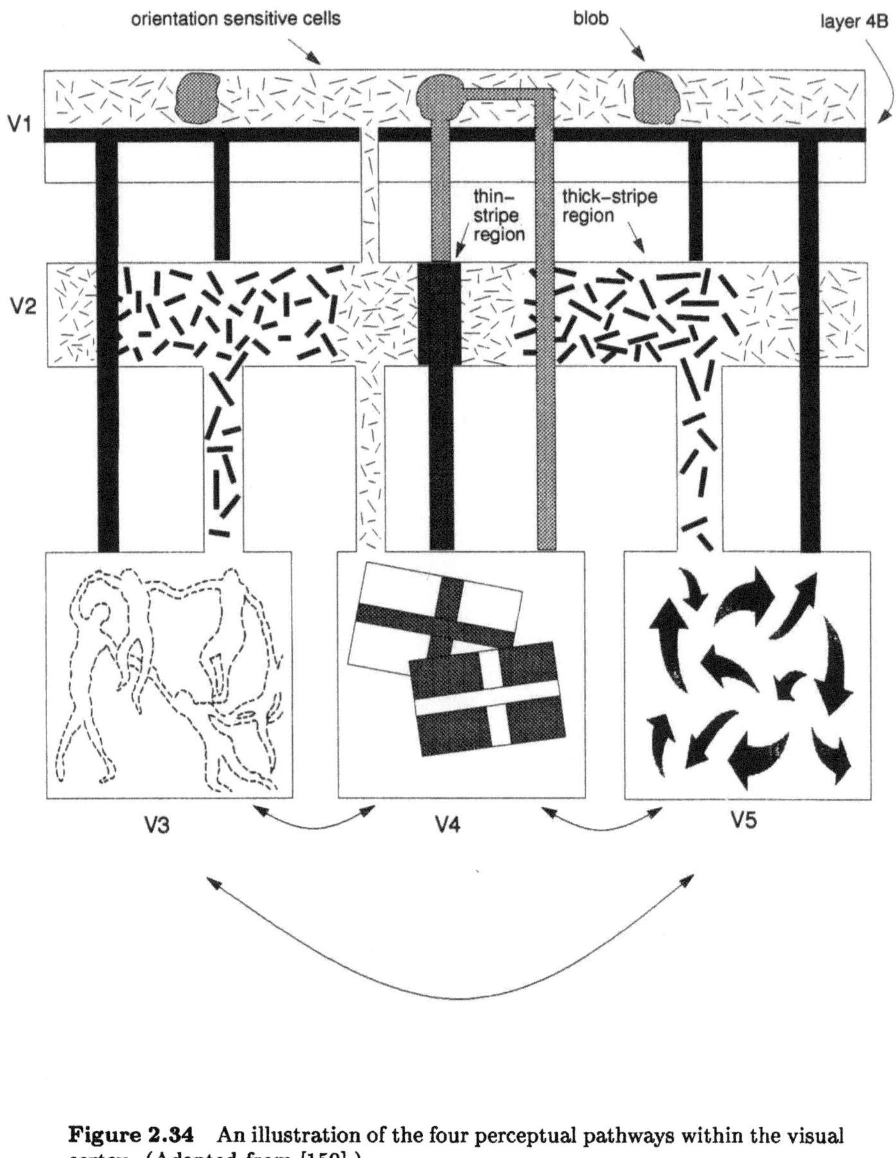

Figure 2.34 An illustration of the four perceptual pathways within the visual cortex. (Adapted from [150].)

tion. The segregation becomes very apparent in patients with lesions or other forms of damage to any of the visual areas. A lesion in area V4 may, for example, lead to a case where a patient sees only shades of gray. This syndrome is different from the usual varieties of color blindness, in that patients not only fail to see the world in color but they cannot even recall the impression of color from a time before the lesion formed. This leads one to believe that area V4 is not only involved in the processing of color from the eyes but is also involved in the visualization of recalled situations from memory. However, the patient's knowledge of form, depth and motion remains intact, unless other areas are damaged as well.

In the same manner, a lesion in area V5 may result in the loss of appreciation of motion; objects at rest are perfectly visible, but motion causes them to vanish. Again, other attributes of vision remain unaffected. This is another indication of the specificity of perceptual pathways.

It is difficult to understand how the integration of information from all these pathways is achieved. It appears likely, however, that the higher level areas, V3, V4 and V5, isolate properties that are deemed to be of major importance in the visual field. These may, for example, be regions of consistent motion corresponding to a particular object. These specialized areas signal the appearance of an event of interest to the more detailed representations located in areas V1 and V2. This is accomplished by the use of *reentrant*, or feedback, paths.

The reentrant inputs to areas V1 and V2 are structurally different from the forward connections leading from V1 and V2 to the different specialized areas. The forward projections are convergent, with inputs from rather wide regions ending up in smaller regions. The return, or feedback, projections are diffuse and fairly non-specific. They contact not only the cells that send forward projections to their regions, but also other cells in the vicinity. As an example, the return input from V5 to layer 4B of V1 is diffuse, and reaches all types of cells in that layer, including the ones that project into area V3.

Reentrant, or feedback, connections allowing information to flow both ways between different areas, are considered essential for resolving conflicts between cells with different capabilities that respond to the same stimulus. An example of the resolution of such conflicts has been studied using the *Kanitza triangle* (see Figure 2.35) as stimulus, and measuring the corresponding responses of cells in V1 and V2. In this illusion one can observe a triangle defined by an incomplete structure. The brain will then create lines defining the central, white triangle, where none are existent.

Figure 2.35 The Kanitza triangle containing illusory contours.

It has been demonstrated that the form selective cells in area V1 do not respond to the illusion, and do not indicate that a line is present. The V2 cells, on the other hand, do respond to the illusion indicating the presence of a line. The V2 cells receive their inputs from V1, but the V2 cells have larger receptive fields and more specialized functions.

The integration quite likely proceeds in multiple stages. It appears not to require the convergence of output into any particular master area; this would imply another, more sophisticated system overseeing the output of this area. The integration seems to be a gradual process encompassing perception as well as comprehension of the visual world, rather than being postponed until all the visual areas have completed their individual operations.

The reentrant, or feedback, system is assumed to serve three purposes. It unites and synchronizes the signals for form and motion found in different visual pathways. It can isolate and refer the coarse information about motion back to an area with a more detailed map. It integrates form and dynamics-in-form together with color information.

A major observation regarding the organization of the visual system is its use of different pathways for different types of information. Although details are not known, the system appears to produce various relatively "free-floating" attribute maps with some suitable resolution. It then relies on some ability to combine features that relate to the same phenomenon. This is different from our usual conceptual view in computer vision, where we like to make an early, comprehensive description of the phenomenon, and label or associate it with various attributes such as orientation, color, length, etc.

One reason for the strategy of the visual system may be that linkage is a combinatorially and computationally very expensive process. While the generation of features is a relatively straightforward procedure, the generation of linkages is not. Consequently the linkage may be done on demand, driven by suitable focus-of-attention mechanisms.

Beyond the areas mentioned above, the visual system seems to divide into two major projections [103], one upon the *inferior temporal cortex* (IT), the other upon the *posterior parietal cortex* (PP), see Figure 2.16. It is found that area V4 sends axons mainly, but not at all exclusively, to IT, whereas axons from areas V3 and V5 favor PP but do not exclude IT.

The PP is, as opposed to the IT and the earlier visual areas, *multimodal.* This means that it receives information from different sensory pathways, in this case the visual and somatosensory systems. Measurements of cell responses and experiments with patients with lesions in PP areas show that it is involved in the representation of *spatial position* of objects. Patients with lesions in PP may be perfectly able to describe and name objects in their visual field, but unable to determine their positions. PP is consequently a likely locus for the WHERE-function.

Studies show that IT is crucial for *object recognition*, or the WHAT-function. Perett, Rolls and Caan [110] found cells in IT areas of monkeys with a preference for *faces* of people or monkeys. Responses were unaffected by changes in scale and color. Some cells were sensitive to faces of specific individuals, in the sense that their responses were different for different faces. It is known that lesions in the IT area may lead to an inability to distinguish between individual faces, between different car models, etc.

IT and PP have strong connections with the frontal lobe of the cerebral cortex. This part of the brain is responsible for the highest cognitive functions and for the generation of high-level motor programs. Much is known about the connection to and implementation of motor functions [25], but this is outside the scope of this book.

3

LOW LEVEL OPERATIONS

3.1 INFORMATION REPRESENTATION

Image computers store and process large chunks of data representing image information in various forms. This is usually done in the form of arrays, which are contiguous areas of memory with data elements laid out one after the other. The storage layout of arrays simplifies data access when all items are elements in regular arrays. Then the program merely needs to take the position of the element and multiply the y-coordinate by the row length, and to add the x-coordinate and the start address of the array. For images or data arrays with a fixed size, this is a convenient representation.

In most cases of processing of structural or spatial information, such as images, we describe features in relation to the spatial position in the image. It has been well established that spatial position is a necessary parameter for successful interpretation of image information. This is the reason why descriptors such as global Fourier transforms never have been of much interest in image analysis.

Representation of features in terms of spatial coordinates is sensible at low levels, such as for information about local orientation. A hierarchical processing structure will have a succession of representations, going from low level to higher level features. One property which seems to characterize higher level features is that they become increasingly sparse. This implies that we cannot expect most of the points in the feature plane to have a non-zero value. This in turn implies that most of the storage space allocated for high level features will be empty. We have a problem if a program needs to store data having a variable number of elements, such as might describe events that do not represent an even quantisation of physical space or any other real or conceptual space. For example, we may not want to waste storage space on events of type `Nothing happening` in every element over a large region. The problem of how to allocate space on demand is a crucial issue, and is of major interest in the field of Neural Networks.

While position is an important property for features at lower levels, this may be less so for higher levels, or, at least, there may be no reason to arrange feature elements in terms of spatial coordinates. There are indications in the visual system that relations between events may only be taken into account within a limited size window. This implies, for example, that we do not tend to relate or link points within the same class if they are located far away from each other. The distance over which such a linkage is made, appears to be dependent upon the level of the feature in question. At higher levels features tend to be more global, which means that they are linked or associated over greater distances of the image.

In the coming sections we will discuss different types of information representation for low level processing. There are many aspects of the choice of information representation, which are crucial to the design of processing systems. In that pursuit we will also look at biological systems for hints of useful mechanisms. Although the goal is not primarily to establish visual system models, the discussion may occasionally deal with the interpretation of hints from biological systems.

3.1.1 Scalar representation

A common type of image is the *monochrome image*, which is a two-dimensional light intensity function $f(\xi_1, \xi_2)$, typically representating an ordinary gray scale photographic image. We will here as well as in the remaining sections use ξ_1 and ξ_2, rather than x and y, to denote global spatial coordinates. The value of f at a point (ξ_1, ξ_2) can be proportional to the brightness or *gray level* of the image at that point. The value of f can also be some function of the density of the image. A pixel is typically represented by a byte, which provides a resolution of 256 levels. For many applications, e.g. X-ray anlysis, it is necessary to have a resolution of up to 12 bits per pixel.

An alternative, simpler representation is obtained by thresholding the gray scale image, to describe pixels above the threshold as $f = 1$ and pixels below the threshold as $f = 0$. This produces a *binary image*. This is a very compact representation which is useful in the representation of topological properties of objects that exhibit a good separation from the background. There is an extensive literature on processing of binary images [123], but we will not consider this topic here.

3.1.2 Representation of multi-component features in 2-D space

In most cases we want to represent features containing several components. A typical example is a color image, which in RGB representation contains three components. In this case we have an image described by a vector field, where each vector has three components. In spite of their dominance in research and literature, scalar images are in fact a special case. Representations must in the general case contain several components, which is why we in this book will deal extensively with the consequences of using multi-component image representations.

The Curse of Multi-dimensionality

Something that makes a profound difference in the representation and use of vector features in comparison with scalar features, is what is often referred to as *The Curse of Multi-dimensionality.* The basis for this is the difficulty of ordering items in a space with dimensionality higher than one. We illustrate this problem in Figure 3.1. In the left part of the figure are three items, t_1, t_2 and t_3, arranged along a single dimension. In this case the ordering is unambiguous. Item t_2 is unambiguously located in the interval between t_1 and t_3. Item t_3 is located outside the interval between t_1 and t_2, etc.

In the right part of Figure 3.1 are three items, s_1, s_2 and s_3, arranged in a two-dimensional space. The ordering between these items is only a matter of convention. If we start at the upper left corner and scan horizontally, we obtain a sequence s_1, s_2 and s_3. This scanning sequence and the ordering derived are not unambiguous, however, but a question of convention. If we instead start at the upper right corner and scan vertically from right to left, we obtain a sequence s_3, s_2, s_1. We can see that the ordering sequence is as dependent upon the rather arbitrary conventions we set up as upon the spatial arrangements of items in the two-dimensional plane.

This situation is of fundamental importance, and it has a number of effects in related fields. It is the reason that polynomials of two or more variables

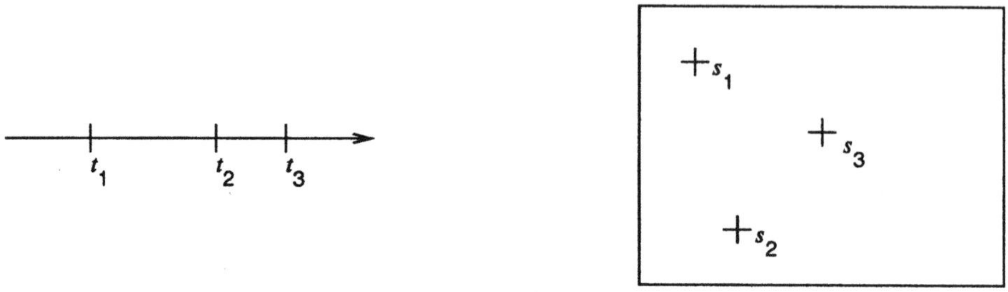

Figure 3.1 Ordering of events along one dimension (left) and along two dimensions (right).

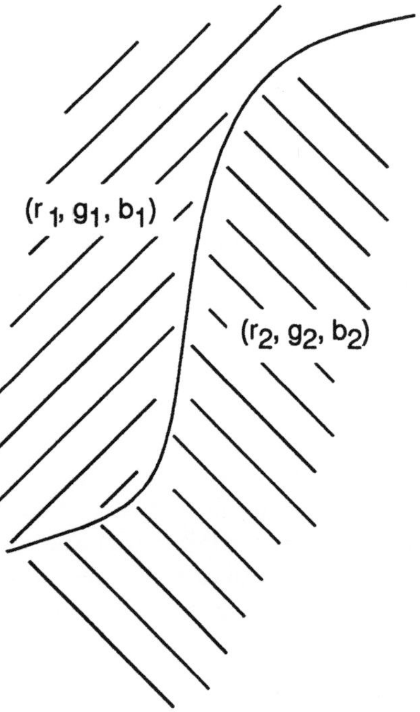

Figure 3.2 Boundary separating two vector fields.

cannot, in general be factored. It is the reason why stability in recursive filters is very difficult to determine for filters with a dimensionality of two or higher. It is also the reason why grammars or rule-based structures are inherently one-dimensional, although their effects may span a higher dimensionality as a result of combination, concatenation or symmetric application.

The center of our interest will be the consequences of the curse of multi-dimensionality to the characterization of boundaries between vector fields. Figure 3.2 shows a boundary section between two vector fields. The vector fields can represent two different colors, in which case the notations easily associate to the color components red, green and blue. In this case it is not possible to assign a vector direction across the boundary, i.e. in the way we can obtain the gradient for a scalar field. For a scalar field (i.e. a gray scale image) we have a one-dimensional variable, and we can easily set up an unambiguous notation for a vector being perpendicular to the separating boundary, going from darker to brighter. In the color case, where we have no fewer than three components and, consequently, a three-dimensional variable space, no such convention is possible. Consequently, we can not find an unambiguous direction of a vector perpendicular to the separating boundary between the two regions. What we can find is a sufficiently useful descriptor of the boundary orientation, such as the double angle descriptor discussed in Chapter 1.

The description of boundaries between vector fields is a much more fundamental issue than the description of boundaries between scalar fields. Scalar fields are rare anomalies, while vector fields are the normal case for description of colors, textures and, in general, the combination of multiple features to be used for reliable description.

3.1.3 Complex valued representation

One very useful case of multi-component representations contains two components. As it has many of the properties common to non-scalar images, but a minimal complexity, it will discussed at some length. It furthermore can be expressed as a complex valued representation, which is often advantageous. A pixel is described by a complex number $\mathbf{p} = pe^{i\alpha}$, where $p = |\mathbf{p}|$ and $\alpha = \arg[\mathbf{p}]$. It is represented in component form as

$$\mathbf{p} = p\cos\alpha + i\,p\sin\alpha \tag{3.1}$$

Although a complex number has two components, it is a different mathematical object from a vector with two components. For complex numbers there is a particular set of computation rules defining different operations, and the relative simplicity of these makes the representation attractive.

For generality of representation a given input image is considered to be complex valued. Ordinary gray scale scalar images can be considered as a special case of the complex valued representation, where the gray scale value is reflected in one of the parameters of the complex number. An obvious variety is that the gray scale is reflected in the magnitude value of the representation, with the argument set to zero. If the gray scale value is f for a pixel at position (ξ_1, ξ_2), the corresponding complex valued pixel is defined

$$\mathbf{p}(\xi_1, \xi_2) = p(\xi_1, \xi_2) e^{i\alpha(\xi_1, \xi_2))} = f(\xi_1, \xi_2) e^{i0}. \tag{3.2}$$

An alternative representation of scalar images in complex valued notation is:

$$\mathbf{p}(\xi_1, \xi_2) = p(\xi_1, \xi_2) e^{i\alpha(\xi_1, \xi_2)} = p(\xi_1, \xi_2) e^{icf(\xi_1, \xi_2)}. \tag{3.3}$$

In this case the gray scale value is reflected in the argument of the variable $\mathbf{p}$. This is more consistent with the rules that have been prescribed for a good representation to use in a hierarchy. It represents the class membership of a phenomenon as an argument or an angle. This allows easily updatable descriptors as we will show later. The constant c is a scale factor giving the correspondence between intensity value and argument. $p(\xi_1, \xi_2)$ reflects the confidence associated with a particular statement of value of the argument $cf(\xi_1, \xi_2)$. $p(\xi_1, \xi_2)$ is usually set to a fixed value at the conversion from a density variable to a complex variable. This value may then be updated and modified in the processing, e.g. as a function of an enhancement.

3.2 COMPLEX VALUED CONVOLUTION FUNCTIONS

Many image processing operations employ differentiation in some form. Typically this is implemented as a convolution. In this section we will discuss the use of a particular set of complex valued weighting functions.

Let us assume a two-dimensional image, containing complex valued image elements

$$\mathbf{p}(\boldsymbol{\xi}) = p(\boldsymbol{\xi})\, e^{i\alpha(\boldsymbol{\xi})}. \tag{3.4}$$

In order to compute descriptors of local structure, the image is convolved with a set of operator kernels. These kernels typically implement line and edge detectors in four different orientations as suggested in Chapter 1. Let N denote the number of complex valued kernels with real and imaginary parts

$$g_k^{re}(\boldsymbol{\xi}) \qquad \text{and} \qquad g_k^{im}(\boldsymbol{\xi}) \tag{3.5}$$

for $k = 1, 2, \ldots, N$. Consequently, we define the complex valued kernels as

$$\mathbf{g}_k(\boldsymbol{\xi}) = g_k^{re}(\boldsymbol{\xi}) + i\, g_k^{im}(\boldsymbol{\xi}). \tag{3.6}$$

By forming a complex kernel $\mathbf{g}_k$ of the real kernels g_k^{re} and g_k^{im} it is possible to detect orientation regardless of whether the input has a structure similar to a line kernel or an edge kernel or any combination thereof.

The convolution between the image described by the pixel values $\mathbf{p}$ and the filter kernel $\mathbf{g}_k$ is written

$$\mathbf{q}_k(\boldsymbol{\xi}) = \sum_{\mathbf{x}} \mathbf{p}(\boldsymbol{\xi} - \mathbf{x})\, \mathbf{g}_k(\mathbf{x}). \tag{3.7}$$

The summation is to be taken over all points of the kernel $\mathbf{g}_k$, which in any practical implementation must be finite. Consequently, for each point $\boldsymbol{\xi}$ we can consider a neighborhood $\mathcal{N}$ that is covered by the kernel $\mathbf{g}_k$ and represent $\mathbf{q}_k$ in that point as a linear combination or scalar product between the elements of $\mathbf{p}(\mathbf{x})$ for $\mathbf{x} \in \mathcal{N}$ and the elements of $\mathbf{g}_k$.

From the complex product sum $\mathbf{q}_k$, we can compute a magnitude and argument representation

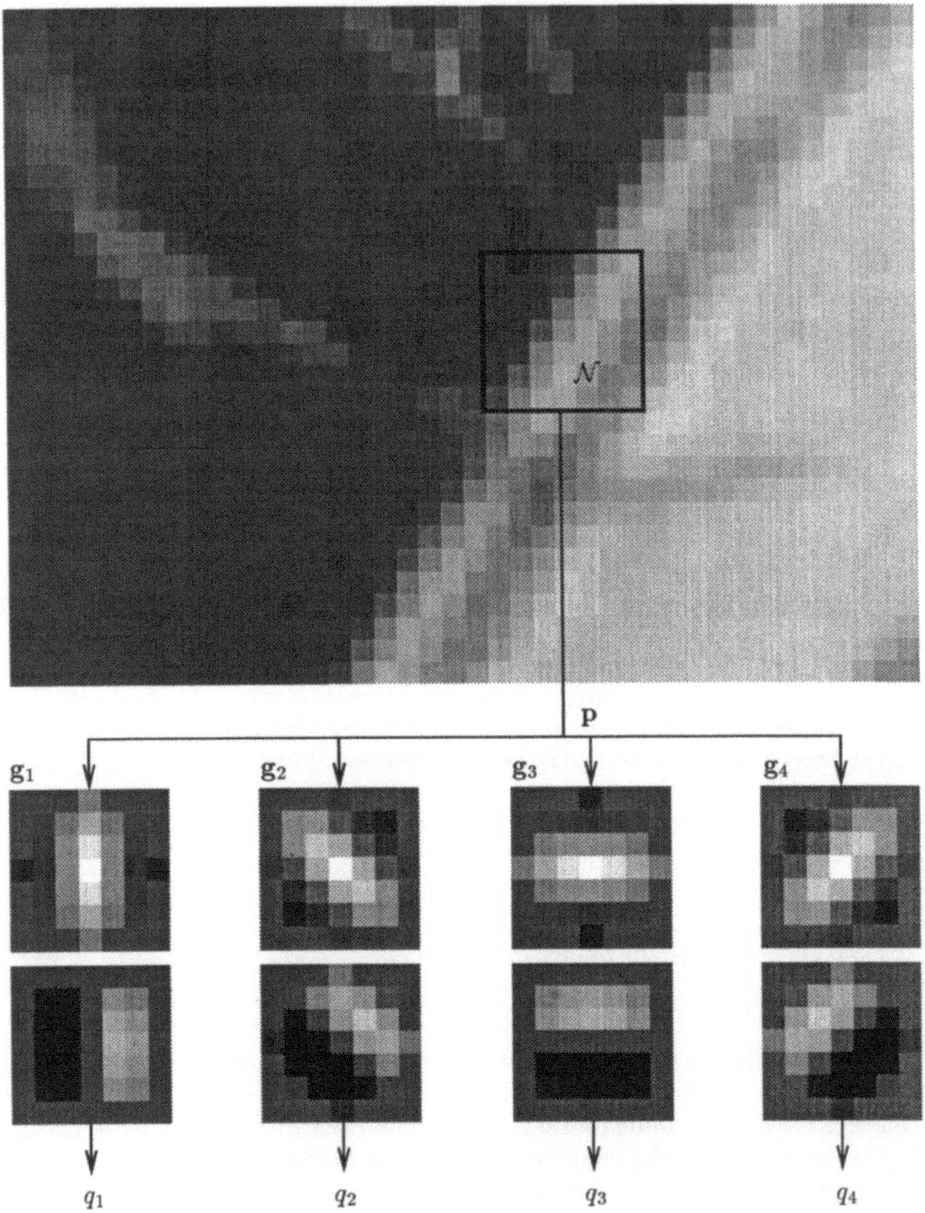

Figure 3.3 Illustration of the generation of feature components representing edge and line content within a neighborhood.

$$\begin{cases} q_k &= |\mathbf{q}_k| \\ \theta_k &= \arg[\mathbf{q}_k] \end{cases} \tag{3.9}$$

The relation between line and edge components can be represented as magnitude and phase according to Figure 1.10 in Chapter 1. The magnitude q_k is an appropriate measure for the combination of line and edge content, as the corresponding kernels are orthogonal and the individual components consequently can be added geometrically. We assume that edge and line components are weighted sine and cosine components.

It should be apparent that from each one of the complex kernels, in a particular orientation labeled k, a feature component, q_k, is produced. For a particular neighborhood a feature vector $\mathbf{q}$ is obtained:

$$\mathbf{q} = \begin{pmatrix} q_1 \\ \vdots \\ q_N \end{pmatrix}. \tag{3.9}$$

For a typical case $N = 4$, as we are considering four different orientations of the otherwise identical kernels.

The situation is illustrated in an intuitive way in Figure 3.3, where a particular neighborhood is matched by a complex valued combination of line and edge detectors in four different orientations producing four different values of q_k. These values q_k represent the magnitude of line and edge content in a particular orientation as indicated.

3.3 COMPACT REPRESENTATION

In general, all of the magnitude components, q_k, derived from a convolution will be nonzero, as the neighborhood $\mathcal{N}$ moves over the image and matching is performed with the kernels. A representation of the convolution result would necessarily include all the components, equivalent to the full n-dimensional feature vector $\mathbf{q}$ of the state within the neighborhood.

It turns out that there is generally a correlation between components such that not all features q_k are not likely to have large values simultaneously within a given neighborhood. This can be shown to be the case if we have a sufficiently simple model. For low-level processing, this model restriction is formulated as

- **Local dominance:** *Within a neighborhood of limited size the content can be represented by the dominant component only, i.e. the variation of gray scale values in the dominant direction.*
- **Local one-dimensionality:** *It is unlikely that lines or edges that are perpendicular will occur simultaneously within a particular small neighborhood.*

The assumption that two features are not likely to occur simultaneously implies that they exhibit a negative correlation. Consequently two such features can be represented along only one dimension. If we represent the occurrence of one of the features with a positive value, then we can represent the other feature with a negative value. In consequence we state that two such features q_1 and q_3 can be sufficiently well represented by one variable z_1:

$$z_1 = q_1 - q_3. \tag{3.10}$$

A vertical edge or line produces a positive value of z_1, while a horizontal edge or line produces a negative z_1.

We can formulate the total output as a complex number, $\mathbf{z}$, as a function of all features q_k for $N = 4$:

$$\mathbf{z} = z_1 + iz_2 = (q_1 - q_3) + i(q_2 - q_4) \tag{3.11}$$

An alternative way of writing this is

$$\mathbf{z} = \sum_{k=1}^{4} q_k e^{i(k-1)\frac{\pi}{2}}. \tag{3.12}$$

This formulation is illustrated in the block diagram in Figure 3.4.

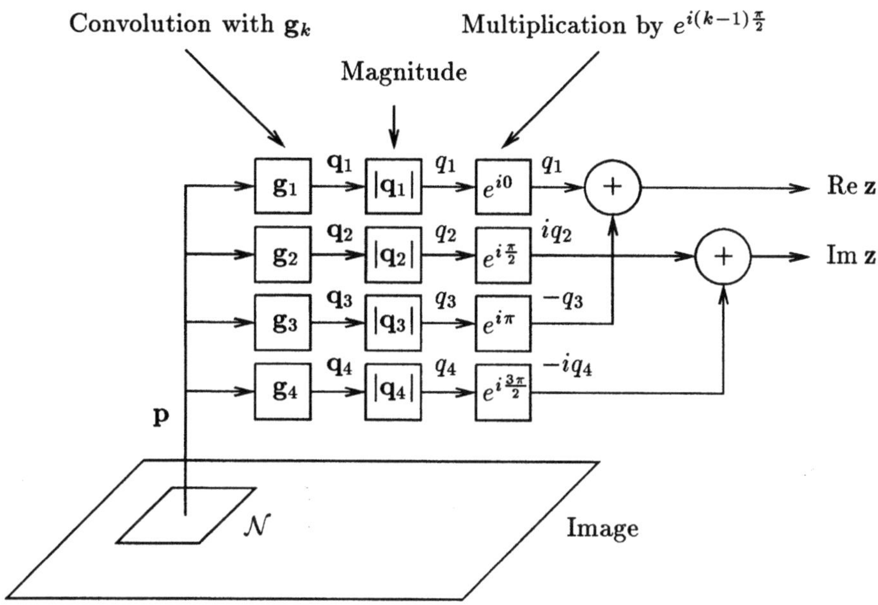

Figure 3.4 Block diagram of the generation of complex valued output $\mathbf{z}$ from feature components q_k.

Figure 3.5 illustrates the relationship between dominant orientation of a structure within a neighborhood, the kernel set optimally tuned for this orientation of structure, and direction of the output complex value produced for such a structure.

FIgure Figure 3.6 gives examples of the outputs from processing of various signals using the filters of in Figure 3.3. Note that input 1, presenting one singular dominant orientation, produces an orientation output with relatively large margnitude. Input 2, which shows one dominant orientation, produces a lower magnitude than input 1 due to noise which decreases the assumption of one-dimensionality. Input 3, presenting two equally strong and perpendicular orientations, produces a zero output which indicates that there is no dominant orientation. This is also the case for input 4, which in fact presents no dominant orientation. The arguments of complex numbers $\mathbf{z}$ indicate the preference orientations of input structures. The arguments themselves should, however, be interpreted as indicated in Figure 3.10.

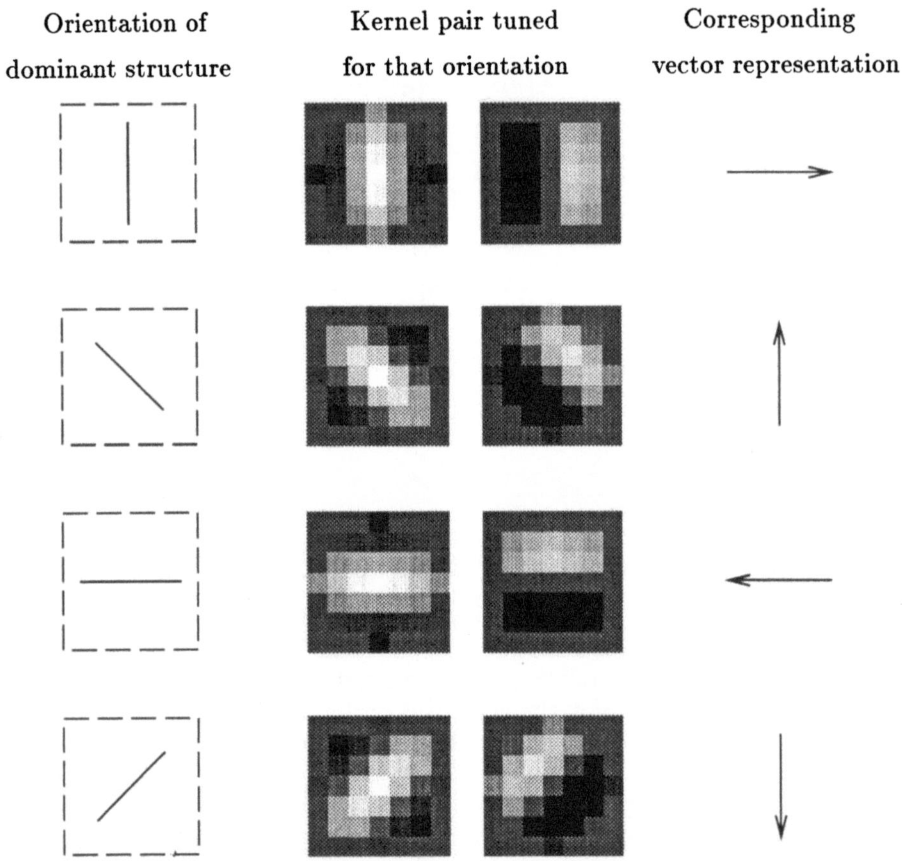

Figure 3.5 Illustration of the correspondence between the orientation of the dominant structure within the neighborhood, kernel set giving maximum output, and the direction of the output vector.

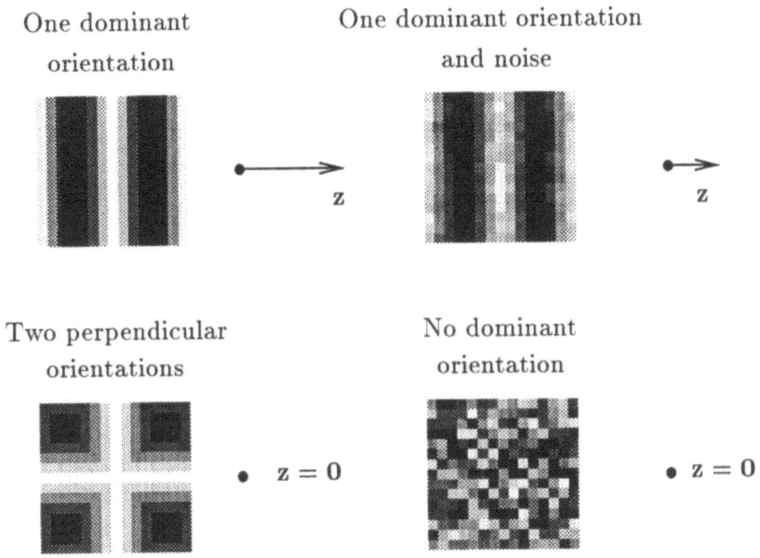

Figure 3.6 Examples of output from line and edge kernels combined for different orientations.

3.4 EXAMPLES OF PROCESSING

We would conceptually consider the two halves of the left image in Figure 3.7 maximally different in terms of line orientation, as the lines are at right angles. According to the preceding discussion, they should be represented by vectors with opposite directions. This is illustrated by the right image, where the two regions have opposing vectors in accordance with the conventions illustrated in Figure 3.5. In Figure 3.8 an original image, containing a circular region, is shown together with a stylized processing result image showing vector directions of the corresponding transform.

Actually computing a local orientation angle estimate can be done with four kernel pairs having evenly distributed symmetry axes. An example of the eight real kernels which constitute four quadrature filter pairs are shown in Figure 3.5. The kernel size is 7×7 elements. The angle estimate is produced by combining the outputs from each kernel pair as described earlier.

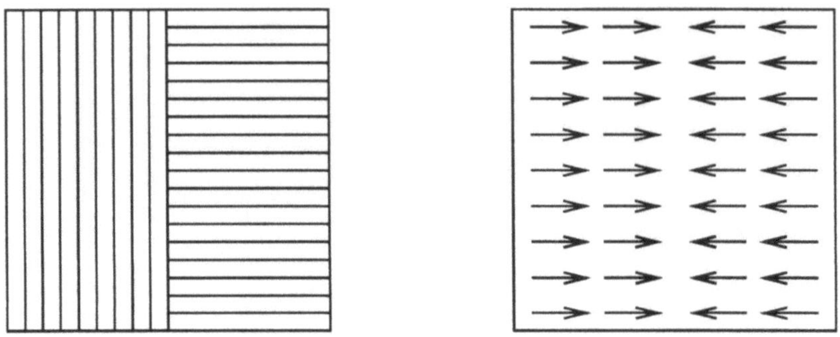

Figure 3.7 Example of the first level transform output upon a stylized image.

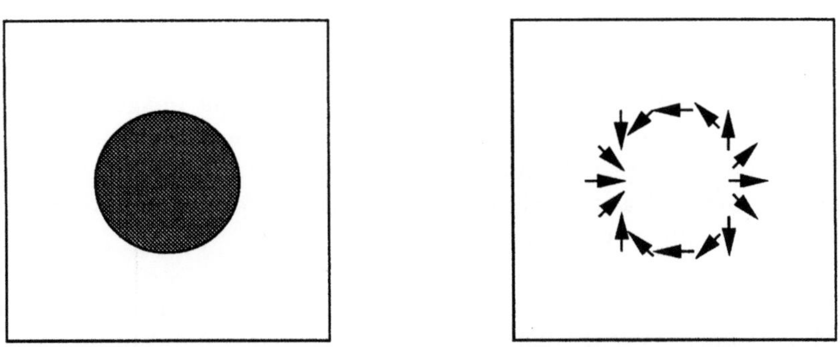

Figure 3.8 Illustration of the relationship between an original image and its vector transform.

To visualize the effect of this type of operation, we use a test pattern that is circularly symmetric, shown in the left part of Figure 3.9. The four kernel pairs operate upon the test pattern, each giving an estimate for the corresponding orientation, shown in the right part of Figure 3.9. The estimates are scalar, i.e. the brightness indicates the magnitude of the kernel outputs. The direction of the vector, on the other hand, is defined by the main orientation of the corresponding kernel pair. Top left of Color image 3 shows the result after a combinaton of the outputs from the four kernel pairs. Here the complex numbers or vectors are represented by colors such that vertical orientation is represented by green, horisontal orientation by red, etc.

3.4.1 Cascaded operations

We have seen in the preceding sections how combinations of differentiating kernels can be used to detect and represent variations in a gray scale image. It is now possible to use the same type of operations upon the vector field resulting from the previous operation. Figure 3.7 shows a stylized image with two different regions. The first level transform illustrated in the right image is complex valued. We can continue with another operation of the same type

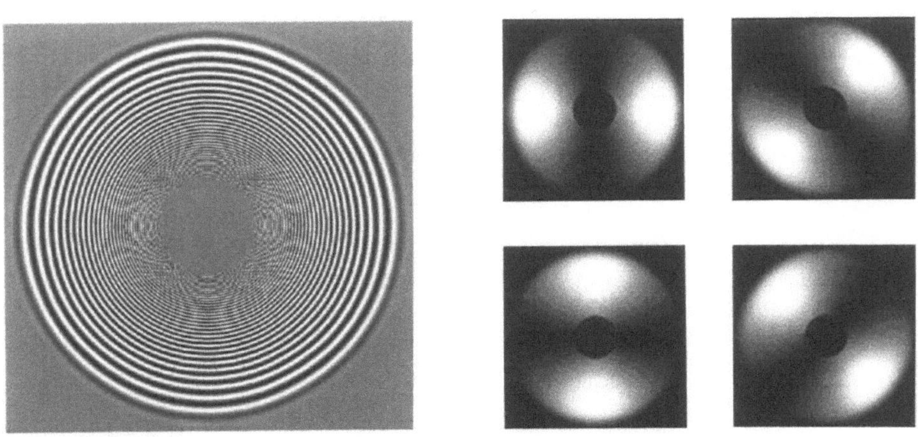

Figure 3.9 Left: test pattern containing spatial frequencies in a range 0.5 – 0.05 cycles per pixel for all orientations. Right: the four filter outputs q_k for the test image.

upon the first-level transform image to obtain a second-level transform. This comes out as shown in Chapter 1, Figure 1.15c.

Color image 5shows an example of two levels of processing upon a texture image, with two regions exhibiting different orientations. The left hand top is the original image of a calf skin from Brodatz' book on textures [24], where a part of the pattern has been cut out and rotated. The right hand top part is the first order transform output. The lower left part displays the same vector angle as the first order transform, but with the magnitude set to 1. The lower right part displays the second order transform delineating the two regions.

This can be seen as a simple, stylized example of texture segmentation. The first level transform gives a description of the texture field properties. The second level transform gives a delineation between the two regions having different properties. We will see more of multi-level operations in the chapters to follow.

3.5 MORE ON COMPATIBILITY REPRESENTATION

We will extend the discussion on compatibility representation that began in Chapter 1, and repeat the very significant Figure 3.10. The complex **z**-plane is represented with the principal directions of q_k indicated. We can see the angular positions of outputs resulting from vertical, horizontal and the two intermediary orientations of structures. There are several observations that can be made from the figure and the expressions given.

1. The representation gives an unambiguous description of the orientation of a boundary, which works as well for vector images. This is accomplished by scaling the argument ot the orientation of a line or edge by a factor of two, to produce the argument of the orientation description vector.

2. The representation described can be viewed as a data compression, where four features are condensed into two. With reference to Figure 3.10 we can view the angle of **z** as pointing out the dominant class membership, and the magnitude of **z** indicating the confidence or certainty in the class membership identified. Class memberships symmetrically located about

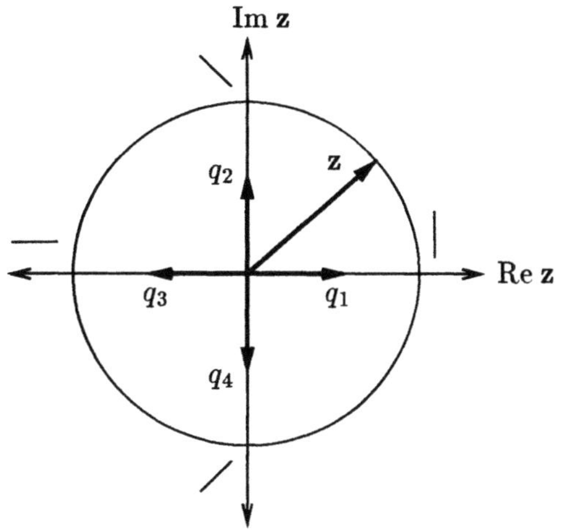

Figure 3.10 Illustration of the direction of output vector **z** as a function of the dominant orientation of structure.

the origin are termed *incompatible*, as they are unlikely to occur simultaneously within a particular neighborhood.

3. Structures that are perpendicular produce vector components that are opposing. This is intuitively appealing as we would say that perpendicular structures are maximally different compared in terms of orientation, which should give them a representation as maximally different vectors. This is an effect of the model restriction to one-dimensionality stated earlier.

4. A particular output value **z** can be produced by a number of combinations of features q_k. We can view the output as given by the dominant component, corrected by a penalty function consisting of the non-dominant component.

5. Two feature measures that have the same magnitude and are incompatible produce a resulting output that is zero. Incompatible pairs of measurements, according to our model, should always act to cancel out each other and produce a zero output. This implies an undecided observation about the state within the neighborhood.

All models will fail for some input situation; the best we can hope for is that they do this in a graceful way. The current model and representation structure do this by proclaiming zero or low confidence in the case of conflicting statements. This is a very important feature, allowing a local measure of reliability. This measure can later be updated and modified by other information available.

The information about the orientation – the information about class membership – is extremely important to retain, as it turns out to be more robust to noise and consequently more useful than the magnitude. The orientation information can be used for very specific consistency operations as will be shown later.

3.6 GENERAL PROPERTIES OF A USEFUL INFORMATION REPRESENTATION

The meaningful application of most signal processing concepts requires that the data represent some aspect of the real world in an orderly way. More precisely, an increased difference between events in the real world must generally result in an increased distance between the data points representing these events. Using representations that violate these requirements makes many signal processing concepts meaningless, and also greatly reduces the possibility of gaining an intuitive feeling for "what is going on". For many real-world problems, however, establishing well behaved representations is a non-trivial task and in these cases a first and crucial step of any analysis is to find such a representation.

In this section we will generalize the problem of finding a suitable vector representation for two-dimensional lines. In the following, a *line* will be taken to mean a *symmetrical* separator of two regions such that no relation between the regions can be inferred. As a contrast, an *edge* will be understood as an *asymmetrical* separator, such that an ordering, simple or ambiguous, of the two regions is always implied. Note that a consequence of this definition is that lines can only be unambigously associated with an angle over an interval of 180 degrees, while edges can be unambiguously associated with angles over an interval of 360 degrees. Another interpretation is to consider that both lines and edges are associated with angles spanning a 180 degree interval and that in addition, edges are *signed* while lines are *unsigned*.

Considering the above, we can argue that a general partitioning element must have the properties of a line, as it is not possible to consistently define an ordering of transitions between regions, if the data constituting the different regions is multi-dimensional.

Considering the fact that a substantial part of the concepts used in image processing are of multi-dimensional nature (color, texture fields, shape, etc.), it becomes apparent that, in any image processing theory claiming generality, a partitioning element must have the above mentioned line properties.

In trying to find a vector representation for line orientation, a two-dimensional half plane is appropriate, the vector angle being identical to the line orientation and the vector length being a positive valued function of some other property of the line, e.g. its energy, probability or some other definition of its *magnitude.* A space resulting from this type of direct mapping will be referred to as the original space. It is obvious in the present case that, for a given vector length, the original space corresponds to a half circle. The problem with this original space is that it leads to an unacceptable distance measure as vectors representing two lines differing by a small angle can end up being represented by vectors that are very far apart, i.e. one line is represented close to one end of the half circle and the other line is represented close to the opposite end.

The method presented in this chapter, implicitely solves this problem for two-dimensional signals. For a discussion of representational issues for higher dimensional signals we will proceed to following chapters.

Figure 3.11 An original image, a portrait of a girl

4

FOURIER TRANSFORMS

This chapter presents an overview of definitions and theorems for Fourier transforms in the context of computer vision. The presentation is made without any claims on being either thorough or mathematically strict. Some results are stated without proofs and the transforms of various functions are assumed to exist without further ado. For a more rigorous treatment of these issues, see, for example, [23]. Furthermore, it is assumed that the reader is familiar with the one-dimensional Fourier transform, both the continuous version and the discrete Fourier transform, as well as some mathematical constructions such as generalized functions or distributions.

The Fourier transform is in practice the name of a collection of similar transforms, each defined in a particular way in order to suit a particular purpose and with the common property of mapping functions to functions. First of all, we may distinguish between the discrete Fourier transform (DFT) which maps functions defined on discrete and finite points, e.g. the pixel values of an image, and the continuous Fourier transform which maps functions defined on continuous and infinite sets, e.g. functions on $\mathbb{R}^2$. Though related, these two transforms are defined for two quite different sets of functions and should therefore not be confused. The functions being transformed may have arbitrary numbers of variables; this means that a Fourier transform which maps functions of one variable is not the same this as a Fourier transform mapping functions of two variables. In the following the *dimensionality* of a Fourier transform will refer to the number of variables of the involved functions. Hence, a one-dimensional Fourier transform maps functions of one variable to functions of one variable, etc. In the case when we consider functions of n variables, where n is some unspecified but positive integer, we refer to the corresponding Fourier transform as multi-dimensional. Also the details of the definitions of the Fourier transform may be different, e.g. the variables or the resulting function may be scaled in different ways. Furthermore, it should be mentioned that term "Fourier transform" is used to refer both to the process of mapping a function to a function and to the resulting function.

In the context of computer vision, the functions to be transformed are often digital images or filter kernels. Typically, such images or filters have dimensionality two or more, are real- or complex-valued, and are defined on a regular discrete point grid with finite extension in all dimensions. Hence, to actually compute the Fourier transform of such a function, e.g. for designing a filter kernel to be used in some image operation, the DFT must be used. As will be apparent in Section 4.8, however, the DFT must be used with care. For example, it identifies the vertical borders as well as the horizontal borders of a two-dimensional image and therefore treats the image as infinitely periodic. The DFT also suffers from being cumbersome to use in theoretical analysis of a signal and, from that point of view, the continuous Fourier transform is advantageous. The DFT can, however, often be seen as a sufficiently good approximation to the continuous transform and we will therefore use the continuous transform when considering an image and its Fourier transform from a theoretical point of view. Furthermore, images of dimension two or more are often separable, in one way or another, into one-dimensional functions. Since the Fourier transform itself is separable as well, this implies that Fourier transforms of multi-dimensional images often can be achieved by with the one-dimensional Fourier transform or variants thereof.

The rest of this chapter is divided into nine sections. Section 4.1 is a short introduction to Fourier transforms in the context of computer vision. Section 4.2 presents the basics of Fourier transforms with which the reader is assumed to be familiar with but which is included here for reference. This section also contains notations and conventions used in the rest of the book. Section 4.3 presents three aspects of the Fourier transform which may not be covered by elementary textbooks on transform theory, most prominently the uncertainty relation. Separability, seen as a relation between the multi-dimensional and the one-dimensional Fourier transform is presented in Section 4.4. The concept of analytical signals will prove an important issue in the following chapters and is therefore treated in both one and multiple dimensions in Section 4.5. Two examples of the use of statistical descriptions of an image in connection with the Fourier transform are presented in Section 4.6. A catalogue of transform pairs is included in Section 4.7. Section 4.8 presents various aspects of using the Fourier transform on signals which are sampled and truncated. Section 4.9 ends the chapter with a presentation of the DFT.

4.1 INTRODUCTION

Historically, computer vision is quite a new branch on the tree of signal processing, an area where Fourier analysis has had one of its most important applications. One-dimensional signals such as audio, radio, television and radar have been investigated for over a century and it is a fact that Fourier analysis often gives a useful representation of a signal since not only the signal properties but also basic operations such as linear filtering and modulations are easily described in the Fourier domain. When the first attempts in computer vision began back in the early 70's, Fourier analysis of the images was therefore a natural approach. These early attempts were often based on transforming the entire image, in the same way as the Fourier transform is applied to the entirety of an audio signal. This approach, called a *global transform* of the image, was, however, soon recognized as inappropriate, at least for more complicated images. To understand why, let us illustrate with an example. Consider an image containing a small number of simple objects as in Figure 4.1. We consider the objects to be distinct as well as different in both shape and position. The Fourier transform of this image, however, contains in each point information about all of these objects, and in a non-trivial form. This illustrates one of the basic properties of the Fourier transform, namely the *uncertainty principle* which implies that information which is well localized in space is distributed in frequency and vice versa. For images in general, often far more complicated than our example, descriptions of local features and objects are therefore smeared out and not easily identified in the transform domain.

The problem with global Fourier transforms is thus that if we consider a specific region in an image and find that it contains some very simple feature, e.g. a line segment, then neither the feature itself nor its position in the image is clearly described by the global Fourier transform. One solution to this problem is to consider Fourier transforms of image regions rather than of the entire image. This approach is referred to as a *local transform*. Conceptually we may think of this operation as producing a set of transforms, one for each region. As an example, for each image point we can define a neighborhood of some appropriate size and transform each such neighborhood. The result will then be a set of Fourier transforms, one for each point in the image, each of which describes a region around a specific image point, as illustrated in Figure 4.2. The advantage with this approach is twofold. First, if the region size is small enough, we can expect each region to contain very few and simple features, resulting in a simple transform. Second, the position of each feature described by a local transform is given explicitly by the position of of the corresponding region in the image.

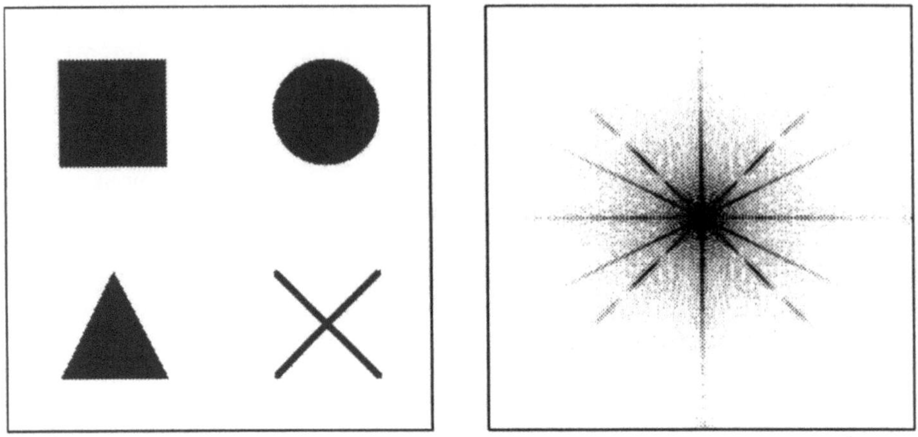

Figure 4.1 An image containing some objects (left) and the magnitude of the corresponding Fourier transform (right).

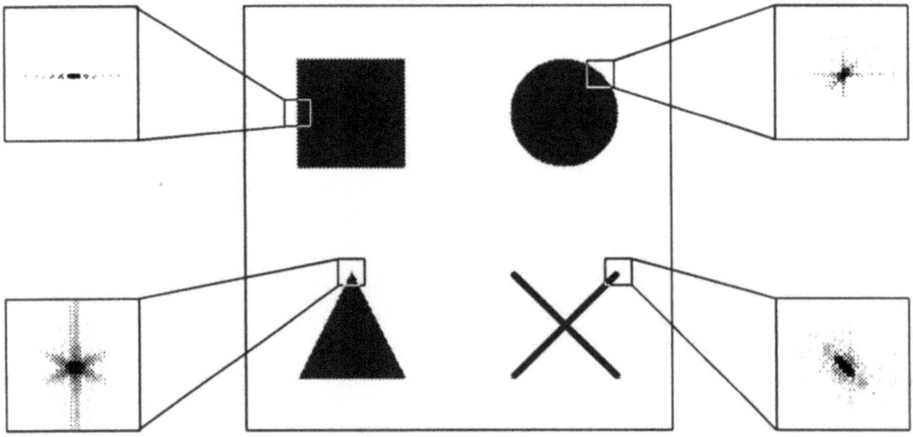

Figure 4.2 Local Fourier transforms of some image regions

Using local Fourier transforms is, however, not enough. As previously mentioned, actually computing a Fourier transform numerically, as was done for the illustrations, requires the use of the DFT. As a consequence of its construction it identifies the vertical borders as well as the horizontal borders of a two-dimensional image. Hence, the resulting function, which describes the frequency content of an image region, will not only describe the image structures within the region but also artifacts which arise from the wrap-around of the borders. As an example, consider the local Fourier transform of the top of the triangle in Figure 4.2. The transform contains two oblique and one vertical linear structures. Without going into any details on why the transform has this appearance, is seems reasonable to assume that the two oblique lines are related to the oblique borders of the triangle. The vertical line in the transform is, however, not at all related to the triangle. The region does not cover the horizontal border of the triangle which would, in fact, show up as a vertical line in the transform. The reason for the vertical line in the transform is that the image region being transformed has one intensity at the lower horizontal border and a different intensity at the upper horizontal border creating an artificial horizontal edge in the transformed region.

The artifacts that arise from using the DFT on images can be avoided by weighting each image region to be transformed. The concept of weighting implies that the pixel values of the region are multiplied with some suitable function that has unit value in the centre of the region and vanishes at the borders. For reasons which will become apparent in the following sections, this weighting function should make a smooth transition from unity to zero. We will not go into detail here on how to define suitable weighting functions, but merely point out that the weighting, apart from suppressing artifacts, will also affect those parts of the transform that are of interest. Figure 4.3 illustrates the result of using a weighted local Fourier transform. As seen, artifacts due to the border wrap-around of the DFT are suppressed. Using weighted local Fourier transforms for analysis of images is today a widely accepted strategy, and a reference to a Fourier transform in the context of image processing almost always means a weighted local transform.

One may ask why the Fourier transform is such a commonly used tool in computer vision. The immediate answer is that most features believed to be important for solving image processing tasks are often more easily described in the Fourier domain than in terms of the variation of the image itself. Examples of such features are orientation, phase, frequency and curvature. As a consequence, the algorithms used to detect and represent such features are often described in terms of Fourier transforms. This, however, does not imply that the actual implementations of these algorithms use Fourier transforms as

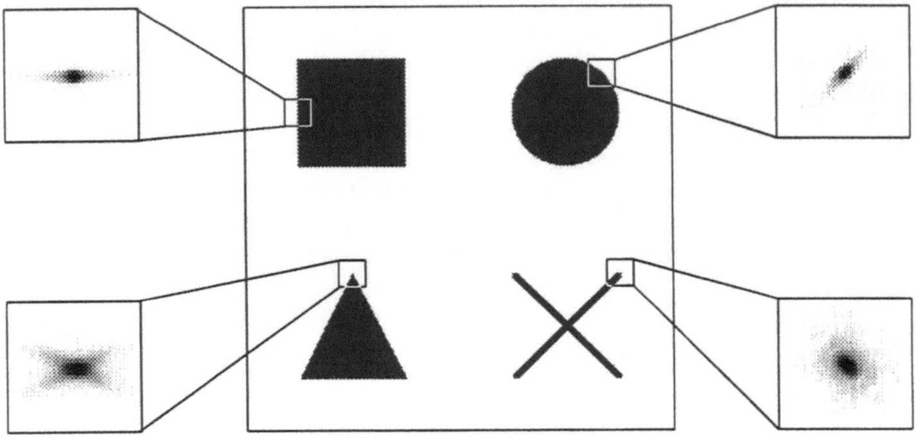

Figure 4.3 Local Fourier transforms of the same regions as in Figure 4.2 using a window function

a means to compute feature descriptions. Furthermore, it should be recognized that Fourier analysis may not always provide the only or the most efficient means for feature description and extraction.

4.2 BASICS

This section presents the basics of the continuous Fourier transform. The reader is assumed to be familiar with the results which are included here for reference. We begin with a short presentation of the notations and conventions used here followed by definitions of some useful functions. We include a short treatment of impulse functions for use in the following sections. Then we present the actual definition of the continuous Fourier transform along with a discussion of its implications. Following this comes the definition of the inverse Fourier transform, a short list of the most basic theorems related to Fourier transforms and, finally, a discussion of even, odd and Hermitian functions and their Fourier transforms.

4.2.1 Notations and conventions

In the rest of this chapter we deal with functions of multiple variables, using n to denote the number of variables, and it will prove practical to view the argument of such functions either as n real numbers or as a vector in $\mathbb{R}^n$. We use the convention of writing such vectors with lower case boldface letters and represent them as $n \times 1$ matrices. The elements of such a vector are written in lower case italics with an index. As an example, a vector $\mathbf{x} \in \mathbb{R}^4$ with the components x_1, x_2, x_3 and x_4 is represented with a 4×1 matrix,

$$\mathbf{x} = \begin{pmatrix} x_1 \\ x_2 \\ x_3 \\ x_4 \end{pmatrix}. \tag{4.1}$$

If f is a function defined on, for example, $\mathbb{R}^4$, we write either $f(x_1, x_2, x_3, x_4)$ or $f(\mathbf{x})$, thus meaning the same thing. We denote transposition of a vector with a T-sign, i.e. if $\mathbf{x}$ is defined as in Equation (4.1), then

$$\mathbf{x}^T = (x_1 \; x_2 \; x_3 \; x_4) \tag{4.2}$$

The scalar product between the two vectors $\mathbf{x}, \mathbf{y} \in \mathbb{R}^n$ is written

$$\mathbf{x} \cdot \mathbf{y} = \mathbf{x}^T \mathbf{y} = \sum_{k=1}^{n} x_k \, y_k. \tag{4.3}$$

In multiple dimensions, Cartesian coordinates may not be the most natural to use. For example, if considering a function which varies only with the distance to the origin, then directional coordinates or spherical coordinates will often prove more practical than Cartesian coordinates. *Directional coordinates* for a point $\mathbf{x} \in \mathbb{R}^n$ are defined by the relation

$$\mathbf{x} = r \, \hat{\mathbf{x}}, \tag{4.4}$$

where

$$\begin{aligned} r &= |\mathbf{x}|, \\ \hat{\mathbf{x}} &= \frac{\mathbf{x}}{r}. \end{aligned} \tag{4.5}$$

Hence, instead of the n Cartesian coordinates $x_1, \ldots, x_n$, the real number r and the unit vector $\hat{\mathbf{x}}$ can be used to define a specific point. Instead of the unit vector $\hat{\mathbf{x}}$, it is sometimes convenient to describe the direction of $\mathbf{x}$ with angles. The number r together with $n-1$ angles are referred to as *spherical coordinates*. It should be noted that the term spherical coordinates does not imply that the space under consideration is three-dimensional, as may be the standard definition. In two dimensions, spherical coordinates are the same thing as the *polar coordinates* (r, φ), where

$$\begin{aligned} x_1 &= r \cos\varphi, \\ x_2 &= r \sin\varphi, \qquad 0 \le \varphi < 2\pi. \end{aligned} \tag{4.6}$$

In three dimensions we may use the usual spherical coordinates $(r, \varphi_1, \varphi_2)$, where

$$\begin{aligned} x_1 &= r \cos\varphi_1 \sin\varphi_2, \\ x_2 &= r \sin\varphi_1 \sin\varphi_2, \\ x_3 &= r \cos\varphi_2, \qquad 0 \le \varphi_1 < 2\pi, \qquad 0 \le \varphi_2 < \pi. \end{aligned} \tag{4.7}$$

The concept of spherical coordinates can be extended to arbitrary dimensions, even though two or three dimensions will be the standard examples in this presentation.

Depending on the type of problem considered, we will use either Cartesian, directional or spherical coordinates to describe a point $\mathbf{x}$. This approach is of course applicable to points in the Fourier domain. To distinguish between spherical coordinates in the spatial and in the Fourier domain respectively we use the following convention.

- The *radial coordinate in the spatial domain* is denoted r, i.e. $r^2 = x_1^2 + x_2^2 + \ldots + x_n^2$.
- The *radial coordinate in the Fourier domain* is denoted ρ, i.e. $\rho^2 = u_1^2 + u_2^2 + \ldots + u_n^2$.
- The *angular coordinates in the spatial domain* are denoted φ_k, where $k = 1, \ldots, n-1$.
- The *angular coordinates in the Fourier domain* are denoted ψ_k, where $k = 1, \ldots, n-1$.

Sometimes it will be necessary to consider functions of Cartesian and spherical coordinates at the same time. As an example, if $f(x_1, x_2)$ is some two-dimensional function, then we may consider also "the same function" but defined with spherical coordinates

$$f_1(r(x_1, x_2), \varphi(x_1, x_2)) = f(x_1, x_2) \tag{4.8}$$

The two functions f_1 and f are or course different functions, but since they are related by a specific change of coordinates, i.e. Cartesian to spherical, we will treat them as two different descriptions of one and the same function and simply write

$$f(r, \varphi) = f_1(r, \varphi), \tag{4.9}$$

relying on the context and the variables used to indicate whether we use the Cartesian or spherical description of a particular function. Hence, when going from Cartesian to spherical coordinates we may replace $f(x, y)$ with $f(r, \varphi)$.

Unit vectors, i.e. vectors of unit length, are normally denoted with a circumflex, e.g. $\hat{\mathbf{e}}$. Note that a unit vector may be an element of both the spatial and the Fourier domain. The reason is that these vectors do not have any physical dimensionality, e.g. spatial position or frequency, but carry only directional information. As an example, if considering a unit vector $\hat{\mathbf{e}}$ in the spatial domain which describes a particular direction therein, we may consider the same direction in the Fourier domain as also described by $\hat{\mathbf{e}}$.

4.2.2 Some useful functions

The following is a short list of functions used in this book. The graphs of the functions are illustrated in Figure 4.4. It should be noted that some of the functions, e.g. the sinc function, may be defined in different ways in other textbooks.

The rectangular function.

$$\operatorname{rect}(x) = \begin{cases} 1 & |x| < \frac{1}{2} \\ 0 & |x| > \frac{1}{2} \end{cases} \tag{4.10}$$

The sinc function.

$$\operatorname{sinc}(x) = \begin{cases} \frac{\sin \pi x}{\pi x} & x \neq 0 \\ 1 & x = 0 \end{cases} \tag{4.11}$$

The sign function.

$$\operatorname{sign}(x) = \begin{cases} -1 & x < 0 \\ 1 & x \geq 0 \end{cases} \tag{4.12}$$

The step function or the Heaviside function.

$$\operatorname{step}(x) = \begin{cases} 0 & x < 0 \\ 1 & x \geq 0 \end{cases} \tag{4.13}$$

Note that $\operatorname{sign}(x) = 2\operatorname{step}(x) - 1$.

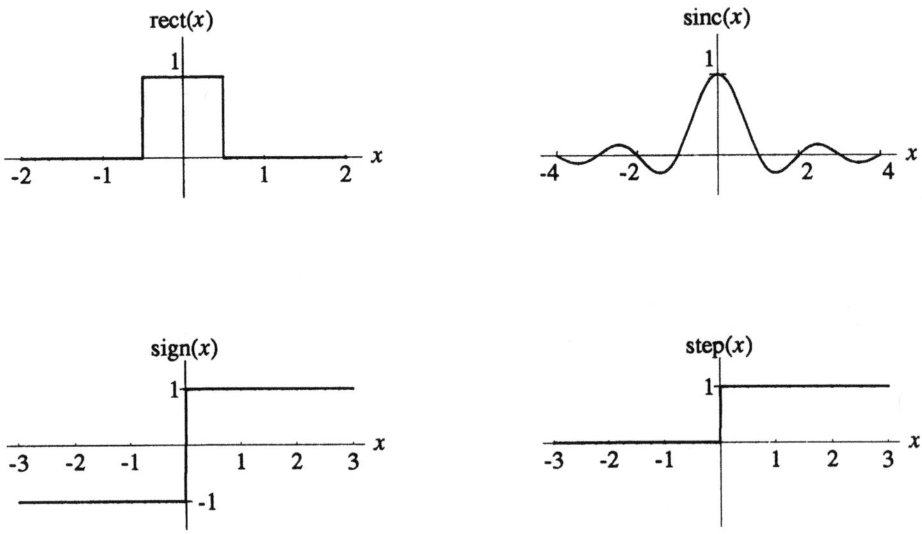

Figure 4.4 Graphs of the functions defined in Section 4.2.2

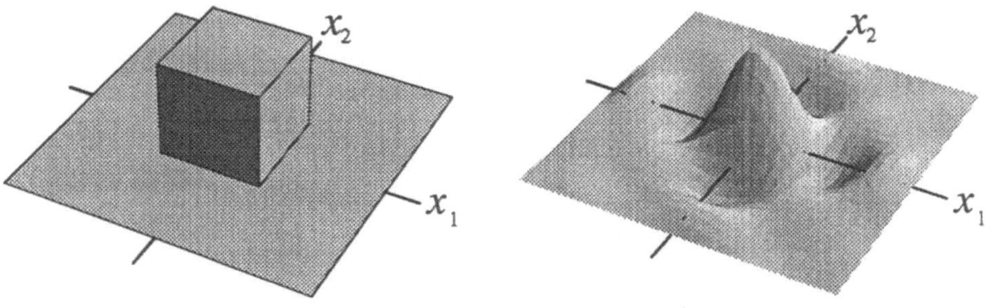

Figure 4.5 Graphs of the functions $\mathrm{rect}^2(\mathbf{x})$ and $\mathrm{sinc}^2(\mathbf{x})$.

Some of these function will, in the following, prove useful when defined for multiple variables. For $\mathbf{x} \in \mathbb{R}^n$ we therefore define

$$\mathrm{rect}^n(\mathbf{x}) = \mathrm{rect}(x_1) \cdot \mathrm{rect}(x_2) \ldots \mathrm{rect}(x_n), \tag{4.14}$$

which corresponds to an n-dimensinal box which is centred at the origin, and of unit width in along all coordinates. We also define

$$\mathrm{sinc}^n(\mathbf{x}) = \mathrm{sinc}(x_1) \cdot \mathrm{sinc}(x_2) \ldots \mathrm{sinc}(x_n). \tag{4.15}$$

The last two functions are illustrated in Figure 4.5 for the case $n = 2$.

4.2.3 Impulse functions

The generalized function called the impulse function, the delta function, or the Dirac function, can be defined in several ways. Here we will present two useful definitions of the one-dimensional impulse function. Impulse functions in multiple dimensions are then defined either similarly or as a product of one-dimensional impulse functions.

Figure 4.6 The one-dimensional impulse function (left) and the shah function (right)

The one-dimensional impulse function

The impulse function $\delta(x)$ is defined uniquely by the following property. Let f be a function, then

$$\int_{-\infty}^{\infty} f(x)\,\delta(x_0 - x)\,dx = f(x_0) \tag{4.16}$$

From this definition of the impulse function it follows immediately that for $a \neq 0$ it is the case that

$$\delta(ax) = \tfrac{1}{|a|}\,\delta(x) \tag{4.17}$$

It is sometimes also convenient to consider the impulse function as the derivative of the step function, i.e.

$$\delta(x) = \tfrac{d}{dx}\,\mathrm{step}(x). \tag{4.18}$$

It is noteworthy that a shift of a function $f(x)$ to $f(x - x_0)$ can be defined in terms of a convolution with an impulse function according to

$$f(x - x_0) = f(x) * \delta(x - x_0) \tag{4.19}$$

The shah function or the sample function

It will prove practical, especially when dealing with discrete functions, to define a function that is an infinite sum of impulses with unit spacing. This is the shah function, denoted III(x), and defined as

$$\mathrm{III}(x) = \sum_{k=-\infty}^{\infty} \delta(x - k) \tag{4.20}$$

Figure 4.7 Left: an impulse line in three dimensions. Right: an impulse plane in three dimensions.

Multi-dimensional impulse functions

In n dimensions we define a generalized function $\delta^n(\mathbf{x})$ such that

$$\int_{\mathbb{R}^n} f(\mathbf{x})\, \delta^n(\mathbf{y}-\mathbf{x}) d^n x = f(\mathbf{y}). \tag{4.21}$$

for all functions on $\mathbb{R}^n$. This δ^n is then the natural generalization of the one-dimensional function δ. Evidently it must be the case that

$$\delta^n(\mathbf{x}) = \delta(x_1)\, \delta(x_2) \ldots \delta(x_n), \tag{4.22}$$

which is easily proved by integrating the right hand side of Equation (4.21) one variable at a time.

In n dimensions we will sometimes encounter products of $n-1$ one-dimensional delta functions rather than n as above. Let $\hat{\mathbf{e}} \in \mathbb{R}^n$ be a unit vector and let $\{\hat{\mathbf{e}}_k,\ k = 1, \ldots, n\}$ be a basis of orthonormal vectors where $\hat{\mathbf{e}}_1 = \hat{\mathbf{e}}$, and define

$$\delta_{\hat{\mathbf{e}}}^{\text{line}}(\mathbf{x}) = \delta(\mathbf{x}^T\hat{\mathbf{e}}_2)\, \delta(\mathbf{x}^T\hat{\mathbf{e}}_3) \ldots \delta(\mathbf{x}^T\hat{\mathbf{e}}_n). \tag{4.23}$$

This generalized function, a product of $n-1$ impulse functions, will then depend on the direction of $\hat{\mathbf{e}}$. We may think of this function as an impulse line that passes through the origin and is parallel to the vector $\hat{\mathbf{e}}$, see Figure 4.7.

Let Γ denote this line. This implies

$$\int_{\mathbb{R}^n} f(\mathbf{x})\, \delta_{\hat{\mathbf{e}}}^{\text{line}}(\mathbf{x})\, d^n x = \int_{\Gamma} f\, ds, \tag{4.24}$$

where s is a parameterization of Γ and, hence, the right hand side of Equation (4.24) is the integral of f along Γ.

In n dimensions we may also consider generalized functions which are simply one-dimensional impulse functions. Let $\hat{\mathbf{e}}$ denote a unit vector in $\mathbb{R}^n$ and define

$$\delta_{\hat{\mathbf{e}}}^{\text{plane}}(\mathbf{x}) = \delta(\mathbf{x}^T \hat{\mathbf{e}}) \tag{4.25}$$

This function may be thought of as an impulse hyperplane that is perpendicular to $\hat{\mathbf{e}}$, see Figure 4.7. Let P denote this hyperplane. Then

$$\int_{\mathbb{R}^n} f(\mathbf{x})\, \delta_{\hat{\mathbf{e}}}^{\text{plane}}(\mathbf{x})\, d^n x = \int_{P} f\, d\Omega, \tag{4.26}$$

where Ω is a parameterization of the hyperplane P and, hence, the right hand side of Equation (4.26) is the integral of f over P.

Sample function, multidimensional Also the III function is extended to multiple dimensions (Figure 4.8) as

$$\text{III}^n(\mathbf{x}) = \text{III}(x_1)\, \text{III}(x_2) \ldots \text{III}(x_n). \tag{4.27}$$

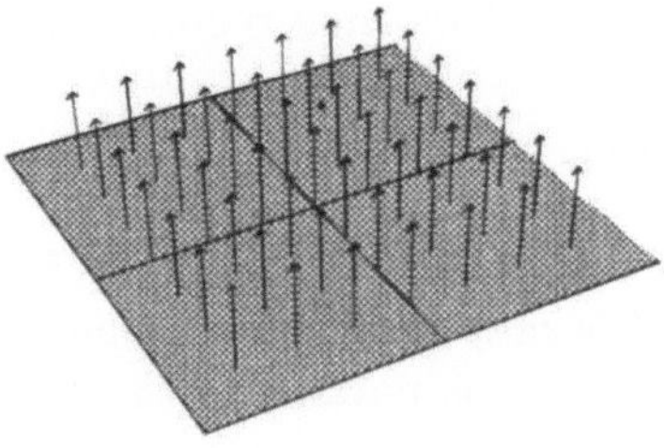

Figure 4.8 The two-dimensional function III^2.

4.2.4 Definition

Let f be a function, $f : \mathbb{R}^n \to \mathbb{C}$. We then define a new function $F : \mathbb{R}^n \to \mathbb{C}$, as

$$F(\mathbf{u}) = \int_{\mathbb{R}^n} f(\mathbf{x})\, e^{-i\mathbf{u}^T\mathbf{x}}\, d^n x. \tag{4.28}$$

The function F is referred to as the *Fourier transform* of f or corresponding to f. Given this definition we will spend some time discussing its implications.

First, f can not be chosen arbitrarily and we may assert that only functions f for which the corresponding transform F is well-defined will be of interest here. Therefore, we must restrict f to some set S_1 of functions that have well-defined Fourier transforms. The nature of S_1, i.e. what functions have well-defined Fourier transforms, is a subject completely outside the scope of this presentation and we will simply take the engineer's point of view and assert that most functions of interest have a well-defined Fourier transform. Some care must be observed when applying this proposition, however, since there are a number of common functions that do not have a well-defined Fourier transform, e.g. the exponential function. The relation between f and F, according to Equation (4.28), defines a map from S_1 to a new set of functions, S_2. In the following, we will refer to maps between sets of functions as *transforms*. Therefore, we may describe the right hand side of Equation (4.28) as a transform $\mathcal{F} : S_1 \to S_2$ and write

$$F = \mathcal{F}\{\, f \,\}. \tag{4.29}$$

Note the use of calligraphic letters and curly braces to distinguish a transform from a function on $\mathbb{R}^n$. Note, furthermore, that also $\mathcal{F}$ is referred to as a Fourier transform implying that the word *Fourier transform* may refer either to the process of transforming f to F or to the result of that process, i.e. F. In the following, however, there will be little or no risk for ambiguities. According to the foregoing, we may define S_2 as the image of S_1 under $\mathcal{F}$, denoted

$$S_2 = \mathcal{F}\{\, S_1 \,\}, \tag{4.30}$$

or, in other words, S_2 contains precisely those functions that are Fourier transforms of functions in S_1. In practice, the two sets S_1 and S_2 must contain objects other than functions, namely, generalized functions or distributions. Examples of distributions are the impulse functions and its derivatives. At certain points it is necessary to distinguish between proper and generalized functions, but we will in general refer to the elements of S_1 and S_2 simply as functions.

In the following we will assume that $S_1 = S_2$. This may seem strange in view of how these two sets have been introduced above. The reason to distinguish between the two sets is purely practical and is related to how we interpret the elements of S_1 and S_2, respectively. In the context of computer vision, the functions to be Fourier transformed are functions of one or several variables where the variables describe a point in space and/or time. Hence, these variables have the dimension of a length or time unit and it will prove practical to define one such unit as the distance between two adjacent image pixels. Despite the fact that some of these variables may be related to time, we will simply call them *spatial variable* and the corresponding functions *spatial functions*. Hence, S_1 is a set of spatial functions and is therefore called the *spatial domain*, a term that is used also for the space spanned by the variables of a spatial function. With this interpretation of the elements in S_1, the Fourier transform will give a description of a spatial function in terms of spatial (or temporal) frequency. Hence, the elements of S_2 are called *frequency functions* or simply Fourier transforms and their variables are *frequency variables*. Both S_2 and the space spanned by frequency variables are called the *Fourier domain* or the *frequency domain*. The dimension of frequency variables is radians per length unit. As an example, a function in S_1 may describe the intensity of an image along a specific line. The corresponding Fourier transform will then describe the image intensity in terms of frequency components. We will employ the convention that elements of S_1 are written with lower case Roman letters and corresponding elements of S_2 with the corresponding upper case letter. Note that from a mathematical point of view we need not distinguish between the two sets S_1 and S_2, implying that any property stated to be valid for the elements of one set is automatically also is valid for the elements of the other. Note also, that there is one distinct Fourier transform $\mathcal{F}$ as well as one pair S_1 and S_2 for each value of n.

The definition of the Fourier transform in Equation (4.28) is valid for any dimensionality of the function f. In the one-dimensional case, where $n = 1$, we get the familiar expression

$$F(u) = \int_{-\infty}^{\infty} f(x)\, e^{-iux}\, dx. \tag{4.31}$$

In higher dimensions, the Fourier transform can also be defined using spherical coordinates. In two dimensions, $n = 2$, Equation (4.28) corresponds to

$$F(u_1, u_2) = \int_0^{2\pi} \int_0^{\infty} r\, f(r, \varphi)\, e^{-ir(u_1 \cos\varphi + u_2 \sin\varphi)}\, dr d\varphi, \tag{4.32}$$

or

$$F(\rho, \psi) = \int_0^{2\pi} \int_0^{\infty} r\, f(r, \varphi)\, e^{-i\rho r \cos(\psi - \varphi)}\, dr d\varphi, \tag{4.33}$$

or

$$F(\rho, \psi) = \int_{-\infty}^{\infty} \int_{-\infty}^{\infty} f(x_1, x_2)\, e^{-i\rho(x_1 \cos\psi + x_2 \sin\psi)}\, dx_1 dx_2. \tag{4.34}$$

For $n = 3$ we can, for example, write

$$F(u_1, u_2, u_3) = \tag{4.35}$$

$$\int_0^{\pi} \int_0^{2\pi} \int_0^{\infty} f(r, \varphi_1, \varphi_2)\, e^{-ir([u_1 \cos\varphi_1 + u_2 \sin\varphi_2] \sin\varphi_2 + u_3 \cos\varphi_2)}\, r^2\, \sin\varphi_2\, dr d\varphi_1 d\varphi_2,$$

and so on.

It should be recognized that there are definitions of the Fourier transform other than that of Equation (4.28). As an example, we may want to use frequency variables that have a physical dimension of periods per length or time unit (Hertz). In one dimension this implies the following definition of the Fourier transform,

$$F(s) = \int_{-\infty}^{\infty} f(x)\, e^{-2\pi i s x}\, dx. \tag{4.36}$$

Another example, from physics, is

$$F(k) = \frac{1}{\sqrt{2\pi}} \int_{-\infty}^{\infty} f(x)\, e^{-ikx}\, dx, \tag{4.37}$$

The only difference between Equations (4.31), (4.36) and (4.37) is the scaling of the spatial or frequency variables and of the resulting function, an operation that has no consequence for the general properties of the Fourier transform.

4.2.5 Inverse

One important property of the Fourier transform is that it maps one-to-one from S_1 to S_2, implying that there exists an inverse. If F is defined from f according to Equation (4.28), it is the case that

$$f(\mathbf{x}) = \frac{1}{(2\pi)^n} \int_{\mathbb{R}^n} F(\mathbf{u})\, e^{i\mathbf{u}^T\mathbf{x}}\, d^n u. \tag{4.38}$$

This relation defines a new transform, the *inverse Fourier transform* denoted $\mathcal{F}^{-1}$. Hence, we can write

$$f = \mathcal{F}^{-1}\{\, F \,\} \qquad \text{and} \qquad S_1 = \mathcal{F}^{-1}\{\, S_2 \,\}. \tag{4.39}$$

Note that there is one distinct $\mathcal{F}^{-1}$ for each value of n.

The general formulation of the multi-dimensional case above reduces to

$$f(x) = \frac{1}{2\pi} \int_{-\infty}^{\infty} F(u)\, e^{iux}\, du. \tag{4.40}$$

for the one-dimensional case. In higher dimensions, spherical coordinates can be used to define the inverse. As an example, for $n = 2$ we have

$$f(r,\varphi) = \frac{1}{(2\pi)^2} \int_0^{2\pi} \int_0^{\infty} \rho\, F(\rho,\psi)\, e^{i\rho r\cos(\psi-\varphi)}\, d\rho d\psi. \tag{4.41}$$

In the following we will establish a number of relations between the spatial and the Fourier domains. Because of the relative symmetry of the Fourier transform and its inverse these relations are valid in both directions, even though we occasionally have to introduce some multiples of 2π or a change of sign. As an example, let f be a spatial function and F its corresponding Fourier transform. Given this f, we can define a function F_1 according to

$$F_1 = \mathcal{F}^{-1}\{f\}, \tag{4.42}$$

i.e. we use the inverse Fourier transform in the "wrong" direction. Since $S_1 = S_2$, however, this is not a severe violation. It is easy to prove that F and F_1 must be related by

$$F_1(u) = \frac{1}{(2\pi)^n} F(-u) \tag{4.43}$$

The consequence of this relation is that we can use tables of Fourier transforms in order to obtain inverse transforms as well. If we know a function f and its corresponding Fourier transform F, then the inverse Fourier transform of f is given by F according to Equation (4.43).

4.2.6 Convolution

Consider two n-dimensional functions $f, g \in S_1$. The *convolution* between f and g, denoted $f * g$, is a function $h \in S_1$, defined as

$$h(\mathbf{x}) = \int_{\mathbb{R}^n} f(\mathbf{y})\, g(\mathbf{x} - \mathbf{y})\, d^n y. \tag{4.44}$$

In one dimension, this corresponds to the familiar expression

$$h(x) = \int_{-\infty}^{\infty} f(y)\, g(x - y)\, dy. \tag{4.45}$$

It is easy to prove that the following relations are valid

$$\begin{aligned} f * g &= g * f, \\ (\alpha f + \beta g) * h &= \alpha(f * h) + \beta(g * h). \end{aligned} \tag{4.46}$$

Since $S_1 = S_2$, the definition of convolution and related properties are also valid for functions in S_2.

Convolution in the spatial domain has the interesting property of corresponding to multiplication in the Fourier domain and vice versa. Consider two functions $f, g \in S_1$ with corresponding Fourier transforms $F, G \in S_2$. Then,

$$\begin{aligned} \mathcal{F}\{ f \cdot g \} &= \frac{1}{(2\pi)^n} F * G \\ \mathcal{F}\{ f * g \} &= F \cdot G \end{aligned} \qquad (4.47)$$

4.2.7 Affine coordinate changes

An affine coordinate changes means that the coordinate system under consideration is subjected to a translation of the origin followed by a linear operator. In n dimensions this means that a coordinate $\mathbf{x}$ is changed to $\mathbf{y}(\mathbf{x}) = \mathbf{A}[\mathbf{x}+\mathbf{b}]$, where $\mathbf{A}$ in an invertible matrix. Note that the multiplication by $\mathbf{A}$ is the same as a change of basis. It is often the case that functions in one domain are related by affine coordinate changes and we would of course want to know what this relation looks like in the other domain.

Let f and g be two n-dimensional functions such that

$$g(\mathbf{x}) = f(\mathbf{y}(\mathbf{x})) = f(\mathbf{A}[\mathbf{x} + \mathbf{b}]). \qquad (4.48)$$

It is simple to show that this implies that F and G, their corresponding Fourier transforms, must be related by

$$G(\mathbf{u}) = \frac{1}{|\det \mathbf{A}|} e^{i\mathbf{u}^T\mathbf{b}} F([\mathbf{A}^T]^{-1}\mathbf{u}). \qquad (4.49)$$

Apparently, the translation of the coordinate system, described by the vector $\mathbf{b}$, causes a complex modulation of F. The modulation is made in the direction described by $\hat{\mathbf{b}}$ and the number of cycles per frequency unit of the modulation is proportional to the norm of $\mathbf{b}$, i.e. proprtional to the length of the shift. Since this is the only effect of the shift, we can in the following study how the change of basis, specified by the matrix $\mathbf{A}$, affects F independent of the shift. This is illustrated in the following two examples.

Example. Let f and g be a pair of two-dimensional functions that are related according to Equation (4.48), where $\mathbf{b} = \mathbf{0}$ and $\mathbf{A}$ is given by

$$\mathbf{A} = \begin{pmatrix} a & 0 \\ 0 & b \end{pmatrix}. \tag{4.50}$$

This implies that g and f are related according to a scaling of the spatial coordinates, x_1 and x_2, with factors a and b respectively. According to Equation (4.49), F and G are then related by

$$G(\mathbf{u}) = \tfrac{1}{ab} F(\mathbf{Bu}) = \tfrac{1}{ab} F(\tfrac{u_1}{a}, \tfrac{u_2}{b}), \tag{4.51}$$

since

$$\mathbf{B} = [\mathbf{A}^T]^{-1} = \begin{pmatrix} \frac{1}{a} & 0 \\ 0 & \frac{1}{b} \end{pmatrix}. \tag{4.52}$$

Hence, G is derived from F primarily by a scaling of the u_1 and u_2 frequency coordinates with the factors $\frac{1}{a}$ and $\frac{1}{b}$ respectively. If we consider a function f and its corresponding Fourier transform F, the above discussion implies that if one of the spatial variables of f is scaled, then the corresponding frequency variable of F is scaled in the opposite way. Figure 4.9 illustrates how scaling affects a two-dimensional function and its Fourier transform. In this case the scaling takes place along axes which make a 45° angle to the (x_1, x_2)-axes. □

Example. Another interesting coordinate transformation is rotation. Again, assume f to be two-dimensional, $\mathbf{b} = \mathbf{0}$ and define $\mathbf{A}$ to be

$$\mathbf{A} = \begin{pmatrix} \cos\alpha & -\sin\alpha \\ \sin\alpha & \cos\alpha \end{pmatrix}, \tag{4.53}$$

which implies that g is derived from f by rotating the spatial coordinate system an angle α clockwise around the origin. We note that $\det(\mathbf{A}) = 1$ and also that $[\mathbf{A}^T]^{-1} = \mathbf{A}$, which are natural consequences of the fact that $\mathbf{A}$ is an orthogonal matrix. F and G are then related as

$$G(\mathbf{u}) = F(\mathbf{Au}). \tag{4.54}$$

Hence, also in the Fourier domain is G derived from F according to a rotation of the frequency coordinate system an angle α clockwise around the origin. From this discussion we infer that an orthogonal change of basis in one of the two domains corresponds in the same orthogonal change of basis in the other domain. □

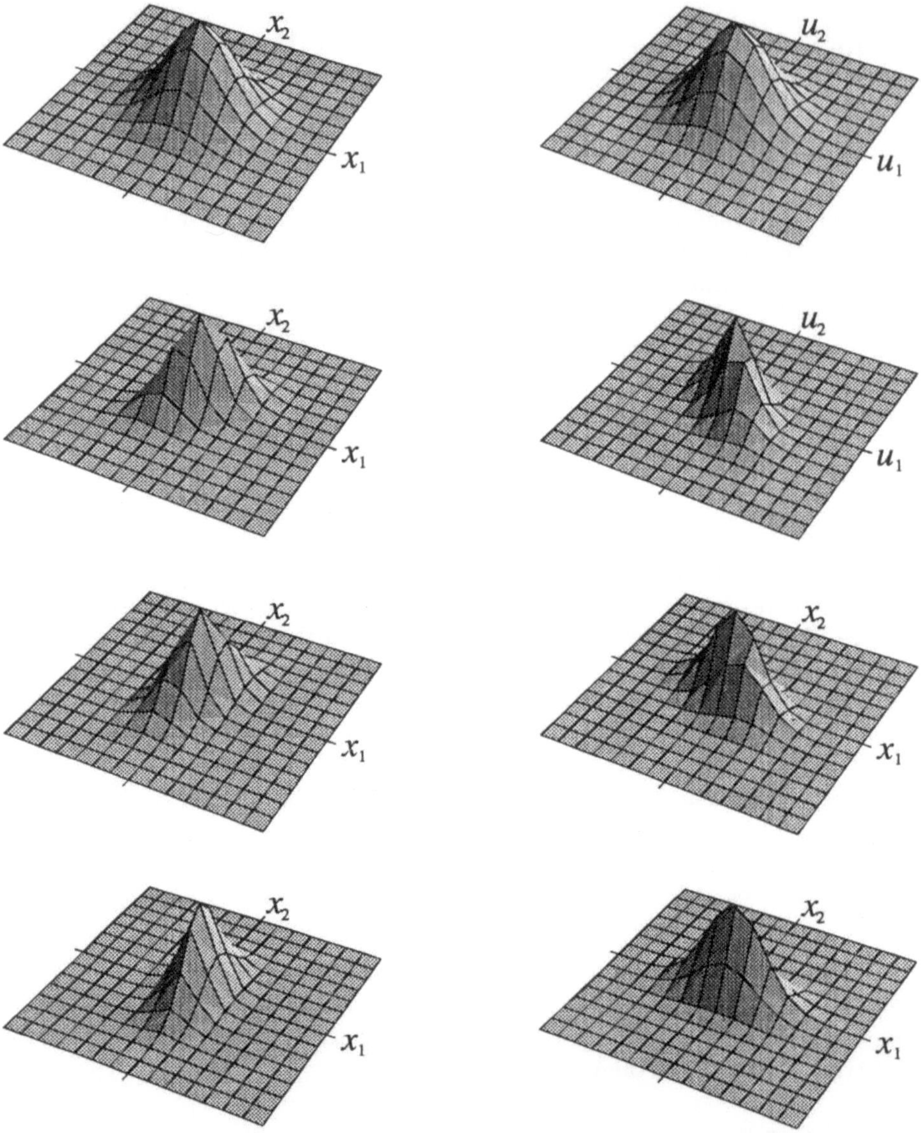

Figure 4.9 Pairs of spatial functions (left) and their corresponding Fourier transforms (right). Top pair: a circular symmetric Gaussian. Second pair: same Gaussian but scaled along an oblique axis. Third and fourth pairs: the scaled Gaussian rotates anti-clockwise in both domains.

4.2.8 Basic theorems

The following is a short list of the most fundamental properties of the Fourier transform. In this presentation f and g are spatial functions with corresponding Fourier transforms F and G.

Linearity.
The Fourier transform is linear, i.e. for $\alpha, \beta \in \mathbb{C}$ it is the case that

$$\mathcal{F}\{\alpha f + \beta g\} = \alpha F + \beta G. \tag{4.55}$$

This is of course also true for the inverse Fourier transform.

Derivatives.
Let us begin with the one-dimensional case and set

$$g(x) = \left(\tfrac{d}{dx}\right)^n f(x). \tag{4.56}$$

It follows that G then is given by

$$G(u) = (iu)^n F(u). \tag{4.57}$$

In the n-dimensional case, we may differentiate with respect to any of the coordinates arbitrary number of times. Let D be a differential operator which differentiates p_1 times with respect to x_1, p_2 times with respect to x_2, etc. Hence,

$$D = \left(\tfrac{\partial}{\partial x_1}\right)^{p_1} \left(\tfrac{\partial}{\partial x_2}\right)^{p_2} \cdots \left(\tfrac{\partial}{\partial x_n}\right)^{p_n}. \tag{4.58}$$

If g and f are related by

$$g(\mathbf{x}) = D f(\mathbf{x}), \tag{4.59}$$

then

$$G(\mathbf{u}) = i^{p_1+p_2+\ldots+p_n}\, u_1^{p_1}\, u_2^{p_2} \ldots u_n^{p_n}\, F(\mathbf{u}) \tag{4.60}$$

Moment.
Let us begin with the one-dimensional case. The p-th moment of f is a scalar m_p, defined as

$$m_p = \int_{-\infty}^{\infty} x^p f(x)\, dx \qquad p \geq 0. \tag{4.61}$$

This moment, m_p, can also be expressed in terms of F according to

$$m_p = i^p \left[\tfrac{d^p}{du^p} F(u)\right]_{u=0}. \tag{4.62}$$

As an example, it follows immediately that

$$m_0 = \int_{-\infty}^{\infty} f(x)\, dx = F(0). \tag{4.63}$$

In the n-dimensional case we can consider integrals of f multiplied by an arbitrary product of the coordinates. Hence, the moment m must have n indices which each indicates the power of the corresponding coordinate, i.e.

$$m_{p_1 p_2 \dots p_n} = \int_{\mathbb{R}^n} x_1^{p_1} x_2^{p_2} \dots x_n^{p_n} f(\mathbf{x}) d^n x. \tag{4.64}$$

As a generalization of Equation (4.63), this moment is also given by

$$m_{p_1 p_2 \dots p_n} = i^{p_1+p_2+\dots+p_n} \left[(\frac{\partial}{\partial u_1})^{p_1} (\frac{\partial}{\partial u_2})^{p_2} \dots (\frac{\partial}{\partial u_n})^{p_n} F(\mathbf{u}) \right]_{\mathbf{u}=\mathbf{0}} \tag{4.65}$$

Power theorem.
This useful result, sometimes also referred to as Parseval's theorem, implies that the integral of products of functions is basically invariant to the Fourier transform

$$\int_{\mathbb{R}^n} f^\star(\mathbf{x})\, g(\mathbf{x})\, d^n x = \frac{1}{(2\pi)^n} \int_{\mathbb{R}^n} F^\star(\mathbf{u})\, G(\mathbf{u})\, d^n u. \tag{4.66}$$

Without going into any details, it can be shown that the two integrals of Equation (4.66) are feasible as a scalar product on the vector space $S_1 = S_2$ of functions. Furthermore, the Fourier transform is only a change of basis in this vector space and is, in addition, basically an orthogonal change of basis with respect to the above scalar product. Sinee orthogonal changes of basis leave scalar products invariant, the power theorem follows directly. From the power theorem it follows directly that

$$\int_{\mathbb{R}^n} |f(\mathbf{x})|^2 \, d^n x = \frac{1}{(2\pi)^n} \int_{\mathbb{R}^n} |F(\mathbf{u})|^2 \, d^n u. \tag{4.67}$$

4.2.9 Even, odd and Hermitian functions

Here, we will distinguish between real functions, i.e. $f : \mathbb{R}^n \to \mathbb{R}$, and complex functions, i.e. $f : \mathbb{R}^n \to \mathbb{C}$. A real function f is called *even* if

$$f(-\mathbf{x}) = f(\mathbf{x}) \qquad \text{for all } \mathbf{x} \in \mathbb{R}^n, \tag{4.68}$$

and *odd* if

$$f(-\mathbf{x}) = -f(\mathbf{x}) \qquad \text{for all } \mathbf{x} \in \mathbb{R}^n. \tag{4.69}$$

A real function f is in general neither even nor odd but it can always be written as a sum of two unique functions f_e and f_o that are even and odd, respectively. To see this, consider the following identity

$$f(\mathbf{x}) = \tfrac{1}{2}[f(\mathbf{x}) + f(-\mathbf{x})] + \tfrac{1}{2}[f(\mathbf{x}) - f(-\mathbf{x})], \tag{4.70}$$

and set

$$\begin{aligned} f_e(\mathbf{x}) &= \tfrac{1}{2}[f(\mathbf{x}) + f(-\mathbf{x})], \\ f_o(\mathbf{x}) &= \tfrac{1}{2}[f(\mathbf{x}) - f(-\mathbf{x})]. \end{aligned} \tag{4.71}$$

A complex function f is called *Hermitian* if

$$f(-\mathbf{x}) = f^{\star}(\mathbf{x}) \qquad \text{for all } \mathbf{x} \in \mathbb{R}^n, \tag{4.72}$$

and *anti-Hermitian* if

$$f(-\mathbf{x}) = -f^{\star}(\mathbf{x}) \qquad \text{for all } \mathbf{x} \in \mathbb{R}^n. \tag{4.73}$$

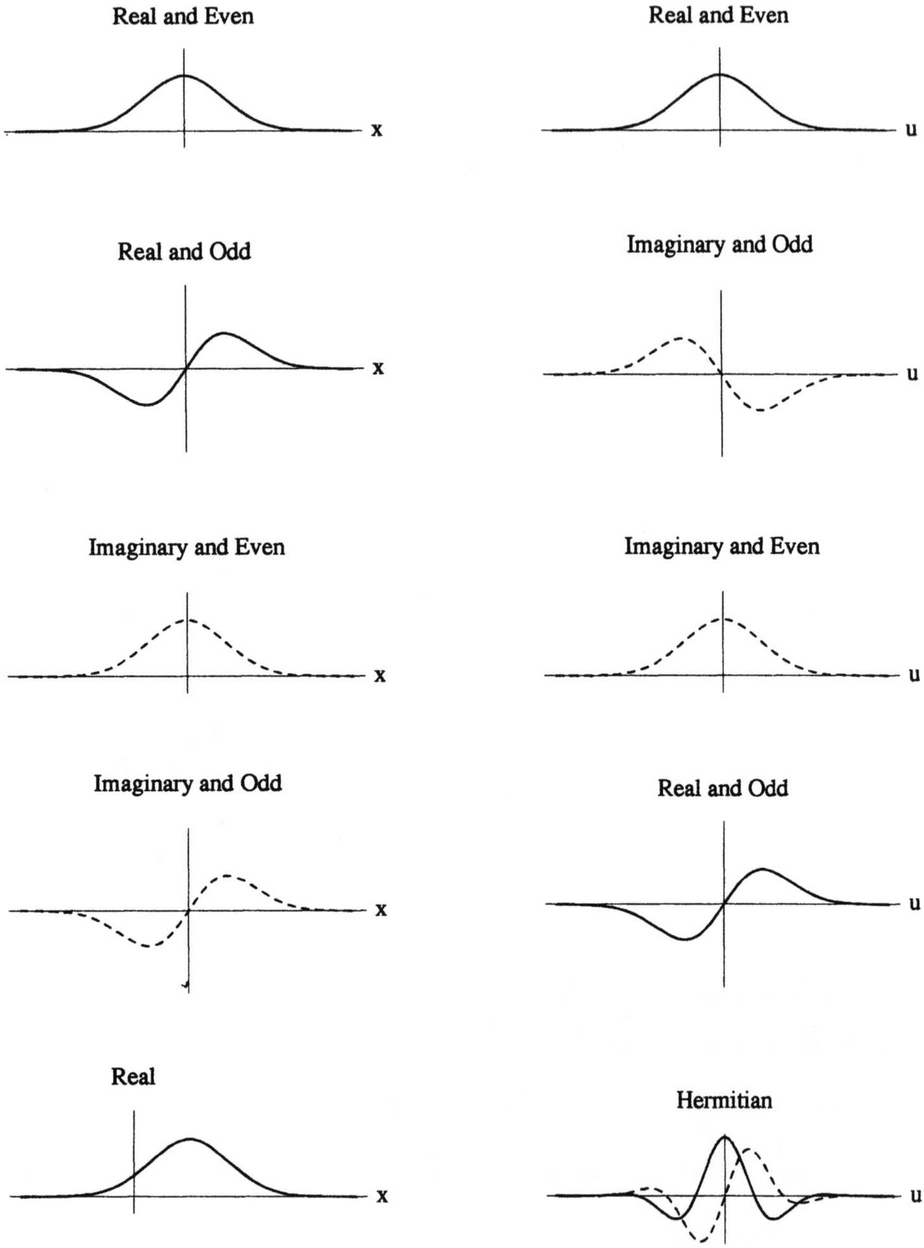

Figure 4.10 Some real and imaginary functions (left) and their corresponding Fourier transforms (right). Imaginary parts are indicated with dashed curves.

It follows immediately that for a Hermitian function the real part is even and the imaginary part is odd. For an anti-Hermitian function the real part is odd and the imaginary part is even.

The Fourier transform has the interesting property of preserving even and oddness. Let f_e and f_o be two real functions that are even and odd respectively. Then

$$\begin{aligned} F_e &= \mathcal{F}\{\, f_e \,\} \text{ is real and even} \\ F_o &= \mathcal{F}\{\, f_o \,\} \text{ is imaginary and odd} \end{aligned} \tag{4.74}$$

Here, imaginary means complex with the real part equal to zero. As a consequence of this relation, it follows that the Fourier transform of a real function is Hermitian and, conversely, that the Fourier transform of a Hermitian function is real. See Figure 4.10.

4.3 THREE ASPECTS OF THE FOURIER TRANSFORM

This section presents three different aspects of the Fourier transform which may not be covered by elementary textbooks on Fourier transforms. The first one relates asymptotic behaviour in one domain to discontinuities of derivatives in the other domain. The second is the famous uncertainty relation that will haunt us for the rest of this book. The third aspect is the projection slice theorem which describes how projections of multi-dimensional functions into lower dimensions can be related using the Fourier transform. A more thorough treatment of the first two aspects is found in [23], Chapter 8.

4.3.1 Asymptotic and impulsive behavior

In the following, we will sometimes need to consider discontinuous functions. If these functions can be assumed to be piece-wise continuous and bounded, Equation (4.18) suggests that their derivatives are impulsive at the points of discontinuity, see Figure 4.11. In other cases, the first order derivative of a function may be continuous but the second order derivative impulsive, etc. In the following, when referring to a function for which the n-th order derivative

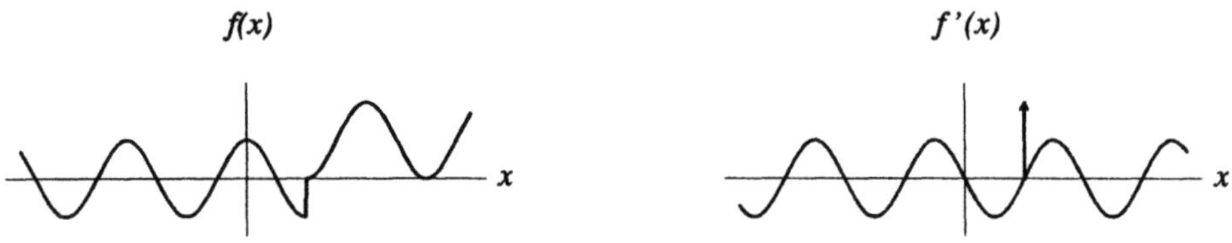

Figure 4.11 A discontinuous function (left) and its first derivative (right).

is impulsive it is understood that lower order derivatives, i.e. of order less than n, are *not impulsive.*

We may in other cases be interested in functions that decay like x^{-n} or u^{-n} for some integer n. The following proposition states that the Fourier transform makes a correspondence between these two classes of functions. To begin with let us assume the one-dimensional case.

Proposition 4.1. Let f be a proper function such that its n-th derivative is impulsive. Then F, the Fourier transform corresponding to f, decays like u^{-n}. The latter statement is the same as saying that the function

$$u^n F(u) \tag{4.75}$$

is bounded and does not go to zero when $u \to \pm\infty$. The proposition is valid for all integers $n \geq 0$, where the 0-th order derivative of f is f itself. □

Because of the relative symmetry between $\mathcal{F}$ and $\mathcal{F}^{-1}$, Proposition 4.1 also works the other way around, i.e. if the n-th derivative of F is impulsive, then and only then will f decay like x^{-n}. Evidently, F itself is impulsive if and only if f contains a constant term.

Example. Consider the function f defined as

$$f(x) = \begin{cases} 0 & x < 0 \\ 1 & 0 < x < 1 \\ 2 - x & 1 < x < 2 \\ 0 & 2 < x \end{cases} \tag{4.76}$$

This function is obviously discontinuous at $x = 0$ and, hence, its derivative is impulsive at that point. From Proposition 4.1 it follows that $F(u)$ decays like u^{-1}. Note, that the second order derivative of f is also impulsive. The decay of $F(u)$ is, however, determined by the derivative of lowest order that is impulsive. The function f, and its derivative, are shown in Figure 4.12 which also illustrates that F decays like u^{-1} by showing the magnitude of F multiplyed by u, resulting in a function that tend to be constant for large values of $|u|$. □

Proposition 4.1 is also applicable in multiple dimensions. The derivatives of various orders of f which have been considered above must then be specified to be taken with respect to a certain variable x_k. The proposition then describes the asymptotic behaviour of F with respect to the corresponding frequency variable u_k.

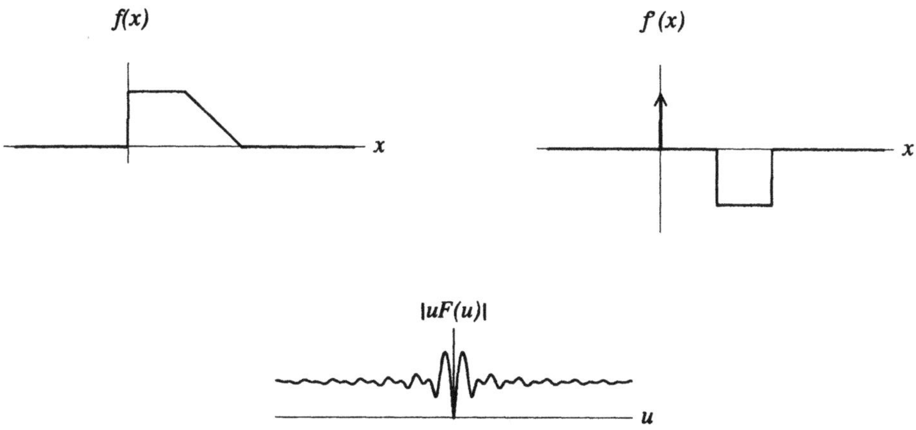

Figure 4.12 A function f (top left), its first order derivative (top right) and the magnitude of $u \cdot F$ (bottom).

4.3.2 The uncertainty principle

Given a one-dimensional function f we define the mean and the variance of $|f(x)|^2$, denoted $\bar{x}$ and σ_f^2, as

$$\bar{x} = \frac{\int_{-\infty}^{\infty} x\, f^\star(x)\, f(x)\, dx}{\int_{-\infty}^{\infty} f^\star(x)\, f(x)\, dx}, \qquad \sigma_f^2 = \frac{\int_{-\infty}^{\infty} (x-\bar{x})^2\, f^\star(x)\, f(x)\, dx}{\int_{-\infty}^{\infty} f^\star(x)\, f(x)\, dx}. \tag{4.77}$$

Let F be the Fourier transform corresponding to f. The mean and variance of $|F(u)|^2$, denoted $\bar{u}$ and σ_F^2, are then defined as

$$\bar{u} = \frac{\int_{-\infty}^{\infty} u\, F^\star(u)\, F(u)\, du}{\int_{-\infty}^{\infty} F^\star(u)\, F(u)\, du}, \qquad \sigma_F^2 = \frac{\int_{-\infty}^{\infty} (u-\bar{u})^2\, F^\star(u)\, F(u)\, du}{\int_{-\infty}^{\infty} F^\star(u)\, F(u)\, du}. \tag{4.78}$$

With some effort it is possible to prove the following inequality [1],

$$\sigma_f^2 \cdot \sigma_F^2 \geq \tfrac{1}{4}. \tag{4.79}$$

In terms of standard deviations, σ_f and σ_F, Equation (4.79) corresponds to

$$\sigma_f \cdot \sigma_F \geq \tfrac{1}{2} \tag{4.80}$$

This inequality is known as the *uncertainty principle* or the *uncertainty relation.* It should be noted that some functions do not have a well-defined, i.e. finite, variance. The importance of this principle cannot be overestimated and we will therefore illustrate the implications of Equation (4.79) with some examples.

[1] See for example Bracewell [23], page 160.

Figure 4.13 The function defined in Equation (4.81) and its corresponding Fourier transform defined in to Equation (4.82).

Example. Consider the function

$$f(x) = e^{-a|x|}, \tag{4.81}$$

where $a > 0$, and its corresponding Fourier transform

$$F(u) = \frac{2a}{a^2+u^2}. \tag{4.82}$$

See Figure 4.13. We compute the standard deviations of these functions

$$\begin{aligned} \sigma_f &= \frac{1}{\sqrt{2}a}, \\ \sigma_F &= a. \end{aligned} \tag{4.83}$$

This gives

$$\sigma_f \cdot \sigma_F = \frac{1}{\sqrt{2}} > \frac{1}{2}, \tag{4.84}$$

which is in agreement with the uncertainty principle. □

The standard deviation can be given an interpretation in terms of the width of a function. In the above example, both f and F have a distinct "bumps" and it is the widths of these bumps that are of interest here. The parameter a serves in this example as a measure of the width of F and of the inverse width of f. If a is made larger, the bump of f will become narrower and the bump of F will become broader, and conversely if a is made smaller.

It is often convenient to define the *energy functions* corresponding to f and F, respectively, as

$$\begin{aligned} e(x) &= |f(x)|^2 = f^\star(x)f(x) \\ E(u) &= |F(u)|^2 = F^\star(u)F(u) \end{aligned} \tag{4.85}$$

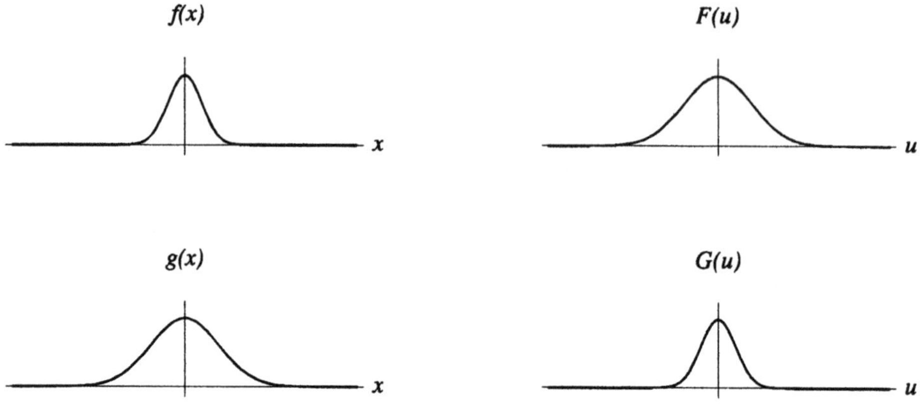

Figure 4.14 Top: A narrow function (left) and its corresponding Fourier transform which is broad (right). Bottom: A broad function (left) and its corresponding Fourier transform which is narrow (right).

These two functions are always real and are in fact the square of the magnitude of f or F, respectively; this implies that they may be used in order to obtain an intuitive description of the magnitudes of f and F. The uncertainty principle asserts that the standard deviations of the two energy functions are related and that they cannot both be made arbitrary small. In intuitive terms this implies that the energy of f and F can not be confined to intervals of arbitrary size both in the spatial domain and in the Fourier domain, respectively. As an example, consider a function f which is zero everywhere except around the point $x = x_0$, resulting in a relatively small value of its variance. The Fourier transform of this function must then have a relatively large value of its variance which corresponds to the Fourier transform being broad. The extreme example of this situation is the spatial impulse function $f(x) = \delta(x)$ and its corresponding Fourier transform $F(u) = 1$. Of course, the situation works the other way around as well. If the Fourier transform of some function, f, is confined within a small region, then the function itself is broad (Figure 4.14).

The uncertainty principle may seem awkward at first sight. Why should we not be able to find a function such that both its energy and the energy of the corresponding Fourier transform are confined within regions that are arbitrary small? The reason is that the Fourier transform of a function is not independent of the function itself. In fact, it is rather the opposite situation. Since the

Fourier transform, $\mathcal{F}$, is invertible, the Fourier transform corresponding to a specific function f contains all information necessary to describe the function itself. If we change the function somehow, then the corresponding Fourier transform will change as well. This is the same type of relation that we have between a function and its derivative. We can not choose a function such that both the function itself and its derivative are arbitrary, for example a function that is growing and a corresponding derivative that is negative.

It should be noted that the uncertainty principle is an inequality, implying that the left hand side of Equation (4.80) is, in general, larger than the right hand side. In fact, for some common functions such as the rectangular function, the left hand side is infinite. The are, however, functions that satisfy the uncertainty principle with equality. This is illustrated with the following example.

Example. Consider the function

$$f(x) = e^{-\frac{1}{2}(\frac{x}{a})^2}, \tag{4.86}$$

with corresponding Fourier transform

$$F(u) = \sqrt{2\pi}\, a\, e^{-\frac{1}{2}(au)^2} \tag{4.87}$$

From these functions we compute the standard deviations.

$$\begin{aligned} \sigma_f &= \frac{a}{\sqrt{2}} \\ \sigma_F &= \frac{1}{\sqrt{2}a} \end{aligned} \tag{4.88}$$

which gives

$$\sigma_f \cdot \sigma_F = \frac{1}{2}. \tag{4.89}$$

Hence, for arbitrary values of a the uncertainty principle is satisfied with equality. For this reason, Gaussian functions are often referred to as optimal with respect to the uncertainty principle. □

A situation where the uncertainty principle is of main importance is in the design of filters for digital signal processing, e.g. image processing. In practical implementation these filters must often have finite spatial extension, i.e. the filter functions must be zero outside an interval $a \leq x \leq b$. It is easy to prove that for any such function it is the case that $\sigma_f \leq b-a$. Hence, if the filter size $\Delta x = b - a$ is specified at the outset, then we are restricted to consider only filters for which

$$\sigma_F \geq \frac{1}{2\Delta x}. \tag{4.90}$$

This implies that F, the Fourier transform corresponding to f, can not be zero outside an interval smaller than $(2\Delta x)^{-1}$. Conversely, if we want to implement a filter with a specified frequency function F, i.e. with a given σ_F, the uncertainty principle states that

$$\Delta x \geq \frac{1}{2\,\sigma_F}. \tag{4.91}$$

In the context of image processing, filter outputs are often related to descriptions of local image properties, such as spatial frequency or orientation. If the filter size is Δx, then the filter output at $x = x_0$ is typically related to the behaviour of the signal in the interval $x_0 - \frac{\Delta x}{2} < x < x_0 + \frac{\Delta x}{2}$. Hence, Δx is a measure of uncertainty with respect to where the property is positioned. Therefore, we often want Δx to be small in order to give the filter output a well-defined spatial position. On the other hand, σ_F is a measure of uncertainty with respect to property. We want often to employ filter functions that are well localized in the Fourier domain, e.g. band-pass filters, in order to assign specific properties to the filter output. This implies that σ_F should be small. According to the uncertainty principle, however, both Δx and σ_F can not both be made arbitrarily small.

With the above result in mind, we may define filter design as a process of finding a filter function that to some acceptable degree is well localized in both domains. Depending on the application this may imply that the localization is relatively better in one domain than the other, but often we are interested in filters with good localization in both domains. In this case, filters for which

$$\sigma_f \cdot \sigma_F = \frac{1}{2} \tag{4.92}$$

are then of interest and are sometimes referred to as *optimal* with respect to the uncertainty relation. According to the previous example, Gaussian filters are optimal in this sense.

We end this discussion on uncertainty by stating a proposition that relates the variances of two functions to the variance of their convolution.

Proposition 4.2. Let f and g be two one-dimensional functions with variances σ_f^2 and σ_g^2, respectively. Let $h = f * g$. Then

$$\sigma_h^2 = \sigma_f^2 + \sigma_g^2, \tag{4.93}$$

i.e. variances add under convolution. □

There is no straightforward extension of the uncertainty principle to multiple dimensions. In [97], this issue is treated by considering variances of f and F along the n axes in the spatial and the Fourier domain, respectively. It is then proved that the uncertainty principle is valid for each of the n pairs of such variances. This requires, however, dealing with n uncertainty products which will additionally depend, in the general case, on the choice of coordinate systems in the two domains.

4.3.3 The projection slice theorem

Let f be an n-dimensional function with corresponding Fourier transform F and let $\hat{\mathbf{e}} \in \mathbb{R}^n$ be a unit vector. This vector then describes a line that passes through the origin according to $\mathbf{x} = \mathbf{x}(y) = y\,\hat{\mathbf{e}}$. Each value of y thus corresponds to a point on the line. A one-dimensional function, $f_1(y)$, is then defined by integrating f along all directions perpendicular to $\hat{\mathbf{e}}$. This operation may also be seen as projecting all values of f orthogonally onto the line. Hence,

$$f_1(x) = \int_{\mathbb{R}^n} f(\mathbf{x})\,\delta(x - \mathbf{x}^T\hat{\mathbf{e}})\,d^n x. \tag{4.94}$$

F_1, the Fourier transform corresponding to f, is then given by

$$\begin{aligned} F_1(u) = \int_{-\infty}^{\infty} f_1(y)\,e^{-iuy}\,dy = \int_{-\infty}^{\infty}\int_{\mathbb{R}^n} f(\mathbf{x})\,\delta(y - \mathbf{x}^T\hat{\mathbf{e}})\,e^{-iuy}\,d^n x\,dy = \\ \int_{\mathbb{R}^n} f(\mathbf{x}) \int_{-\infty}^{\infty} \delta(y - \mathbf{x}^T\hat{\mathbf{e}})\,e^{-iuy}\,dy\,d^n x = \int_{\mathbb{R}^n} f(\mathbf{x})\,e^{-iu\mathbf{x}^T\hat{\mathbf{e}}}\,d^n x = F(u\,\hat{\mathbf{e}}). \end{aligned} \tag{4.95}$$

Hence, F_1 assumes the values of F along a line through the origin which is parallel to $\hat{\mathbf{e}}$. This relation between the Fourier transforms of a function, f, and a projected version thereof, f_1, is called the *projection slice theorem* or the *Fourier slice theorem.*

Example. Consider a simple X-ray tomography device (Figure 4.15). A linear X-ray source, T, sends parallel rays that are detected at a linear sensor, R, and the tissue to be examined is placed between the source and the sensor. The density of the tissue to X-rays is described by a two-dimensional function, $f(x_1, x_2)$ and the goal is to reconstruct this function. Let the X-rays move in the direction perpendicular to a unit vector $\hat{\mathbf{e}}$. At point x at the sensor array we will get the density value $f_1(x)$, as described by Equation (4.94), since the densities along each ray are integrated. For each direction of $\hat{\mathbf{e}}$, we measure the projected values of f which are given by the function f_1. The one-dimensional function f_1, which depends on $\hat{\mathbf{e}}$, has a corresponding Fourier transform, F_1. According to the projection slice theorem, F_1 must have the same values as F, the Fourier transform corresponding to f, along the line parallel to $\hat{\mathbf{e}}$. This means that we may assign values for F at all points along this line. By changing the direction of the X-rays, i.e. by rotating the source and sensor devices, we may assign values to F along all possible lines through the origin. Since, the projection of f in the directions $\hat{\mathbf{e}}$ and $-\hat{\mathbf{e}}$ are mirror images, all values of F are known after half a turn. The function f can then be obtained simply by an inverse Fourier transform of F. Hence, using the above procedure we can reconstruct the density of a slice of the tissue which then can be used for further analysis, e.g. for medical purposes. This procedure is the general outline for all types of tomography methods, even though most equipment for practical purposes uses more sophisticated schemes in order to gain speed and accuracy of the reconstruction. □

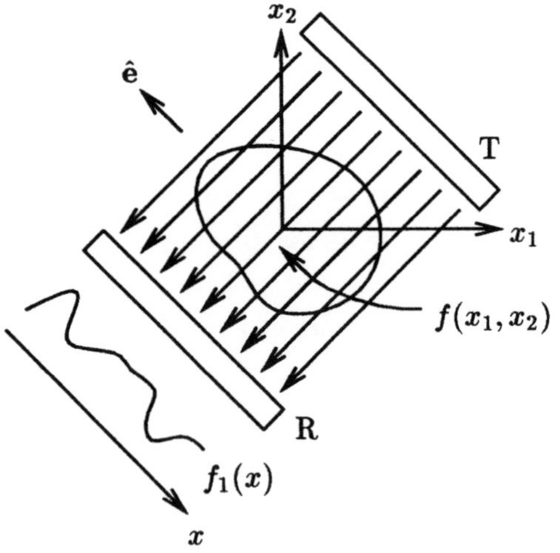

Figure 4.15 A schematic illustration of a tomography device.

4.4 SEPARABILITY

It turns out that many multi-dimensional functions which are of interest are separable. In this context, separable implies that such a function can be written as a product of functions, each of lower dimensionality than the original. In the extreme case, the separation or decomposition can be made using only one-dimensional functions. In some cases the variables of the resulting functions will be the same as those of the original function; in other cases some variable transformation will be done. The former case is referred to as Cartesian separability; an example of the latter is spherical separability.

The fact that certain multi-dimensional functions are somehow separable into one-dimensional functions will have a tremendous impact on the efforts needed to compute their Fourier transforms, both analytically and numerically. As will be demonstrated in the following subsections it will imply that the Fourier transform can be obtained by means of one-dimensional Fourier transforms or relatives thereof. We begin with Cartesian separability, a property that applies both to certain functions and to the Fourier transform itself. We then introduce the concept of simple functions and functions of rank p, and, finally, present a treatment of spherical symmetry and spherical separability.

4.4.1 Cartesian separability

Here, we discuss the concept of Cartesian separability both in terms of the Fourier transform $\mathcal{F}$ and in terms of spatial or frequency functions.

According to its definition in Equation (4.28), it follows that the n-dimensional Fourier transform $\mathcal{F}$ is separable into n one-dimensional Fourier transforms. As an example, consider the two-dimensional case where

$$F(u_1, u_2) = \int_{-\infty}^{\infty} \int_{-\infty}^{\infty} f(x_1, x_2)\, e^{-i(u_1 x_1 + u_2 x_2)}\, dx_1 dx_2. \qquad (4.96)$$

This equation may be rewritten as

$$F(u_1, u_2) = \int_{-\infty}^{\infty} \left[\int_{-\infty}^{\infty} f(x_1, x_2)\, e^{-iu_1 x_1} dx_1 \right] e^{-iu_2 x_2}\, dx_2. \qquad (4.97)$$

Hence, F may be computed by integrating first with respect to x_1 and then with respect to x_2 (or vice versa). During the first integration, x_2 is considered to be a constant but the result will of course depend on the value of x_2. If we define a function f_1, such that

$$f_1(u_1, x_2) = \int_{-\infty}^{\infty} f(x_1, x_2)\, e^{-iu_1x_1}\, dx_1, \tag{4.98}$$

it is evident that f_1 is defined as a one-dimensional Fourier transform of f while keeping x_2 fixed. The function f_1 will of course depend on what fixed value we choose for x_2 and is therefore a function of both u_1 and x_2. Note that f_1 is not an element of either S_1 or S_2, but is rather some sort of half-way transform. By inserting Equation (4.98) into Equation (4.97), we see that

$$F(u_1, u_2) = \int_{-\infty}^{\infty} f_1(u_1, x_2)\, e^{-iu_2x_2}\, dx_2, \tag{4.99}$$

which implies that F may be obtained by a one-dimensional Fourier transform of f_1. Hence, a two-dimensional Fourier transform may be computed by employing two one-dimensional transforms in succession. Note, that the two transforms may be taken in arbitrary order. In general, for n-dimensional functions f, the corresponding Fourier transform F can be computed by taking n one-dimensional Fourier transforms in succession. This property is referred to as the *Cartesian separability* of the Fourier transform $\mathcal{F}$. This discussion applies of course also to the inverse Fourier transform $\mathcal{F}^{-1}$.

Let f be an n-dimensional function which can be written as a product of n one-dimensional functions, i.e.

$$f(x_1, x_2, \ldots, x_n) = f_1(x_1)\, f_2(x_2) \ldots f_n(x_n). \tag{4.100}$$

A function f with this property is referred to as being *Cartesian separable.* As a direct consequence of the Cartesian separability of $\mathcal{F}$, the Fourier transform corresponding to f is given by the following expression

$$F(u_1, u_2, \ldots, u_n) = F_1(u_1)\, F_2(u_2) \ldots F_n(x_n), \tag{4.101}$$

where F_k is the Fourier transform corresponding to f_k. Cartesian separable functions tend to appear quite often in practical problems, and a lot of effort may thus be saved by computing the Fourier transforms of such functions by means of a suitable number of one-dimensional Fourier transforms.

Example. Consider a two-dimensional Gaussian function

$$f(x_1, x_2) = e^{-\frac{1}{2}[(\frac{x_1}{a})^2 + (\frac{x_2}{b})^2]} \qquad a, b > 0. \tag{4.102}$$

This function is Cartesian separable since it is possible to write

$$f(x_1, x_2) = e^{-\frac{1}{2}(\frac{x_1}{a})^2} \cdot e^{-\frac{1}{2}(\frac{x_2}{b})^2}. \tag{4.103}$$

In accordance with the transform pair in Section 4.7, the Fourier transform corresponding to f is then given by

$$F(u_1, u_2) = \sqrt{2\pi}\, a\, e^{-\frac{1}{2}(au_1)^2} \cdot \sqrt{2\pi}\, b\, e^{\frac{1}{2}(bu_2)^2}, \tag{4.104}$$

which may be rewritten as

$$F(u_1, u_2) = 2\pi ab\, e^{-\frac{1}{2}[(au_1)^2 + (bu_2)^2]}. \tag{4.105}$$

We infer from this example that an n-dimensional Gaussian, which is Cartesian separable into n one-dimensional Gaussians, has a corresponding Fourier transform that also is an n-dimensional Gaussian. □

Example. The multi-dimensional function III^n is defined as

$$\text{III}^n(\mathbf{x}) = \text{III}(x_1) \ldots \text{III}(x_n), \tag{4.106}$$

and is apparently Cartesian separable. According to the transform pair in Section 4.7, it must therefore be the case that

$$\mathcal{F}\{\, \text{III}^n(\mathbf{x})\,\} = \text{III}(\tfrac{u_1}{2\pi}) \ldots \text{III}(\tfrac{u_n}{2\pi}) = \text{III}^n(\tfrac{\mathbf{u}}{2\pi}), \tag{4.107}$$

from which we infer that the shah-function is its own Fourier transform, regardless of the dimensionality. □

Sometimes a change of basis must be employed before a function is separable. Assume that $g(\mathbf{x}) = f(\mathbf{A}\mathbf{x})$, where f is Cartesian separable. To obtain, G, the Fourier transform corresponding to g, we simply use Equation (4.49), where F is the Fourier transform corresponding to f. Since f is Cartesian separable, F is given by Equation (4.101).

4.4.2 Simple functions and functions of rank p

A general n-dimensional function may have any type of variation. In this section, however, we will restrict our attention to functions that vary according to the following. Let $\{\hat{\mathbf{e}}_k,\ k = 1,\ldots,p\}$ be a set of p orthonormal vectors in $\mathbb{R}^n$ and let g be a p-dimensional function, $p \leq n$. If an n-dimensional function f can be written as

$$f(\mathbf{x}) = g(\mathbf{x}^T\hat{\mathbf{e}}_1, \mathbf{x}^T\hat{\mathbf{e}}_2, \ldots, \mathbf{x}^T\hat{\mathbf{e}}_p), \tag{4.108}$$

then f is said to have ***rank p***. In intuitive terms this implies that a rank p function varies in p directions and is constant in $n-p$ directions. This follows immediately from the fact that the subspace spanned by vectors orthogonal to all $\hat{\mathbf{e}}_k$ is of dimension $n-p$ and from that there is no variation of f around a point $\mathbf{x} = \mathbf{x}_0$ along a vector of this subspace. It should be noted that in order to make the above definition meaningful we must assume that g can not be written as a rank q function where $q < p$.

Of particular interest are functions of rank one, i.e. those functions that vary in one direction only, which in the following are referred to as *simple functions*. Let $\hat{\mathbf{e}}$ be a unit vector in $\mathbb{R}^n$ and let f be an n-dimensional function such that

$$f(\mathbf{x}) = g(\mathbf{x}^T\hat{\mathbf{e}}), \tag{4.109}$$

where g is a one-dimensional function. Then, f is a simple function. A line parallel to $\hat{\mathbf{e}}$ which passes through the point $\mathbf{x}_0$ is then described by the expression $\mathbf{x} = \mathbf{x}_0 + t\hat{\mathbf{e}},\ t \in \mathbb{R}$. The value of f along this line, as a function of t, is then described by the following expression.

$$f(\mathbf{x}) = f(\mathbf{x}_0 + t\hat{\mathbf{e}}) = g(\mathbf{x}_0^T\hat{\mathbf{e}} + t), \tag{4.110}$$

i.e. the value of f along this line is a shifted version of g, the shift being determined by $\mathbf{x}_0^T\hat{\mathbf{e}}$. If we instead consider the value of f along lines perpendicular to $\hat{\mathbf{e}}$, i.e. when $\mathbf{x} = \mathbf{x}_0 + t\hat{\mathbf{f}},\ \hat{\mathbf{f}}^T\hat{\mathbf{e}} = 0$, then f is given by

$$f(\mathbf{x}) = f(\mathbf{x}_0 + t\hat{\mathbf{f}}) = g(\mathbf{x}_0^T\hat{\mathbf{e}}), \tag{4.111}$$

i.e. f is constant along these lines. Note that for each simple function f there is a unique direction of variation, represented either by $\hat{\mathbf{e}}$ or by $-\hat{\mathbf{e}}$, and having established this direction there is a unique one-dimensional function g which validates Equation (4.109).

Evidently, simple functions are special cases of Cartesian separable functions. Let $\mathbf{A}$ be an orthogonal matrix which transforms $\hat{\mathbf{e}}$ to a unit vector pointing in the direction of the x_1-axes. Then

$$h(\mathbf{x}) = f(\mathbf{A}\mathbf{x}) = g(x_1). \tag{4.112}$$

According to Section 4.4.1, the Fourier transform corresponding to h is

$$\begin{aligned} H(\mathbf{u}) &= (2\pi)^{n-1}\, G(u_1)\cdot\delta(u_2)\,\delta(u_3)\ldots\delta(u_n) = \\ &= (2\pi)^{n-1}\, G(u_1)\,\delta^{\text{line}}_{\hat{\mathbf{u}}_1}(\mathbf{u}), \end{aligned} \tag{4.113}$$

where $\hat{\mathbf{u}}_1$ is a unit vector in the direction of the u_1-axes. Hence, H is zero everywhere except on the u_1-axes on which it varies like G, the Fourier transform of g. To be more precise, apart from this variation, H will also contain the product of $n-1$ impulses on this axes. According to Section 4.2.8, an orthogonal transformation of coordinates in one domain corresponds to the same orthogonal transformation in the other domain. Here, we have made an orthogonal change of coordinates, from f to h, and proved that H has all its energy concentrated along the u_1-axes in the Fourier domain. To obtain F we simply have to transform the coordinates back again which results in the u_1-axes being transformed to the $\hat{\mathbf{e}}$ vector. Hence, we can write

$$F(\mathbf{u}) = (2\pi)^{n-1}\, G(\mathbf{u}^T\hat{\mathbf{e}})\,\delta^{\text{line}}_{\hat{\mathbf{e}}}(\mathbf{u}) \tag{4.114}$$

Example. Let the two-dimensional function f be given by

$$f(x_1,x_2) = e^{-\frac{1}{2}[(\frac{x_1+x_2}{\sqrt{2}})^2]}. \tag{4.115}$$

It is immediately clear that

$$f(\mathbf{x}) = e^{-\frac{1}{2}(\mathbf{x}^T\hat{\mathbf{e}})^2}, \tag{4.116}$$

for the unit vector

$$\hat{\mathbf{e}} = \begin{pmatrix} \frac{1}{\sqrt{2}} \\ \frac{1}{\sqrt{2}} \end{pmatrix}. \tag{4.117}$$

This may also be described as

$$f(\mathbf{x}) = g(\mathbf{x}^T\hat{\mathbf{e}}) \tag{4.118}$$

where

$$g(x) = e^{-\frac{1}{2}x^2}. \tag{4.119}$$

Hence, f is simple and varies only in the $\hat{\mathbf{e}}$-direction and, therefore, its Fourier transform is zero everywhere except on a line through the origin parallel to $\hat{\mathbf{e}}$. On this line F will, apart from an impulse factor, vary like

$$G(u) = \sqrt{2\pi}e^{-\frac{1}{2}u^2}, \tag{4.120}$$

the Fourier transform corresponding to g. □

Example. Consider a three-dimensional function f, defined as

$$f(\mathbf{x}) = \delta_{\hat{\mathbf{e}}}^{\text{plane}}(\mathbf{x}) = \delta(\mathbf{x}^T\hat{\mathbf{e}}), \tag{4.121}$$

for some unit vector $\hat{\mathbf{e}}$. This function is zero everywhere except on the plane through the origin which is perpendicular to $\hat{\mathbf{e}}$. Along all lines parallel to $\hat{\mathbf{e}}$ f will look like the impulse function δ. The Fourier transform of f, F, will thus be zero everywhere except on a line through the origin parallel to $\hat{\mathbf{e}}$. On this line will F vary like the Fourier transform of the impulse function that is equal to 1. Hence, F is a constant impulse line through the origin and parallel to $\hat{\mathbf{e}}$. To summarize, the Fourier transform of an impulse plane is an impulse line, where the line and the plane are perpendicular (Figure 4.16). Because of the relative symmetry of $\mathcal{F}$ and $\mathcal{F}^{-1}$, it must also be the case that the Fourier transform of an impulse line is an impulse plane. □

Figure 4.16 An impulse plane in three dimensions (left) and its corresponding Fourier transform (right), an impulse line.

4.4.3 Spherically symmetric functions

Let f be an n-dimensional function such that

$$f(\mathbf{x}) = g(r) \qquad r^2 = x_1^2 + x_2^2 + \ldots + x_n^2, \tag{4.122}$$

for some one-dimensional function g. A function f of this type is called *spherically symmetric* since its value in a point $\mathbf{x}$ depends only on the radial distance r to the origin. It turns out that the Fourier transform corresponding to a spherically symmetric function also is spherically symmetric. To see this, we simply compute the Fourier transform of the function f, defined as above, i.e.

$$F(\mathbf{u}) = \int_{\mathbb{R}^n} f(\mathbf{x})\, e^{-i\mathbf{u}^T\mathbf{x}}\, d^n x. \tag{4.123}$$

To prove that F is a function only of the radial distance to the origin, here denoted ρ, let $\mathbf{u}_1$ and $\mathbf{u}_2$ be two frequency coordinates such that $|\mathbf{u}_1| = |\mathbf{u}_2|$ and let $\mathbf{A}$ be an orthogonal matrix such that $\mathbf{u}_2 = \mathbf{A}\mathbf{u}_1$. This gives

$$\begin{aligned} F(\mathbf{u}_2) &= \int_{\mathbb{R}^n} f(\mathbf{x})\, e^{-i\mathbf{u}_2^T\mathbf{x}}\, d\mathbf{x} = \int_{\mathbb{R}^n} f(\mathbf{x})\, e^{-i\mathbf{u}_1^T\mathbf{A}^T\mathbf{x}}\, d\mathbf{x} = \left[\mathbf{y} = \mathbf{A}^T\mathbf{x}\right] = \\ &= \frac{1}{|\det \mathbf{A}|} \int_{\mathbb{R}^n} f(\mathbf{A}\mathbf{y})\, e^{-i\mathbf{u}_1^T\mathbf{y}}\, d\mathbf{y} = \int_{\mathbb{R}^n} f(\mathbf{y})\, e^{-i\mathbf{u}_1^T\mathbf{y}}\, d\mathbf{y} = F(\mathbf{u}_1) \end{aligned} \tag{4.124}$$

Another proof is supplied by the projection slice theorem. This theorem states that if f is projected along all directions orthogonal to a unit vector $\hat{\mathbf{e}}$, the Fourier transform of the resulting function f_1 is found in F along the line parallel to $\hat{\mathbf{e}}$. Since f is spherically symmetric, however, this implies that f_1 does not depend on $\hat{\mathbf{e}}$ which therefore must be the case also for F. We will now exemplify this phenomena in two dimensions.

Example. Let f be a two-dimensional function which is spherically symmetric (maybe circularly symmetric is more appropriate in this case). This means that

$$f(x_1, x_2) = g(r), \qquad r^2 = x_1^2 + x_2^2. \tag{4.125}$$

Instead of computing the Fourier transform of f as a function of the two Cartesian frequency coordinates u_1 and u_2, we may here use the polar version of the Fourier transform, described by Equation (4.33), and write

$$F(\rho,\phi) = \int_0^{2\pi}\int_0^{\infty} g(r)\, e^{-i\rho r\cos(\varphi-\phi)}\, r\, dr d\varphi = [\alpha = \varphi - \phi] =$$

$$= \int_0^{\infty} g(r) \left[\int_0^{2\pi} e^{-i\rho r\cos\alpha} d\alpha \right] r\, dr. \tag{4.126}$$

The integral within the brackets is sometimes used to define the Bessel function J_0, i.e.

$$J_0(z) = \frac{1}{2\pi} \int_0^{2\pi} e^{-iz\cos\varphi}\, d\varphi. \tag{4.127}$$

Hence, Equation (4.126) may be rewritten as

$$F(\rho,\phi) = 2\pi \int_0^{\infty} g(r)\, J_0(\rho r)\, r\, dr. \tag{4.128}$$

From this equation it is evident that F is a function of only ρ. □

As demonstrated in the above example, the radial variation of F is given by a transformation of the radial variation of f. This transform is, however, not a Fourier transform since it employs a Bessel function rather than an exponential function in the integrand. Let S_1^+ and S_2^+ denote the sets of radial functions in the spatial and frequency domains, respectively. For each $g \in S_1^+$ we define a function $G_\nu \in S_2^+$ as

$$G_\nu(\rho) = 2\pi \int_0^{\infty} g(r)\, J_\nu(\rho r)\, r\, dr, \tag{4.129}$$

where J_ν is the Bessel function of order $\nu \in \mathbb{R}$. The transform of g defined by the right hand side of Equation (4.129), and also the resulting function G_ν, are called the *Hankel transform* of order ν. This transform, seen as a map from S_1^+ to S_2^+, is denoted $\mathcal{H}_\nu$, i.e. we can write

$$G_\nu = \mathcal{H}_\nu\{\, g\,\}. \tag{4.130}$$

Evidently, the radial variation of F in the previous example is given by a Hankel transform of order zero.

For arbitrary dimensionality n, the radial variation of F is given by a Hankel transform of order $\frac{n}{2}-1$. Let f be an n-dimensional and spherically symmetric function, i.e. it can be written as in Equation (4.122) for some one-dimensional function g. The Fourier transform of f is then given by

$$F(\rho) = \left(\frac{2\pi}{\rho}\right)^{\frac{n}{2}-1} H(\rho), \tag{4.131}$$

where

$$H = \mathcal{H}_{\frac{n}{2}-1}\{r^{\frac{n}{2}-1}\, g(r)\} \tag{4.132}$$

Example. In one dimension, a spherically symmetric function f is the same as an even function. As demonstrated in Section 4.2.9, the Fourier transform of an even function is again even, i.e. spherically symmetric, which is in accordance with the above results. Furthermore, for $n = 1$, the radial variation of F is given by

$$F(u) = \sqrt{2\pi u} \int_0^\infty f(x)\, J_{-\frac{1}{2}}(ux)\, \sqrt{x}\, dx, \tag{4.133}$$

where

$$J_{-\frac{1}{2}}(z) = \sqrt{\frac{2}{\pi z}}\, \cos z. \tag{4.134}$$

Hence,

$$F(u) = 2\int_0^\infty f(x)\, \cos ux\, dx = \int_{-\infty}^\infty f(x)\, e^{-iux}\, dx, \tag{4.135}$$

which is precisely the Fourier transform of f. □

Example. Consider an n-dimensional function f, defined as

$$f(\mathbf{x}) = r^\alpha, \tag{4.136}$$

for $-n < \alpha < 0$. The radial variation of F is then given by

$$F(\rho) = (\tfrac{2\pi}{\rho})^{\frac{n}{2}-1}\, 2\pi \int_0^\infty r^{\alpha+\frac{n}{2}}\, J_{\frac{n}{2}-1}(\rho r)\, dr = [q = \rho r] =$$

$$= (\tfrac{2\pi}{\rho})^{\frac{n}{2}-1}\, 2\pi \int_0^\infty \left(\frac{q}{\rho}\right)^{\alpha+\frac{n}{2}} J_{\frac{n}{2}-1}(q)\, \rho^{-1}\, dq = \tag{4.137}$$

$$= \rho^{-\alpha-n}\, (2\pi)^{\frac{n}{2}-1}\, G(1),$$

where

$$G = \mathcal{H}_{\frac{n}{2}-1}\{\, r^{\alpha+\frac{n}{2}-1}\,\}. \tag{4.138}$$

Hence, we see that

$$\mathcal{F}\{r^\alpha\} \propto \rho^{-\alpha-n}. \tag{4.139}$$

The restriction of α to the interval $-n < \alpha < 0$ guarantees the Hankel transform in Equation (4.138) to be well-defined. □

4.4.4 Spherical separability

An arbitrary vector $\mathbf{x} \in \mathbb{R}^n$ can be represented as

$$\mathbf{x} = r\,\hat{\mathbf{x}}, \tag{4.140}$$

where r is the norm of $\mathbf{x}$ and $\hat{\mathbf{x}}$ is a unit vector. We will sometimes have reason to study functions, f, that have the following property

$$f(\mathbf{x}) = f_r(r) \cdot f_\varphi(\hat{\mathbf{x}}), \tag{4.141}$$

i.e. they are written as a product of one radial function, f_r, and one directional function, f_φ. Such functions, f, are called *spherically separable*. As proved in Section 4.4.1, the Fourier transform of a Cartesian separable function is again Cartesian separable. Unfortunately, however, this is not the case for spherically separable functions, at least not for general directional functions. Spherically symmetric functions are a special case of spherically separable functions ($f_\varphi =$ constant) and, as demonstrated in Section 4.4.3, their Fourier transforms are again spherically separable with constant directional function. There are, however, directional functions other than a constant which result in spherically separable functions in both domains. A thorough investigation of this topic and for arbitrary dimensionality is beyond the scope of this presentation and we will merely observe the following example in two dimensions.

Example. Let f be a two-dimensional function that is spherically separable, i.e.

$$f(r,\varphi) = f_r(r) \cdot f_\varphi(\varphi), \tag{4.142}$$

where

$$f_\varphi(\varphi) = e^{im\varphi}, \qquad m \in \mathbb{Z}. \tag{4.143}$$

The Fourier transform of f is then given by

$$\begin{aligned} F(\rho,\psi) &= \int_0^\infty \int_0^{2\pi} f_r(r)\, f_\varphi(\varphi)\, e^{-i\rho r\cos(\psi-\varphi)}\, r\, d\varphi\, dr = \\ &= \int_0^\infty f_r(r) \int_0^{2\pi} e^{im\varphi}\, e^{-i\rho r\cos(\psi-\varphi)}\, d\varphi\, r\, dr = [\alpha = \varphi - \psi - \pi] = \\ &= (-1)^m\, e^{im\psi} \int_0^\infty f_r(r) \left[\int_0^{2\pi} e^{i(\rho r\cos\alpha + m\alpha)}\, d\alpha \right] r\, dr = \\ &= (-1)^m\, e^{im\psi} \int_0^\infty f_r(r)\, [\, 2\pi\, i^m\, J_m(\rho r)\,]\, r\, dr = \\ &= (-i)^m\, e^{im\psi}\, 2\pi \int_0^\infty f_r(r)\, J_m(\rho r)\, r\, dr = (-i)^m\, e^{im\phi}\, G_m(\rho), \end{aligned} \tag{4.144}$$

where G_m is the Hankel transform of order m of the radial function f_r. Hence, a spatial function that is spherically separable, with directional function $e^{im\phi}$, has a Fourier transform that is also spherically separable and whose directional function is the same. □.

We will now develop the above example a little further. First of all, it is the case that

$$G_{-m}(\rho) = (-1)^m \, G_m(\rho), \tag{4.145}$$

where G_k is the Hankel transform of order k of the radial function f. Hence, the same radial variation of F will result for $m = \pm k$. Furthermore, we note that for even values of m the spatial function f is an even function and for odd values of m f is an odd function. The Fourier transform corresponding to a real and even function is again real and even and real. The Fourier transform of a real and odd function is odd and imaginary. This is in complete agreement with the above results. Apparently, the directional function, f_ϕ, cannot be a linear combination of two or more functions of the type $e^{im\phi}$ unless $m = \pm k$ for one integer k. Different absolute values of m will result in different radial functions and, hence, the corresponding Fourier transform is no longer spherically separable.

For each dimensionality, n, it is possible to find directional functions $f(\hat{\mathbf{x}})$ that can be used to construct spatial functions that are spherically separable in both domains. In general these are not of the above type, for instance in three dimensions they are so called *spherical harmonics*. A general theory for how to construct these functions is beyond the scope of this presentation and we will merely assert that spherical symmetry in both domains is possible only for some special types of directional functions that are characteristic to each dimensionality n.

4.5 ANALYTIC SIGNALS

This section introduces the concept of analytic signals. These are defined first in one dimension which in fact is the only way to define analytic signals properly. The second subsection extends the concept to multiple dimensions in an arbitrary but practical way.

4.5.1 Analytic signals in one dimension

The Hilbert transform

The Hilbert transform is, in the same way as the Fourier transform, a mapping between two sets of functions. In the context of this presentation, we define the Hilbert transform, $\mathcal{H}i : S_1 \rightarrow S_1$, as

$$f_{\mathcal{H}i}(x) = \frac{1}{\pi} \int_{-\infty}^{\infty} \frac{f(\tau)}{\tau - x} \, d\tau, \tag{4.146}$$

and write

$$f_{\mathcal{H}i} = \mathcal{H}i\{ f \}. \tag{4.147}$$

Hence, the Hilbert transform maps one spatial function to another spatial function. Since $S_1 = S_2$, however, we may of course also consider Hilbert transforms in the Fourier domain. In view of Equation (4.45), Equation (4.146) may be rewritten as

$$f_{\mathcal{H}i} = f * \tfrac{-1}{\pi x}, \tag{4.148}$$

i.e. the Hilbert transform of f is obtained by convolving f with the function $\frac{-1}{\pi x}$. From this it follows immediately that the Hilbert transform is a *linear* mapping from S_1 to S_1. Let $F_{\mathcal{H}i}$ denote the Fourier transform corresponding to $f_{\mathcal{H}i}$. According to Equation (4.148) it must then be the case that

$$F_{\mathcal{H}i}(u) = F(u) \cdot i \, \mathrm{sign}(u), \tag{4.149}$$

since convolution in the spatial domain corresponds to multiplication in the Fourier domain. Hence, the Fourier transform of $f_{\mathcal{H}i}$ is obtained by multiplying F with the imaginary unit i and then changing the sign of the result for negative frequencies. Another way of describing this process is to rotate the argument of the frequency components an angle $\frac{\pi}{2}$ in the positive direction for positive frequencies and in the negative direction for negative frequencies, (Figure 4.17). From this it follows immediately that two Hilbert transforms in succession will simply change the sign of a function. Furthermore, it should be pointed out that the Hilbert transform is not well-defined for all functions in S_1. The unit function is one example.

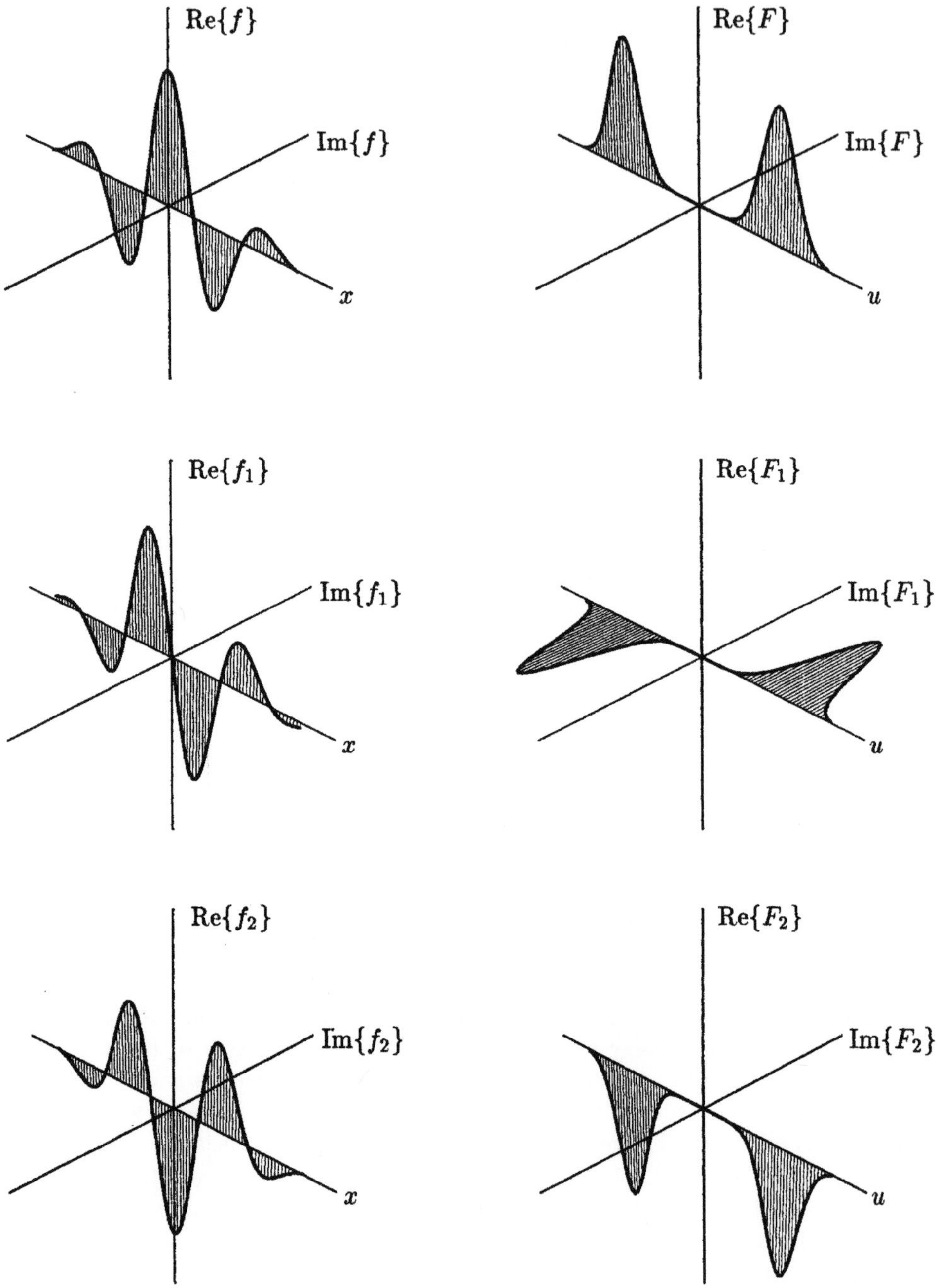

Figure 4.17 Top row: A real function f and its corresponding Fourier transform F. Middle row: The function $f_1 = \mathcal{H}i\{f\}$ and its corresponding Fourier transform F_1. Bottom row: The function $f_2 = \mathcal{H}i\{f_1\} = -f$ and its corresponding Fourier transform $F_2 = -F$.

The following is a short list of functions and their corresponding Hilbert transforms.

Function	Hilbert transform
$\cos(x)$	$-\sin x$
$\sin(x)$	$\cos x$
$\delta(x)$	$\dfrac{-1}{\pi x}$
$\mathrm{rect}(x)$	$\dfrac{1}{\pi}\ln\left\|\dfrac{2\mathrm{x}-1}{2\mathrm{x}+1}\right\|$

The analytic signal

Having defined the Hilbert transform, we continue and introduce the so-called *analytic function* or *analytic signal*, f_A, that corresponds to a function f. The analytic function corresponding to f is defined as

$$f_A = f - i f_{\mathcal{H}i}. \tag{4.150}$$

From this definition, it is clear that the construction of the analytic function is also a linear mapping from S_1 to itself.

Given the above transform table, we get the following correspondences

$$\begin{aligned} &\text{the analytic signal corresponding to } \cos x \text{ is } e^{ix}, \\ &\text{the analytic signal corresponding to } \sin x \text{ is } -ie^{ix}. \end{aligned} \tag{4.151}$$

Hence, the construction of the analytic function is not a one-to-one map since both $\cos x$ and $i \sin x$ have the same analytic signal, i.e. e^{ix}. In the light of

Equation (4.148) it is clear that we can rewrite Equation (4.150) as

$$f_A = f * [\delta(x) + \tfrac{i}{\pi x}], \tag{4.152}$$

which in the Fourier domain corresponds to

$$F_A = F \cdot [1 + \text{sign}(u)] = 2\,F \cdot \text{step}(u). \tag{4.153}$$

Hence, the analytic signal corresponding to f is obtained by suppressing all its negative frequencies and multiply by two. Note that this implies that an analytic function cannot be real and non-zero. This is a result from Section 4.2.9, where it is shown that the Fourier transform of a real function is always Hermitian, which F_A fails to be. However, this also means that the Hilbert transform of a real function f is indeed one-to-one, since the values of F for $u < 0$ may be derived from F for $u > 0$.

To summarize the above, we see that the analytic signal is of interest only for real signals. For such a signal, its corresponding analytic signal is complex, with the real part being the original signal itself and the imaginary part being its Hilbert transform. This is illustrated in Figure 4.18.

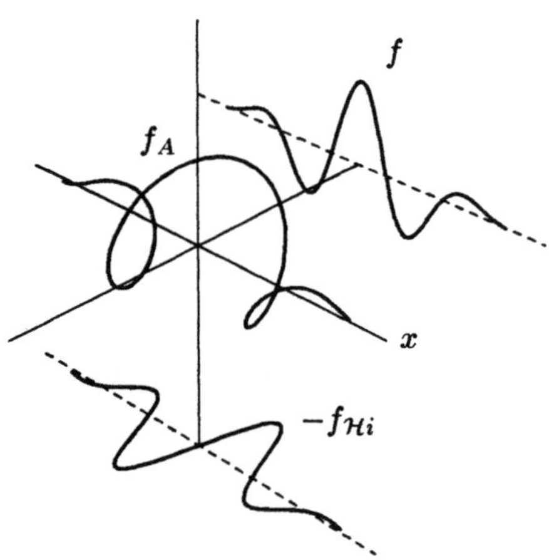

Figure 4.18 The analytic signal f_A corresponding to a real function f. The Hilbert transform $f_{\mathcal{H}i}$ of f is shown with reversed sign in the imaginary domain.

Applications of the analytic signal

We will now try to motivate the introduction of the Hilbert transform and the analytic signal. First of all, we assume that f is real and has a well-defined frequency. As an example, consider

$$f(x) = A\cos(\omega x) \qquad A > 0. \tag{4.154}$$

In other words, the signal f is a cosine function with amplitude A and angular frequency ω. From the previous discussion it follows that

$$f_{\mathcal{H}i}(x) = -A\sin(\omega x), \tag{4.155}$$

and

$$f_A(x) = A\left[\cos(\omega x) + i\sin(\omega x)\right] = A\, e^{i\omega x}. \tag{4.156}$$

This means that the amplitude A is given directly by

$$A = |f_A(x)| = \sqrt{[f(x)]^2 + [f_{\mathcal{H}i}(x)]^2}. \tag{4.157}$$

Note that a measurement of the instantaneous value of f at some point $x = x_0$ will result in any value between $-A$ and A, depending on x_0; this means that A cannot be obtained from the signal f in any simple way.

Correspondingly, the value of f at $x = x_0$ gives no information regarding whether the signal is at a local maximum or minimum or whether f is increasing or decreasing in a neighborhood around x_0. The analytic signal, on the other hand, can provide a measure of the local behaviour of the signal. This information is given by $\arg[f_A]$, which in this case is a linear function of x. In the following, the argument of f_A is referred to as the *phase* of f. Below is a short list demonstrating how the phase can be used to infer the local behaviour of the signal in some typical cases (Figure 4.19).

- If $\arg[f_A] = 2\pi k$ then f assumes its maximal value A at $x = x_0$.
- If $\arg[f_A] = \pi + 2\pi k$, then f assumes its minimal value $-A$ at $x = x_0$.
- If $\arg[f_A] = \frac{\pi}{2} + 2\pi k$, then f is just passing the zero value going from negative values to positive at $x = x_0$.
- If $\arg[f_A] = -\frac{\pi}{2} + 2\pi k$, then f is just passing the zero value going from positive values to negative at $x = x_0$.

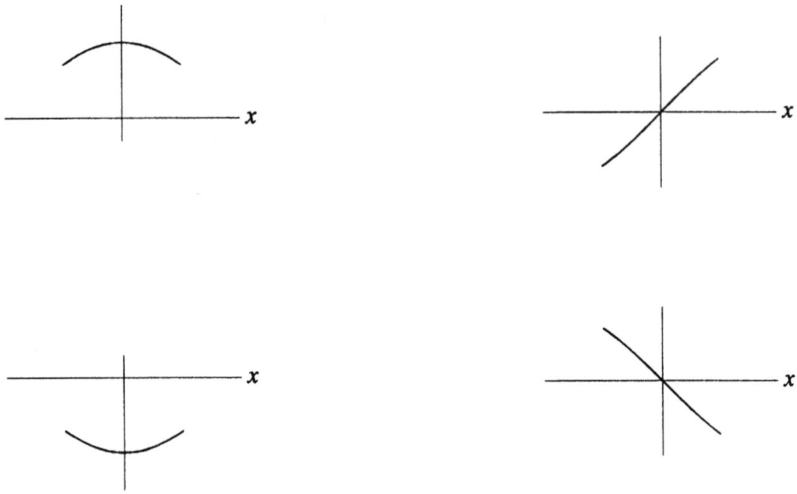

Figure 4.19 Four different phase positions at $x = x_0$ represented by, left to right and top to bottom, $\arg[f_A] = 0, \frac{\pi}{2}, \pi$ and $\frac{3\pi}{2}$, respectively.

Within certain limits, the analytic signal can also give a measure of the angular frequency ω. Consider the analytic signal $f_A(x)$ and a delayed version of the same signal

$$f_D(x) = f_A(x - d), \tag{4.158}$$

for some fixed value of d. Assuming that f is defined by Equation (4.154), this gives

$$f_D(x) = Ae^{i\omega(x-d)}. \tag{4.159}$$

By combining Equations (4.156) and (4.159) in different ways, we obtain a measure of ω. Two examples are

$$f_1(x) = f_A(x) \cdot f_D^{\star}(x) = A^2 e^{i\omega d}, \tag{4.160}$$

and

$$f_2(x) = \frac{f_A(x)}{f_D(x)} = e^{i\omega d}. \tag{4.161}$$

Since d is fixed, the argument of both these signals are linear functions of ω and does not depend on x. If d is known, the angular frequency ω can be computed directly from the argument of either f_1 or f_2. However, since the argument can be have a value in an interval of length 2π, a correct measures of ω can only be obtained in the interval $p \leq \omega < 2\pi + p$ for a fixed value of p.

As indicated above, the phase of f is a linear function of x, i.e.

$$\arg[f_A(x)] = \omega x. \tag{4.162}$$

Hence, a third way of obtaining the angular frequency is to differentiate the phase with respect to x, i.e.

$$\omega = \tfrac{d}{dx} \arg[f_A(x)]. \tag{4.163}$$

Though seemingly more elaborate, this is the most general definition of frequency, not only for signals like the one used in the previous examples but for arbitrary one-dimensional real signals.

The previous discussion demonstrates how the analytic signal, if available, can provide both the amplitude, phase and frequency of real signals with well-defined frequencies. In general, of course, real signal are not of this type which means that the above procedures must either be generalized or used with care. However, the analytic signal is well-defined for real signals in general, which means that given a particular real signal, we can consider the absolute value, argument, and the arguments derivative of the corresponding analytic signal. Hence, we make the following definitions

$$\text{the } \textit{instantaneous amplitude} \text{ of } f(x) = |f_A(x)|. \tag{4.164}$$

For a real signal in general, its instantaneous amplitude may not be a constant function as in the above example. Correspondingly, the instantaneous phase is defined as

$$\text{the } \textit{instantaneous phase} \text{ of } f(x) = \arg[f_A(x)]. \tag{4.165}$$

The instantaneous phase will in general *not* describe the local behaviour of f as in the previous discussion unless f has a well-defined angular frequency. Often, however, it is sufficient that f is a function of small frequency bandwidth for the phase to be useful. Finally, for any real function such that its phase has a well-defined derivative with respect to x, we define

$$\text{the } \textit{instantaneous frequency} \text{ of } f(x) = \tfrac{d}{dx} \arg[f_A(x)]. \tag{4.166}$$

As in the case of the instantaneous amplitude, the instantaneous frequency is, in general, also a non-constant function of x.

Local behaviour and spatial frequency have long been recognized as important properties for image processing. Chapters 7 and 8 discuss methods for estimation of instantaneous or local phase and frequency, respectively. These methods are based on convolving the signal f with various filters in order to obtain an estimate of either phase or frequency. According to the definition of the analytic signal f_A, all negative frequencies of f are suppressed. Furthermore, we know that convolution in the spatial domain corresponds to multiplication in the Fourier domain. Hence, if we employ filters for which the corresponding Fourier transform is zero for negative frequencies, these filters can be thought of as convolved with the analytic signal f_A, even though they are in practice convolved with f. Such a filter, which can be seen as the analytic function of a real filter, is complex and is referred to as a *quadrature filter*.

4.5.2 Analytic signals in multiple dimensions

The previous section dsicussed the Hilbert transform for one-dimensional functions. Its effect on a function f was to change the argument of F by an amount of $\frac{\pi}{2}$, but in opposite directions for positive and negative frequencies. A regular generalization of this transformation to multiple dimensions is not possible since the concepts of positive and negative frequencies are undefined in that case. The only way to implement these concepts is to introduce a direction of reference in the Fourier domain, e.g. by employing a unit vector $\hat{\mathbf{e}}$. Given this reference vector, a frequency coordinate, $\mathbf{u}$, is in a natural way labeled positive if $\mathbf{u}^T\hat{\mathbf{e}} > 0$ and negative if $\mathbf{u}^T\hat{\mathbf{e}} < 0$. For this purpose we define a multi-dimensional sign-function, depending also on $\hat{\mathbf{e}}$, as

$$\text{sign}_{\hat{\mathbf{e}}}(\mathbf{u}) = \begin{cases} 1 & \text{if } \mathbf{u}^T\hat{\mathbf{e}} > 0 \\ -1 & \text{if } \mathbf{u}^T\hat{\mathbf{e}} < 0. \end{cases} \tag{4.167}$$

It should be noted that instead of defining the Hilbert transform in the spatial domain, as in the previous section, we may equally well define it in the Fourier domain. Hence, given a direction of reference, described by $\hat{\mathbf{e}}$, the correspondence between a function f and its Hilbert transform $f_{\mathcal{H}i}$ is defined as

$$F_{\mathcal{H}i}(\mathbf{u}) = F(\mathbf{u}) \cdot i \, \text{sign}_{\hat{\mathbf{e}}}(\mathbf{u}), \tag{4.168}$$

where F and $F_{\mathcal{H}i}$ are the Fourier transforms corresponding to f and $f_{\mathcal{H}i}$, respectively. Note the similarity to Equation (4.149). In the spatial domain, this correspondence is defined by a convolution between f and the inverse Fourier transform of $i\,\mathrm{sign}_{\hat{\mathbf{e}}}(\mathbf{u})$. This latter function is simple in the $\hat{\mathbf{e}}$-direction and, consequently, its inverse Fourier transform is

$$\mathcal{F}^{-1}\{i\,\mathrm{sign}_{\hat{\mathbf{e}}}(\mathbf{u})\} = \frac{-1}{\pi \mathbf{x}^T \hat{\mathbf{e}}}\, \delta_{\hat{\mathbf{e}}}^{\mathrm{line}}(\mathbf{x}), \tag{4.169}$$

and we can write

$$f_{Hi} = f * \left[\frac{-1}{\pi \mathbf{x}^T \hat{\mathbf{e}}}\, \delta_{\hat{\mathbf{e}}}^{\mathrm{line}}(\mathbf{x})\right]. \tag{4.170}$$

With this approach, based on the use of a predefined direction of reference, the Hilbert transform $f_{\mathcal{H}i}$ of a function f can be defined in multiple dimensions. However, it should be noted that this is done in an arbitrary way and, from a strict point of view, the result cannot be called a proper multi-dimensional Hilbert transform. Despite this fact, we will refer to $f_{\mathcal{H}i}$, defined according to Equation (4.170), as the Hilbert transform of f. However, it is important to keep in mind that in the multi-dimensional case the Hilbert transform will assume on a specific direction of reference.

Given the Hilbert transform $f_{\mathcal{H}i}$ of a real function f, the analytic signal f_A is defined in the same way as in one dimension, i.e.

$$f_A = f - i f_{\mathcal{H}i}. \tag{4.171}$$

In the previous section, we restricted the use of the analytic signal to the case when f is real. In multiple dimensions, we will restrict its use even further and consider only simple real functions. Note, however, that it is practical use rather than theoretical limitations which call for these restrictions. If f is a simple function, it has direction of variation that is unique as far as orientation is concerned. This direction can be represented by a unit vector $\hat{\mathbf{e}}$ or $-\hat{\mathbf{e}}$, which in a natural way may serve as a direction of reference. Hence, the concept of analytic signal falls into place quite naturally provided that we are dealing with real and simple functions. Furthermore, if the variation of a simple function is approximately a cosine function, as in Equation (4.154), the discussion in Section 4.5.1 suggests that the amplitude, phase and frequency of the variation can be estimated from the analytical signal. If the simple function is not of this type, its instantaneous amplitude, phase and frequency are still well-defined according to Equations (4.164), (4.165) and (4.166).

Even the definition of quadrature filters can be extended to multiple dimensions. As mentioned in Section 4.18, these filters are complex and can be seen as the analytic signal of a real filter. In multiple dimensions, this means that such a filter must have a unique direction of reference, and its Fourier transform must be zero in the "negative part" of the Fourier domain relative to this direction. However, such a filter need not be a simple function: it is sufficient that the functions or signals to be filtered have this characteristic.

4.6 EXAMPLES OF STATISTICAL CALCULATIONS

In this section we study expectation values of the Fourier transforms or, rather, their energy functions. The idea to be explored here is to obtain a statistical description, in this case the expectation value, of the energy function given a probability density function that describes the probability of various events in the signal.

Example. Let us calculate the expectation value of the energy function of F when f is a two-dimensional edge with uniformly distributed orientation. As a schematic representation of an edge function we will use

$$f(\mathbf{x}) = \mathrm{sign}(\mathbf{x}^T\hat{\mathbf{e}}_1), \tag{4.172}$$

where $\hat{\mathbf{e}}_1$ is a unit vector representing the direction of the slope of the edge. The direction of $\hat{\mathbf{e}}_1$ is uniformly distributed the interval $[0, 2\pi[$. However, to make the following calculations less complicated, we will multiply this function by a sinc-function in the direction perpendicular to $\hat{\mathbf{e}}_1$, i.e.

$$f(\mathbf{x}) = \mathrm{sign}(\mathbf{x}^T\hat{\mathbf{e}}_1) \cdot \mathrm{sinc}(\frac{\mathbf{x}^T\hat{\mathbf{e}}_2}{2\pi a}), \tag{4.173}$$

Hence, f looks like a sign-function in the $\hat{\mathbf{e}}_1$-direction and like a sinc-function in the $\hat{\mathbf{e}}_2$-direction. Furthermore, the parameter a will, for larger and larger values, make the variation of f in the $\hat{\mathbf{e}}_2$-direction look more and more like a unit constant in the vicinity of the origin. For the time being, however, we will regard a as having a fixed, positive value. The Fourier transform of f is

$$F(\mathbf{u}) = \frac{2}{i(\mathbf{u}^T\hat{\mathbf{e}}_1)} \cdot \frac{1}{2\pi} \mathrm{rect}(a\, \mathbf{u}^T\hat{\mathbf{e}}_2). \tag{4.174}$$

The energy function of F, denoted E, is defined as $F^{\star}F$, i.e.

$$E(\mathbf{u}) = \frac{a^2}{\pi^2} \frac{1}{(\mathbf{u}^T \hat{\mathbf{e}}_1)^2} \operatorname{rect}(a\, \mathbf{u}^T \hat{\mathbf{e}}_2). \tag{4.175}$$

Let the angle between $\hat{\mathbf{e}}_1$ and the x_1-axis, as well as between $\hat{\mathbf{e}}_2$ and the x_2-axis, be denoted α. Employing spherical coordinates, E is then expressed as

$$E(\rho, \psi) = \frac{1}{\rho^2} \frac{a^2}{\pi^2} \frac{1}{\cos^2(\alpha - \psi)} \operatorname{rect}(a\, \rho\, \sin(\alpha - \psi)). \tag{4.176}$$

Evidently, E depends on $\hat{\mathbf{e}}_1$. However, since E looks the same if $\hat{\mathbf{e}}_1$ changes sign, we need only consider the interval $0 \leq \alpha < \pi$. We will now assume that α is described by a probability density function $p(\alpha)$, i.e. the probability that $\alpha = \alpha_0$ is $p(\alpha_0)\, d\alpha$. The expectation value of E at (ρ, ψ) is then given by

$$\bar{E}(\rho, \psi) = \int_0^{\pi} E(\rho, \psi)\, p(\alpha)\, d\alpha. \tag{4.177}$$

Assuming that all directions of $\hat{\mathbf{e}}_1$ are equally probable, i.e. $p(\alpha) = \frac{1}{\pi}$, the expectation value of E is given by

$$\bar{E}(\rho, \psi) = \frac{1}{\rho^2} \frac{a^2}{\pi^3} \int_0^{\pi} \frac{1}{\cos^2(\alpha - \psi)} \operatorname{rect}(a\, \rho\, \sin(\alpha - \psi))\, d\alpha. \tag{4.178}$$

The rect-function implies that the integral results in different expressions for $a\,\rho < \frac{1}{2}$ and for $a\,\rho > \frac{1}{2}$. To calculate the latter, set $\theta = \arcsin(\frac{1}{2a\rho})$.

$$\bar{E}(\rho, \psi) = \frac{1}{\rho^2} \frac{4a^2}{\pi^3} \int_0^{\theta} \frac{d\beta}{\cos^2 \beta} = \frac{1}{\rho^2} \frac{4a^2}{\pi^3} \Big[\tan \beta \Big]_0^{\theta} = \frac{1}{\rho^2} \frac{4a^2}{\pi^3} \tan \theta =$$

$$= \frac{1}{\rho^3} \frac{2a}{\pi^3} \frac{1}{\sqrt{1 - (\frac{1}{2a\rho})^2}}. \tag{4.179}$$

Let us summarize the discussion so far. We began with the function f, defined by Equation (4.173), meant to resemble a two-dimensional edge function with the edge slope in the $\hat{\mathbf{e}}_1$-direction. To look like a proper edge, however, we must assume that a is sufficiently large to make the sinc-variation perpendicular to the edge look more and more like a constant in the vicinity of the origin. Equation (4.179) is the expression for the expectation value of the energy function E, assuming that all directions of the edge are equally probable. The expression is only valid for $2\,a\,\rho > 1$, but if a is assumed to be sufficiently large, the expression can be said to be valid outside a correspondingly small region in the Fourier domain which contains the origin. In this region, the right hand side of Equation (4.179) will tend asymptotically towards the function ρ^{-3} for large ρ. As a conclusion, we assert that the energy function of a spatial edge, with uniformly distributed orientation, behaves like ρ^{-3} in the Fourier domain. It should be noted that we will get the same result also for edges which do not pass the origin. If the edge is moved out from the origin, this will cause a complex modulation of the corresponding Fourier transform F. However, E is unaffected by this modulation of F. Hence, E behaves like ρ^{-3} for edges with both random orientation and random position, provided that the former is uniformly distributed. □

Example. We will now consider a randomly oriented line, i.e. f is given by

$$f(\mathbf{x}) = \delta(\mathbf{x}^T\hat{\mathbf{e}}_1) \cdot \mathrm{sinc}(\frac{\mathbf{x}^T\hat{\mathbf{e}}_2}{2\pi a}). \tag{4.180}$$

As demonstrated in the previous example, an edge function in the spatial domain results in an energy function that behaves like ρ^{-3} in the Fourier domain. It should be noted that this results comes from a ρ^{-2} factor because of the fact that the Fourier transform of a sign-function squared behaves like ρ^{-2} and from a ρ^{-1} factor because of the integration over the angle α. Hence, if f now is defined according to Equation (4.180), the expectation value of the energy function must behave like the Fourier transform of the impulse function squared, i.e. like a constant, times the factor ρ^{-1} from the integration. Without further ado, we therefore assert that in the case of lines in the spatial domain, with random orientation and random position, where the former is uniformly distributed, the corresponding energy function in the Fourier domain has an expectation value that behaves like ρ^{-1}. □.

The previous two examples have demonstrated that the energy functions of random distributed edges and lines behave like ρ^{-3} and ρ^{-1}, respectively. The reason for this investigation is simply that two-dimensional images in general have the characteristic property of containing edges and lines of different orientations and positions. As a consequence, we expect the energy function in the Fourier domain to behave like a weighted sum of ρ^{-3} and ρ^{-1} functions. The main result of this discussion is simply that if we are dealing with natural images, most of the energy in the Fourier domain is found near the origin and the expected value of the energy will decrease at least as fast as ρ^{-1} does.

4.7 TRANSFORM PAIRS

The following is a short list of one-dimensional functions and their corresponding Fourier transforms, accompanied by figures illustrating these functions. The table is arranged such that for each function in the left column, its corresponding Fourier transform is presented in the right column. The table can of course also be used for obtaining the inverse Fourier transform of the functions in the right column by looking in the left column. In view of the discussion in Section 4.2.5, we may also obtain the inverse Fourier transform of functions in the left column by taking the function in the right column, changing the sign of its variable and dividing by 2π. Correspondingly, the Fourier transform of a function in the right column is obtained by taking the function in the left column, changing the sign of its variable and multiplying by 2π.

As an example, from this table we can deduce that

- the Fourier transform of $\text{sign}(x)$ is $\frac{2}{iu}$,
- the inverse Fourier transform of $\frac{2}{iu}$ is $\text{sign}(x)$.

But it is also the case that

- the inverse Fourier transform of $\text{sign}(u)$ is $\frac{1}{2\pi}\frac{2}{i(-u)} = \frac{i}{\pi u}$.
- the Fourier transform of $\frac{2}{ix}$ is $2\pi\,\text{sign}(-u) = -2\pi\,\text{sign}(u)$,

In the following figures the imaginary parts are shown dashed.

$\cos \omega x$

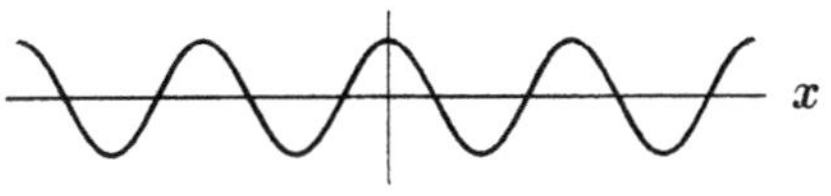

$\pi[\delta(u+\omega) + \delta(u-\omega)]$

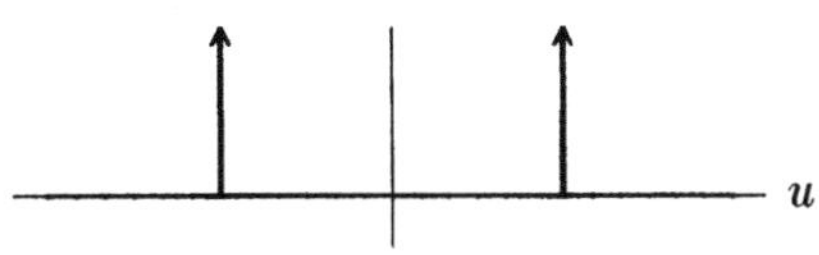

$\sin \omega x$

$i\pi[\delta(u+\omega) - \delta(u-\omega)]$

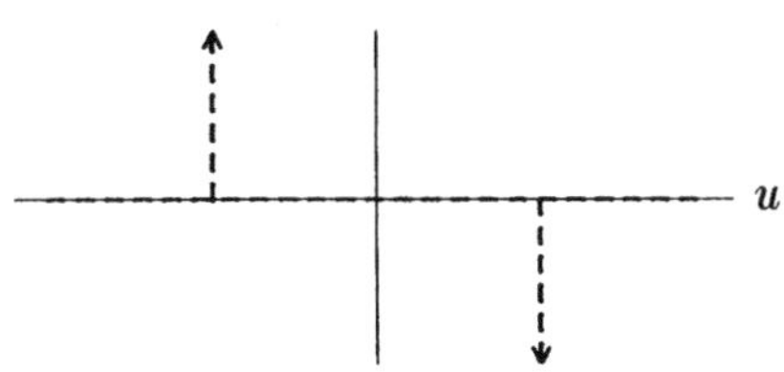

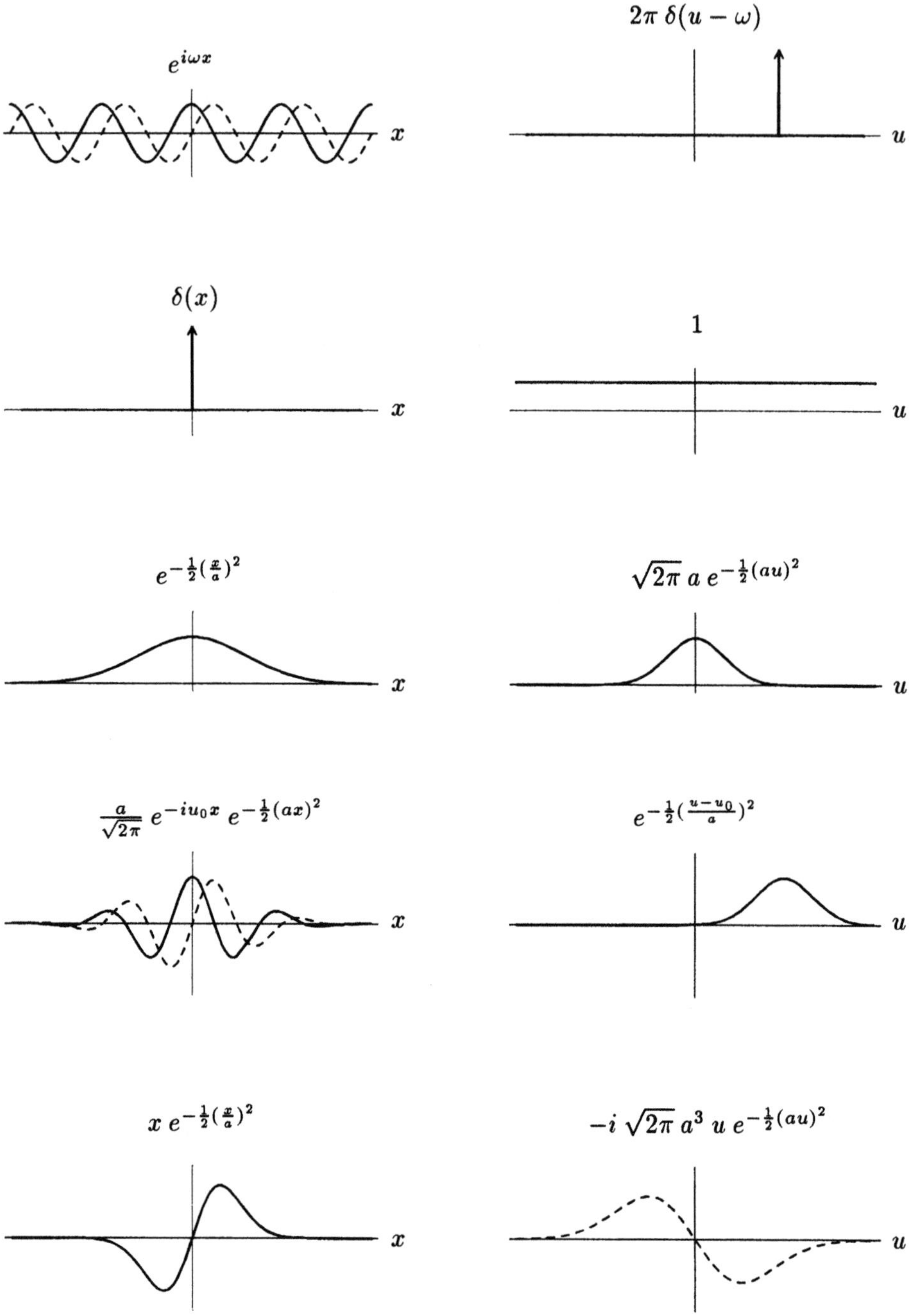

$e^{i\omega x}$
x
$2\pi\,\delta(u-\omega)$
u
$\delta(x)$
x
1
u
$e^{-\frac{1}{2}(\frac{x}{a})^2}$
x
$\sqrt{2\pi}\,a\,e^{-\frac{1}{2}(au)^2}$
u
$\frac{a}{\sqrt{2\pi}}\,e^{-iu_0x}\,e^{-\frac{1}{2}(ax)^2}$
x
$e^{-\frac{1}{2}(\frac{u-u_0}{a})^2}$
u
$x\,e^{-\frac{1}{2}(\frac{x}{a})^2}$
x
$-i\,\sqrt{2\pi}\,a^3\,u\,e^{-\frac{1}{2}(au)^2}$
u

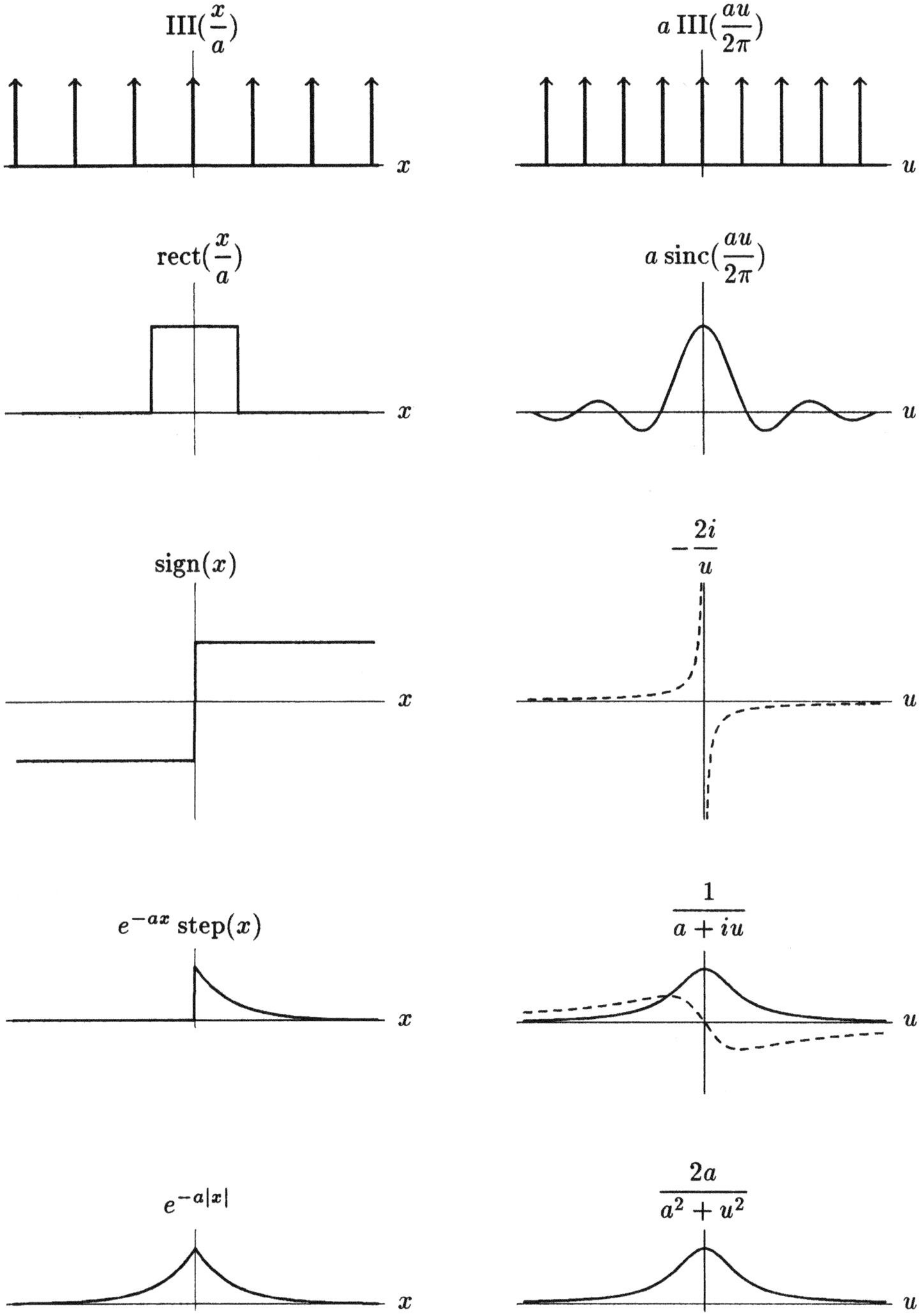
$\mathrm{III}(\frac{x}{a})$
x
$a\,\mathrm{III}(\frac{au}{2\pi})$
u
$\mathrm{rect}(\frac{x}{a})$
x
$a\,\mathrm{sinc}(\frac{au}{2\pi})$
u
$\mathrm{sign}(x)$
x
$-\frac{2i}{u}$
u
$e^{-ax}\,\mathrm{step}(x)$
x
$\frac{1}{a+iu}$
u
$e^{-a|x|}$
x
$\frac{2a}{a^2+u^2}$
u

4.8 THE FOURIER TRANSFORM AND DISCRETE FUNCTIONS

In the following subsections, the terms continuous and discrete are used frequently. In this context, a continuous function refers to a function defined on the entirety of $\mathbb{R}^n$. A discrete function is then the opposite, i.e. it is defined on a set of discrete points where the number of points is either finite or infinite.

The previous sections have presented the Fourier transform for functions that are continuous. In the context of computer vision, however, the spatial functions or signals are always discrete. Note that this does not imply that the corresponding Fourier transforms are discrete as is demonstrated below. Furthermore, the functions are always truncated, i.e. defined only for a finite number of points. As an example, a typical two-dimensional image is defined on a 512×512 point grid. These two restrictions, discreteness and truncation of the functions, imply that some care must be observed when analyzing the functions or designing algorithms for signal processing using Fourier transforms. This section discusses these issues.

4.8.1 Sampled functions

Often, a discrete function can be seen as a *sampled* version of a continuous function. As an example, consider an ordinary two-dimensional digital image. From a physical point of view it is reasonable to believe that the image intensity can be defined for any point (x_1, x_2), where $x_1, x_2 \in \mathbb{R}$. However, a digital image that is supplied by a video camera will only describe the image intensity for discrete values of x_1 and x_2. Let $M \subset \mathbb{R}^n$ denote the set of points for which a discrete function $\tilde{f}$ is defined. The function $\tilde{f}$ may thus be seen as corresponding to a continuous function f according to a sampling process

$$\tilde{f}(\mathbf{x}) = \begin{cases} f(\mathbf{x}) & \mathbf{x} \in M \\ \text{undefined} & \text{otherwise} \end{cases} \tag{4.181}$$

Given the set M, the sampling process may be seen as a linear mapping or transform from continuous functions to discrete functions, the latter defined on M. In reality, the sampling process is often more sophisticated, but the first order effects are well described by the above scheme. In the following, a discrete function is always assumed to be the result of a sampling process, unless stated otherwise.

Figure 4.20 A Cartesian sampling grid (left) and a hexagonal sampling grid (right).

Note that the use of sampling is not restricted to only the spatial domain. In the following we will have cause to consider sampling in the Fourier domain as well. For the moment, however, we concentrate on spatial sampling.

The points in M may be distributed in any possible way. For example, there are some practical implementations that employ a hexagonal sampling grid in two dimensions. For practical reasons, however, the sampling points are more often evenly spaced in each dimension of the spatial domain (Figure 4.20). In the following, it is assumed that the sampling is done on a Cartesian grid, where the samples are d_k length units apart for coordinate x_k. Hence, Equation (4.181) may be rewritten as

$$\tilde{f}(\mathbf{x}) = \begin{cases} f(\mathbf{x}) & x_k = d_k m, \quad m \in \mathbb{Z} \\ \text{undefined} & \text{otherwise} \end{cases} \tag{4.182}$$

For simplicity, all d_k are often assumed to be of unit value, i.e. the unit of length is the distance between two pixels.

The result of the above sampling process, $\tilde{f}$, can not be Fourier transformed. The reason is that the integral of $\tilde{f}$ times an exponential function vanishes, at least if f is bounded. However, this situation can be overcome simply by multiplying $\tilde{f}$ by an impulse at each $\mathbf{x} \in M$ and extending its domain of definition to the entirety of $\mathbb{R}^n$ by assigning a zero value for all $\mathbf{x} \notin M$. This gives a new continuous function, $\hat{f}$,

$$\begin{aligned} \hat{f}(\mathbf{x}) &= f(\mathbf{x}) \cdot \tfrac{1}{d_1}\,\text{III}(\tfrac{x_1}{d_1}) \cdot \tfrac{1}{d_2}\,\text{III}(\tfrac{x_2}{d_2}) \ldots \tfrac{1}{d_n}\,\text{III}(\tfrac{x_n}{d_n}) = \\ &= f(\mathbf{x}) \cdot \frac{1}{\det \mathbf{D}}\,\text{III}^n(\mathbf{D}^{-1}\mathbf{x}), \end{aligned} \tag{4.183}$$

where the two functions III and III^n are defined in Equation (4.20) and Equation (4.27), respectively, and $\mathbf{D}$ is a diagonal matrix with d_k in its k-th diagonal element. Despite the fact that $\hat{f}$ is defined on the entirety of $\mathbb{R}^n$, we will refer to it as a discrete function.

The following theorem asserts that there is a one-to-one correspondence between discrete functions in the spatial domain and periodic functions in the Fourier domain.

Theorem 4.1. $\hat{f}$ is defined according to Equation (4.183) for some continuous function f if and only if $\hat{F}$ is periodic in all coordinates, where the period for coordinate u_k is $\frac{2\pi}{d_k}$. □

Proof of Theorem 4.1. Let us begin by proving that periodicity of $\hat{F}$ implies discreteness of $\hat{f}$. Since $\hat{F}$ is periodic in all coordinates, we may cut out an n-dimensional interval of $\hat{F}$ by multiplying it by an n-dimensional rect-function such that the interval contains precisely one period in all dimensions. $\hat{F}$ can then be reconstructed by putting infinitely many of this interval adjacent to each other along all dimensions. Formally, this corresponds to

$$\hat{F}(\mathbf{u}) = \left[\hat{F}(\mathbf{u}) \cdot \mathrm{rect}^n(\frac{\mathbf{D}\,\mathbf{u}}{2\pi})\right] * \left[\frac{\det \mathbf{D}}{(2\pi)^n}\,\mathrm{III}^n(\frac{\mathbf{D}\mathbf{u}}{2\pi})\right], \tag{4.184}$$

which in the spatial domain gives

$$\hat{f}(\mathbf{x}) = (2\pi)^n \left[\hat{f} * \frac{1}{\det \mathbf{D}}\,\mathrm{sinc}^n(\mathbf{D}^{-1}\,\mathbf{x})\right] \cdot \frac{\det \mathbf{D}}{(2\pi)^n}\,\frac{1}{\det \mathbf{D}}\,\mathrm{III}^n(\mathbf{D}^{-1}\mathbf{x}) =$$

$$= \left[\hat{f} * \mathrm{sinc}^n(\mathbf{D}^{-1}\,\mathbf{x})\right] \cdot \frac{1}{\det \mathbf{D}}\,\mathrm{III}^n(\mathbf{D}^{-1}\mathbf{x}). \tag{4.185}$$

Comparing Equations (4.183) and (4.185) it is evident that $\hat{f}$ is a sampled version of the function in the brackets of Equation (4.185).

Conversely, assume that Equation (4.183) is valid for some continuous function f. In the Fourier domain this corresponds to

$$\hat{F}(\mathbf{u}) = \frac{1}{(2\pi)^n} F(\mathbf{u}) * \mathrm{III}^n(\frac{\mathbf{D}\mathbf{u}}{2\pi}) =$$

$$= \frac{1}{(2\pi)^n} F(\mathbf{u}) * \left[\sum_{\alpha,\beta,\ldots,\nu} \delta(\frac{d_1 u_1}{2\pi} - \alpha)\, \delta(\frac{d_2 u_2}{2\pi} - \beta) \ldots \delta(\frac{d_1 u_1}{2\pi} - \nu) \right] =$$

$$= \frac{1}{\det \mathbf{D}} F(\mathbf{u}) * \left[\sum_{\alpha,\beta,\ldots,\nu} \delta(u_1 - \frac{2\pi\alpha}{d_1})\, \delta(u_2 - \frac{2\pi\beta}{d_2}) \ldots \delta(u_n - \frac{2\pi\nu}{d_n}) \right] =$$

$$= \sum_{\alpha,\beta,\ldots,\nu} F(u_1 - \frac{2\pi\alpha}{d_1}, u_2 - \frac{2\pi\beta}{d_2}, \ldots, u_n - \frac{2\pi\nu}{d_n}), \tag{4.186}$$

where the summation is to be taken over all integers, i.e. $\hat{F}$ is an infinite sum of shifted versions of F, where the shifts in dimension k are $\frac{2\pi}{d_k}$ apart. From this it follows immediately that

$$\hat{F}(u_1, \ldots, u_k + \tfrac{2\pi}{d_k}, \ldots, u_n) =$$

$$= \frac{1}{\det \mathbf{D}} \sum_{\alpha,\beta,\ldots,\nu} F(u_1 - \frac{2\pi\alpha}{d_1}, \ldots, u_k + \frac{2\pi}{d_2} - \frac{2\pi\beta}{d_2}, \ldots, u_n - \frac{2\pi\nu}{d_n}) =$$

$$= \frac{1}{\det \mathbf{D}} \sum_{\alpha,\beta,\ldots,\nu} F(u_1 - \frac{2\pi\alpha}{d_1}, \ldots, u_k - \frac{2\pi[\beta - 1]}{d_2}, \ldots, u_n - \frac{2\pi\nu}{d_n}) =$$

$$= \frac{1}{\det \mathbf{D}} \sum_{\alpha,\beta,\ldots,\nu} F(u_1 - \frac{2\pi\alpha}{d_1}, u_2 - \frac{2\pi\beta}{d_2}, \ldots, u_n - \frac{2\pi\nu}{d_n}) = \hat{F}(\mathbf{u}), \tag{4.187}$$

which proves that $\hat{F}$ is periodic in all coordinates, and that the k-th coordinate has period $\frac{2\pi}{d_k}$. Had there been a smaller value than $\frac{2\pi}{d_k}$ that validates Equation (4.187), the first part of Theorem 4.1 would have been contradicted. □

As will be apparent in the following, not only the theorem itself but also both Equations (4.185) and (4.186) are of major importance. Equation (4.186) is sometimes referred to as *Poisson's summation formula*. The one-dimensional case of this theorem is illustrated in Figures 4.21 and 4.22.

Figure 4.21 A continuous function, f (left), and a discrete version of it, $\hat{f}$ (right).

Figure 4.22 The Fourier transform of the continuous function f (left) and the Fourier transform of its discrete version, $\hat{f}$ (right).

The sampling theorem

We now address a well-known problem when a function is sampled. A desirable property of the sampling process is that it should be invertible, i.e. it should be possible to reconstruct f from $\hat{f}$. Of course, this cannot be done in general since f may assume arbitrary values between the sample points. However, it is possible to state sufficient conditions in order to make the sampling procedure one-to-one. The following theorem, known as the *sampling theorem*, states one such condition.

Theorem 4.2. Let f be a continuous function with corresponding Fourier transform F such that

$$F(\mathbf{u}) = 0 \text{ for } |u_k| \geq \frac{\pi}{d_k}, \tag{4.188}$$

and let

$$\hat{f}(\mathbf{x}) = f(\mathbf{x}) \cdot \frac{1}{\det \mathbf{D}} \, \mathrm{III}^n(\mathbf{D}^{-1}\mathbf{x}). \tag{4.189}$$

Then

$$f(\mathbf{x}) = \hat{f} * \mathrm{sinc}^n(\mathbf{D}^{-1}\,\mathbf{x}). \tag{4.190}$$

Hence, f can be reconstructed from $\hat{f}$ by convolving the latter function with an n-dimensional sinc-function. □

Proof of Theorem 4.2. To prove the sampling theorem, we simply observe that if Equation (4.188) is valid, this implies that $\hat{F}$ in each point contains information from only one shifted version of F according to Equation (4.186). Hence, F is obtained by

$$F(\mathbf{u}) = \hat{F}(\mathbf{u}) \cdot \det \mathbf{D} \cdot \mathrm{rect}^n(\frac{\mathbf{D}\,\mathbf{u}}{2\pi}), \tag{4.191}$$

which in the spatial domain corresponds to Equation (4.190). □

Note that the sampling theorem does not state a necessary condition for enabling exact reconstruction and that other types of conditions are possible. From a practical point of view, however, it describes the easiest way of restricting f.

From Equation (4.186), and unless the conditions of the sampling theorem are valid, $\hat{F}$ may in some points contain a mix of two or more shifted versions of F. Mixing of frequency components in the context of sampling is referred to as *aliasing* and implies that the reconstructed signal, given by the right hand side of Equation (4.190), is not equal to f. However, since the reconstructed signal is a convolution between the sample values and a particular sinc-function, it will equal f in the sample points $\mathbf{x} \in M$. A consequence of the sampling theorem is that prior to sampling, a signal must be band-limited in order to avoid aliasing. The natural way to accomplish this is to filter the signal with an ideal low-pass filter g which has a corresponding Fourier transform

$$G(\mathbf{u}) = \mathrm{rect}^n(\frac{\mathbf{D}\,\mathbf{u}}{2\pi}). \tag{4.192}$$

The following assumes, unless stated otherwise, that to a discrete function $\hat{f}$ there corresponds a continuous function f which satisfies the conditions of the sampling theorem.

4.8.2 Periodic functions

Section 4.2.4 discussed the symmetry between the spatial and the Fourier domains. It should therfore not come as a surprise that the results of the previous section also apply if the two domains are interchanged.

Theorem 4.3. f is periodic in all coordinates, with period p_k for coordinate x_k, if and only if F is discrete and the distance between the samples along coordinate u_k is $\frac{2\pi}{p_k}$. □

Proof of Theorem 4.3. Following the same steps as the proof of Theorem 4.1, it can be shown that the periodicity of f leads to

$$F(\mathbf{u}) = H(\mathbf{u}) \cdot \frac{\det \mathbf{P}}{(2\pi)^n} \, \mathrm{III}^n(\frac{\mathbf{P}\,\mathbf{u}}{2\pi}) \tag{4.193}$$

where $\mathbf{P}$ is a diagonal matrix with p_k in its k-th diagonal element, and

$$H(\mathbf{u}) = F * \mathrm{sinc}^n(\frac{\mathbf{P}\,\mathbf{u}}{2\pi}). \tag{4.194}$$

Hence, F is a sampled version of H, with the sampling done in accordance with Theorem 4.3. Furthermore, also following the proof of Theorem 4.1, the discreteness of of F implies that

$$f(\mathbf{x}) = \frac{\det \mathbf{P}}{(2\pi)^n} \sum_{\alpha,\beta,\ldots,\nu} h(x_1 - p_1\alpha, x_2 - p_2\beta, \ldots x_n - p_n\nu),$$

where h is the inverse Fourier transform of H. The periodicity of f is thus established. □.

Theorem 4.1 can be combined with Theorem 4.3 into the following statement.

- A spatial function is periodic and discrete if and only if its corresponding Fourier transform is periodic and discrete.

We will develop this observation further on.

4.8.3 Truncated functions

As previously mentioned, signals in the context of computer vision are not only discrete but also truncated. Regardless of how an image is created, continuously or discretely, there is no physical device that can generate or store an image of infinite extension. Specially, this must be the case when an image is stored in a computer. Hence, in the context of computer vision, an image is always truncated. If f is continuous, truncation implies that f is defined only within some interval of $\mathbb{R}^n$. For discrete functions, this implies that they are defined only for a *finite* number of points, provided that a Cartesian sampling grid is employed.

Note that the construction of a discrete and truncated signal from a continuous one implies that the continuous function is sampled and truncated but the order of these two operations is irrelevant since they commute. Hence, when er refer to a truncated function, f_T, it may be either continuous or discrete. In the former case it will be subject to sampling and in the latter case it represents a practical implementation of the discrete signal, e.g. a digital image. Furthermore, we will employ the same trick as in Section 4.8.1, and set $f_T(\mathbf{x}) = 0$ for all $\mathbf{x}$ outside the region of definition, rather than to say that f_T is undefined there.

Let us now look at the consequences of truncating a spatial function f. Without loss of generality, we assume that the interval of definition is centered at the origin and of width a_k in the k-th dimension. According to the above description of the truncation process, f_T is then given by

$$f_T(\mathbf{x}) = f(\mathbf{x}) \cdot \mathrm{rect}^n(\mathbf{A}^{-1}\,\mathbf{a}), \tag{4.195}$$

where $\mathbf{A}$ is a diagonal matrix corresponding to the set $\{a_k\}$. In the Fourier domain this corresponds to

$$F_T(\mathbf{u}) = \frac{\det \mathbf{A}}{(2\pi)^n}\, F(u) * \mathrm{sinc}^n(\frac{\mathbf{A}\,\mathbf{u}}{2\pi}). \tag{4.196}$$

Note that Equation (4.196) defines a linear mapping from S_2 to S_2 by mapping an arbitrary function F to F_T. Furthermore, it must be the case that F_T is an *eigenfunction* of this mapping since

$$f_T(\mathbf{x}) \cdot \mathrm{rect}^n(\mathbf{A}^{-1}\,\mathbf{x}) = f_T(\mathbf{x}) \tag{4.197}$$

and hence

$$\frac{\det \mathbf{A}}{(2\pi)^n} F_T(u) * \mathrm{sinc}^n(\frac{\mathbf{A}\,\mathbf{u}}{2\pi}) = F_T(\mathbf{u}) \tag{4.198}$$

This implies that the Fourier transforms of truncated functions must be restricted by the relation described in Equation (4.198).

It was previously argued that a Fourier transform is discrete if and only if the corresponding spatial function is periodic. Hence, regardless of weather f_T is discrete or not, F_T fails to be discrete since f_T is not periodic. This is confirmed by the left hand side of Equation (4.198) which clearly cannot be discrete.

4.8.4 Sampled and truncated functions

From Proposition 4.2 in Section 4.3.2 we know that variances add under convolution. In Section 4.8.3 it was demonstrated that F_T, the Fourier transform corresponding to a truncated function f_T, is the convolution between F and a sinc-function, where F is the Fourier transform corresponding to the untruncated function f. The sinc-function has infinite variance and, hence, this applies to F_T as well. As a consequence, f_T can not be band-limited since this would result in finite variance of F_T, at least if F_T is bounded. This implies that if f is band-limited, as would be required to satisfy the conditions of the sampling theorem, this property is lost after truncation. It does not help to sample the band-limited function prior to truncation. In this case the convolution with the sinc-function casues smearing of frequency components across the periodic intervals in the Fourier domain, causing aliasing. Consequently, it is, in general, not possible to reconstruct f from a sampled and truncated version, $\hat{f}_T$, not even within the interval of truncation. The reconstructed signal will equal the original signal only at the sample points but not necessarily between them.

According to the above, the sampling theorem is not applicable to signals that are both sampled and truncated in the spatial domain. However, this observation should not be taken as a reason to abandon Fourier transforms as a means of analyzing discrete and truncated signals. There is one class of functions which can in fact be sampled and truncated without problems.

Periodic Functions Revisited

Let us assume that f is periodic, with period p_k for coordinate x_k. It was proved in Section 4.8.2 that F then is discrete, and the distance between the samples along coordinate x_k is $\frac{2\pi}{p_k}$, as indicated by Equation (4.193). If f is band-passed, using the filter G of Equation (4.192) for some values of d_k, this does not affect the discreteness of F, it merely implies that F also satisfies Equation (4.188). Hence, f can be sampled at $\mathbf{x} = (x_1, x_2, \ldots, x_n)$, where $x_k = d_k m$ for $m \in \mathbb{Z}$. The result of the sampling process, $\hat{f}_T$, is both discrete and periodic. As mentioned in Section 4.8.2, this means that $\hat{F}$, too, is discrete and periodic, and that it has the same sampling pattern as F, and its period along coordinate u_k is $\frac{2\pi}{d_k}$.

Now, let $\hat{f}$ be truncated, as described by Equation (4.195), such that *the truncation interval is the same size as the period of* f, i.e. $a_k = p_k$, for all k, and let the result be denoted $\hat{f}_T$. According to Equation (4.196), this means that

$$\hat{F}_T = \frac{\det \mathbf{P}}{(2\pi)^n} \hat{F} * \operatorname{sinc}^n(\frac{\mathbf{P}\,\mathbf{u}}{2\pi}). \tag{4.199}$$

As discussed in Section 4.8.3, $\hat{F}_T$ is not discrete. It is, however, periodic with period $\frac{2\pi}{d_k}$ since the periodicity of $\hat{F}$ is preserved under convolution. By combining Equations (4.193), (4.194) and (4.199), we see that

$$\hat{F}(\mathbf{u}) = \hat{F}_T(\mathbf{u}) \, \mathrm{III}^n(\tfrac{\mathbf{P}\,\mathbf{u}}{2\pi}), \tag{4.200}$$

i.e. $\hat{F}$ can be recovered from $\hat{F}_T$ *by a sampling process in the Fourier domain.* In the spatial domain this corresponds to

$$\hat{f} = \hat{f}_T * \frac{1}{\det \mathbf{P}} \, \mathrm{III}^n(\mathbf{P}^{-1}\,\mathbf{x}), \tag{4.201}$$

which is a complicated way of expressing the obvious fact that if a periodic function $\hat{f}$ is truncated in correspondence with its periodicity, then $\hat{f}$ can be reconstructed by repeating $\hat{f}_T$ infinitely many times, each instance shifted p_k length units in dimension k. Once $\hat{f}$ is reconstructed, f can be reconstructed the usual way, i.e. by convolving $\hat{f}$ by a sinc-function.

To summarize the above discussion, we have demonstrated that a continuous and periodic function f, which in addition satisfies the conditions of the sampling theorem, can be reconstructed from a sampled and truncated version provided that the truncation interval corresponds to the periodicity.

4.8.5 Locality aspects of sampling

As previously argued, a signal should be band-limited before sampling in order to avoid aliasing. In the one-dimensional case we should use the filter

$$G(u) = \text{rect}(\tfrac{du}{2\pi}) \tag{4.202}$$

described by Equation (4.192). Hence, the Fourier transform of the band-passed signal would be zero outside the the interval $-\pi < u < \pi$. As a consequence, it must be the case that

$$\sigma_F \leq \tfrac{1}{\sqrt{3}}\pi, \tag{4.203}$$

where the right hand side is the standard deviation of G. According to the uncertainty principle, this implies that

$$\sigma_f \geq \tfrac{1}{2\sigma_F} \geq \tfrac{\sqrt{3}}{2\pi} \approx 0.28. \tag{4.204}$$

This is not to say that there is a band-limited f for which $\sigma_f = \frac{\sqrt{3}}{2\pi}$. In fact, as demonstrated by Gabor [40], the lower bound of σ_f is $\frac{1}{2}$. Hence, regardless of how localized things may be in the spatial domain of an original continuous signal, the band-limitation will introduce a certain amount of uncertainty in the localization.

Another aspect of locality, related to the above discussion, arises when a continuous signal is sampled, and has to do with *how* the sampling is done. We have previously argued that once the continuous signal has been band-passed, using the filter G, it is sampled at $x \in \mathbb{Z}$. In practice, however, the sampling points can seldom be synchronized with the signal, which means that the only statement that can be made about the sample points is that they have unit distance rather than being positioned at specific points. As a consequence, a more realistic formulation of the sampling process than Equation (4.183) would be

$$\hat{f}(x) = \text{III}(x)\, f(x - \Delta x), \tag{4.205}$$

where Δx is an unknown real number such that $0 \leq \Delta x < 1$. When f is sampled, Δ will have some specific, but unknown, value. This is illustrated in the following example.

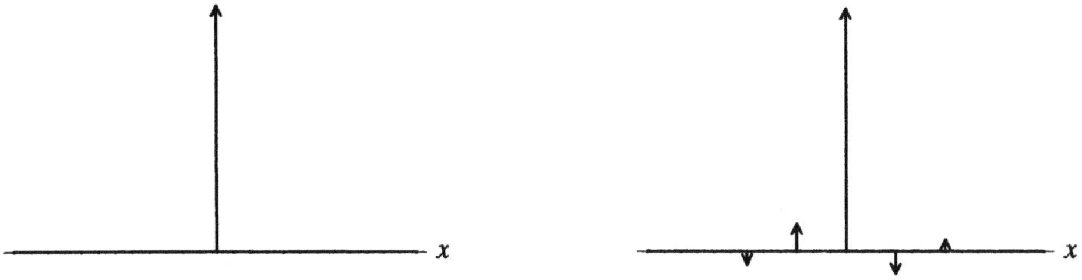

Figure 4.23 An impulse function (left) and a sampled version of it when $\Delta x = 0.1$ (right).

Example. Consider an impulse function

$$h(x) = \delta(x), \tag{4.206}$$

which is to be sampled. First, however, it must be band-limited in order to satisfy the sampling theorem, using the filter G according to Equation (4.192). The result of the band-limitation, denoted f, is

$$f(x) = h * g = \text{sinc}(x). \tag{4.207}$$

If sampled according to Equation (4.205), the corresponding discrete function will be

$$\hat{f}(x) = \text{III}(x)\,\text{sinc}(x - \Delta x) \tag{4.208}$$

for an undetermined Δx. Only in the case $\Delta x = 0$ will $\hat{f}$ look like an impulse. In other cases $\hat{f}$ will, for example, appear as illustrated in Figure 4.23. □

To summarize, we have concluded that band-limiting a signal sets a lower bound of $\frac{1}{2}$ for the spatial uncertainty of a signal which is to be sampled. Furthermore, we can not assume that the sampling is done at specific points relative to the signal, merely that the samples have a specific spacing; this implies that the sampled signal may not appear as expected.

4.9 THE DISCRETE FOURIER TRANSFORM

So far we have considered the Fourier transform which is a mapping from continuous functions to continuous functions. Even in the case of discrete functions, as discussed in Section 4.8, the Fourier transform can be used by employing the trick of describing the sampling process as a multiplication by a III-function, thereby obtaining a Fourier transformable function. From a practical point of view, however, the Fourier transform is inadequate for digital signal processing, such as image processing, which deals only with discrete signals. For example, the images that are to be processed are always discrete in all dimensions, and a practical implementation of a filter must be discrete. Furthermore, the filter may result from a design process which is described in the Fourier domain, and its spatial representation would then be the inverse Fourier transform of its frequency representation, a process that also has to be applied to discrete functions. As a consequence, we must use a discrete version of the Fourier transform which enables transformation of functions that are discrete in both domains. This is the *discrete Fourier transform* or DFT. Even though the DFT shares many of the basic properties of the usual Fourier transform, the two should not be confused since there are some quite important differences, which are discussed below.

We start by defining the DFT and its inverse, then look at some of the basic properties shared by both the Fourier transform and the DFT. Finally, we discuss their differences.

4.9.1 Introduction to the DFT

In Section 4.8.1 we argue that the process of turning a continuous function, f, into a discrete function, $\hat{f}$, should be defined according to Equation (4.183). Note, however, that the multiplication by the III^n-function was introduced merely as a trick to make $\hat{f}$ Fourier transformable. A more intuitive expression is that given for $\tilde{f}$ by Equation (4.182), which simply picks out the sampled values and leaves $\tilde{f}$ undefined elsewhere. In the following, we will consider $\tilde{f}$, and assume that f is sampled at integer values for all coordinates, i.e. all $d_k = 1$. Furthermore, we also assume that $\tilde{f}$ is truncated, i.e. each of its coordinates, x_k, assumes integer values in an interval $a_k \leq x_k \leq b_k$.

Definition of the DFT

Given the above description of $\tilde{f}$, the discrete Fourier transform (DFT) corresponding to $\tilde{f}$, denoted by $\tilde{F}$, is defined as

$$\tilde{F}(\mathbf{u}) = [\Delta_1 \cdot \Delta_2 \ldots \Delta_n]^{-1} \sum_{x_1=a_1}^{b_1} \sum_{x_2=a_2}^{b_2} \cdots \sum_{x_n=a_n}^{b_n} \tilde{f}(\mathbf{x})\, e^{-i\mathbf{u}^T\mathbf{x}}, \tag{4.209}$$

where $\Delta_k = b_k - a_k + 1$ is the number of sample points in the k-th dimension.

Having defined $\tilde{F}$ this way, we make some important observations on the domain of definition of $\tilde{F}$. Clearly, $\tilde{F}$ is well-defined for any $\mathbf{u} \in \mathbb{R}^n$. Furthermore, $\tilde{F}$ is periodic, with period 2π, in each of its coordinates. Hence, $\tilde{F}$ is completely defined by its values on an interval $w_k \leq u_k < w_k + 2\pi$ for each coordinate u_k (the interval may, alternatively, include the other end-point instead). Note that it is of no theoretical consequence where such an interval is positioned, i.e. what value is chosen for each w_k. Instead, practical considerations may suggest that, for each w_k, some value is more appropriate than others.

With each u_k constrained to a half-open interval, we now demonstrate that it is sufficient to calculate $\tilde{F}$ only for multiples of $\frac{2\pi}{\Delta_k}$ for coordinate u_k. The set of spatial functions $\tilde{f}$, defined on a specific interval, can be seen as a vector space of dimensionality $d = \Delta_1 \cdot \Delta_2 \ldots \Delta_n$. With this interpretation, Equation (4.209) is nothing but a change of basis, where the new basis in the spatial domain is represented by discrete and truncated complex exponentials of various frequencies for each coordinate. If each u_k is restricted to only multiples of $\frac{2\pi}{\Delta_k}$, the result will be exactly d basis functions which in addition are orthogonal, i.e. they span the vector space under consideration. If we choose to calculate $\tilde{F}$ for more than d points, the resulting $\tilde{F}$ will not provide more information about the properties of $\tilde{f}$. If other points than multiples of $\frac{2\pi}{\Delta_k}$ are used, the basis functions may not be orthogonal, which as a result will give a more complex formulation of the inverse discrete Fourier transform, to be defined below. The convention of compupting $\tilde{F}$ only in multiples of $\frac{2\pi}{\Delta_k}$ for coordinates u_k is therefore a natural choice.

Let us summarize the discussion so far. We started with a discrete and truncated function $\tilde{f}$. This function is truncated such that it is defined only when coordinate x_k is within the interval $a_k \leq x_k \leq b_k$, where $a_k, b_k \in \mathbb{Z}$. The DFT of $\tilde{f}$, denoted by $\tilde{F}$, is then defined according to Equation (4.209). However, we may restrict the domain of definition of $\tilde{F}$ to multiples of $\frac{2\pi}{\Delta_k}$ for each of its coordinates, u_k, which in addition are confined to a half-open interval of length

2π. These intervals may be arbitrarily positioned for each coordinate. As a consequence, $\tilde{F}$ is also discrete and truncated; this means that the concept of discrete Fourier transform is applicable to $\tilde{F}$ itself as well as to the process of transforming $\tilde{f}$ into $\tilde{F}$ according to Equation (4.209).

The inverse DFT

Provided that $\tilde{F}$ is computed for the points described above, the mapping according to Equation (4.209) is one-to-one. Hence, there is an inverse mapping, the *inverse discrete Fourier transform* or IDFT. It is easy to see that

$$\tilde{f}(\mathbf{x}) = \sum_{u_1} \sum_{u_2} \cdots \sum_{u_n} \tilde{F}(\mathbf{u})\, e^{i\mathbf{u}^T\mathbf{x}}, \tag{4.210}$$

where the summation is to be taken over the prescribed points, constitutes an inversion formula. Note that the inversion formula implies that the previous discussion of restricting the coordinates of $\tilde{F}$ also applies to $\tilde{f}$. Hence, despite the fact that we started with a discrete and truncated $\tilde{f}$, obtained a DFT by Equation (4.209), and returned to $\tilde{f}$ with Equation (4.210), we note that we may in fact calculate $\tilde{f}$ for arbitrary values of $\mathbf{x}$. We develop some consequences of this observation below.

4.9.2 Fourier Transform versus DFT

At first sight it may seem as if the DFT and the Fourier transform are practically the same transformation, the integrals in the definition of the Fourier transform and its inverse having been replaced by finite sums for the DFT and the IDFT. Intuitively, it may seem that the sums should be sufficiently good approximations of the integrals. Is there really more to it than that? It turns out that there is, and the problem is in fact the discreteness of $\tilde{F}$ which implies that $\tilde{f}$ must be periodic. Regardless of the fact that we see $\tilde{f}$ as a sampled and truncated version of some continuous, and in general non-periodic function f, the DFT will describe the features of the Fourier transform of a periodic function, namely the one which is the result of repeating the truncation interval infinitely many times. This is shown by the following.

Let $\hat{f}_P$ denote a version of f which is truncated, sampled and finally made periodic according to

$$\hat{f}_P = \left[f \cdot \text{rect}^n(\boldsymbol{\Delta}^{-1}\mathbf{x}) \cdot \text{III}^n(\mathbf{x})\right] * \frac{1}{\det \boldsymbol{\Delta}} \text{III}^n(\boldsymbol{\Delta}^{-1}\mathbf{x}), \tag{4.211}$$

where $\boldsymbol{\Delta}$ is a diagonal matrix with Δ_k in its k-th diagonal element. Hence, the truncation has the same interval and the sampling is done in the same points as for $\tilde{f}$, and the truncated interval is repeated infinitely many times. According to Section 4.8.2, the discreteness and periodicity of $\hat{f}_P$ implies that its corresponding Fourier transform $\hat{F}_P$, too, is discrete and periodic. Having defined this transform pair, the following theorem is at hand.

Theorem 4.4. The two functions $\tilde{f}$ and $\tilde{F}$ can, for arbitrary values of $\mathbf{x}$ and $\mathbf{u}$, respectively, be expressed as

$$\tilde{f}(\mathbf{x}) = \hat{f}_P(\mathbf{x}) * \text{sinc}^n(\mathbf{x}) \tag{4.212}$$

$$\tilde{F}(\mathbf{u}) = \frac{1}{(2\pi)^n} \hat{F}_P(\mathbf{u}) * \text{sinc}^n(\frac{\boldsymbol{\Delta}\mathbf{u}}{2\pi}) \tag{4.213}$$

□

Proof of Theorem 4.4. It follows directly from the definition of the DFT, Equation (4.209), that

$$\tilde{F}(\mathbf{x}) = \frac{1}{\det \boldsymbol{\Delta}} \sum_{\mathbf{X}} \tilde{f}(\mathbf{x})\, e^{-i\mathbf{u}^T\mathbf{x}} = \frac{1}{\det \boldsymbol{\Delta}} \sum_{\mathbf{X}} f(\mathbf{x})\, e^{-i\mathbf{u}^T\mathbf{x}} =$$

$$= \frac{1}{\det \boldsymbol{\Delta}} \sum_{\mathbf{X}} \int_{\mathbb{R}^n} f(\mathbf{y})\, e^{-i\mathbf{u}^T\mathbf{y}}\, \delta^n(\mathbf{x}-\mathbf{y})\, d\mathbf{y}^n =$$

$$= \frac{1}{\det \boldsymbol{\Delta}} \int_{\mathbb{R}^n} f(\mathbf{y}) \sum_{\mathbf{X}} \delta^n(\mathbf{x}-\mathbf{y})\, e^{-i\mathbf{u}^T\mathbf{y}}\, d\mathbf{y}^n =$$

$$= \frac{1}{\det \boldsymbol{\Delta}} \int_{\mathbb{R}^n} h(\mathbf{y})\, e^{-i\mathbf{u}^T\mathbf{y}}\, d\mathbf{y}^n = \frac{1}{\det \boldsymbol{\Delta}} \mathcal{F}\{\, h \,\}, \tag{4.214}$$

where

$$h(\mathbf{x}) = f(\mathbf{u}) \cdot \mathrm{III}^n(\mathbf{x}) \cdot \mathrm{rect}^n(\boldsymbol{\Delta}^{-1}\mathbf{x}) = \hat{f}_P \cdot \mathrm{rect}^n(\boldsymbol{\Delta}^{-1}\mathbf{x}). \tag{4.215}$$

Hence, $\tilde{F}$, will assume the same values as the Fourier transform of h divided by $\det \boldsymbol{\Delta}$. This is precisely the statement of Equation (4.213).

Similarly, the definition of the IDFT, Equation (4.210), together with Equation (4.213) gives

$$\tilde{f}(\mathbf{x}) = \mathcal{F}^{-1}\{\, K(\mathbf{u}) \,\}, \tag{4.216}$$

where

$$K(\mathbf{u}) = \left[\hat{F}_P * \mathrm{sinc}^n(\frac{\boldsymbol{\Delta}\mathbf{u}}{2\pi})\right] \cdot \frac{\det \boldsymbol{\Delta}}{2\pi} \cdot \mathrm{III}^n(\frac{\boldsymbol{\Delta}\mathbf{u}}{2\pi}) \cdot \mathrm{rect}^n(\frac{\mathbf{u}}{2\pi}). \tag{4.217}$$

Consequently,

$$\begin{aligned} \tilde{f}(\mathbf{x}) &= \left[(2\pi)^n \, \hat{f}_P(\mathbf{x}) \cdot \frac{1}{\det \boldsymbol{\Delta}} \, \mathrm{rect}^n(\boldsymbol{\Delta}^{-1}\mathbf{x})\right] * \\ &* \frac{1}{(2\pi)^n} \, \mathrm{III}^n(\boldsymbol{\Delta}^{-1}\mathbf{x}) * \mathrm{sinc}^n(\mathbf{x}) = \\ &= \hat{f}_P * \mathrm{sinc}^n(\mathbf{x}), \end{aligned} \tag{4.218}$$

which is the statement of Equation (4.212). □

So what, more precisely, does Theorem 4.4 say? An examination of Equations (4.212) and (4.213) shows that $\tilde{f}$ and $\hat{f}_P$ must be equal at the sample points, apart from the multiplicative impulses in $\hat{f}_P$. The same is also true for $\tilde{F}$ and $\hat{F}_P$, disregarding the factor $(2\pi)^{-n}$. Consequently, the DFT is nothing but some sample values of $\hat{F}_p$, and will therefore reflect the properties of $\hat{f}_P$ rather than f. As an example, let f be two-dimensional function, and let the truncation be done such that the values of f on e.g. the corresponding points of the left and right vertical borders are not the same. This implies that a discontinuity will appear in the periodic function $\hat{f}_P$. The discontinuity will somehow show in $\hat{F}_P$, and therefore also in $\tilde{F}$. This was illustrated in the beginning of this chapter, in Figure 4.2, where artifacts due to border wrap-around appear in the local transforms. Consequently, it is incorrect to say that a DFT describes only the features of some local region, represented by $\tilde{f}$; it contains a description of a periodic version of the region, represented by $\hat{f}_P$.

Having realized the problems of the DFT, what can we do about them? There is no unambiguous answer, but at least one approach deserves mention, the *windowed DFT.* This implies that $\tilde{f}(\mathbf{x})$, prior to the computation of the DFT, is multiplied by a window function, $\tilde{w}(\mathbf{x})$, designed to reduce artifacts of border wrap-around. In most cases this implies that $w(\mathbf{x})$ has a value close to or equal to one at the centre, i.e. in the vicinity of the origin, and a value close to or equal to zero at the borders. Precisely how the transition between one and zero is implemented depends on what type of artifacts are to be avoided and how much the window is allowed to effect the DFT. The multiplication by $w(\mathbf{x})$ corresponds to convolution with $W(\mathbf{u})$ in the Fourier domain. Consequently, the reduction of border wrap-around is achieved at the cost of a smearing that causes aliasing of the frequency components. In order to reduce the aliasing as well, the window function should be as smooth as possible.

To summarize the above discussion, we may assert that the DFT must be used with care since it will describe the relation between a spatial and a frequency function, *which both are discrete and periodic.* Hence, if we design a function in the Fourier domain, e.g. a filter function, and then represent it by discrete values $\tilde{F}(\mathbf{u})$, the IDFT will give a function $\tilde{f}$ that represents one period of the periodic function $\hat{f}_P$.

5

KERNEL OPTIMIZATION

In this chapter it is shown how convolution kernels can be implemented in practical situations. The material is based on [65]. The presentation is restricted to the case where the desired ideal filter is given in the Fourier domain. An optimal kernel is the set of coefficients that minimizes some distance measure with respect to the ideal filter. A family of distance measures suitable for multidimensional image signals is given. The dependence of attainable error levels on kernel size is demonstrated and convolution results on test images discussed.

5.1 SPATIAL RESTRICTIONS

The two main restrictions imposed on convolution kernels are both of spatial nature:

1. Kernels are represented by samples in a rectangular fashion.
2. Kernels are of limited size.

(As the effect of limited resolution in kernel coefficients has proved to be of relatively minor importance, this will not be discussed here). The first restriction implies that the Fourier transform of the kernels will be repetitive in all dimensions with a period of 2π, if the sampling distance in all dimensions is taken to be 1, i e:

$$\tilde{F}(\mathbf{u}) = \tilde{F}(\mathbf{u} + \sum_i 2k_i\pi\hat{\mathbf{u}}_i) \tag{5.1}$$

where $\hat{\mathbf{u}}_i$ is a unit frequency vector in the direction of the i-th dimension and the k_i-s are integers. The limited kernel size will impose a smoothness restriction on the realizable filter functions. (Taking the symmetry of the Fourier transform into account the effect of a limited kernel size can also be seen as a low pass filter acting in the frequency domain, i.e. the uncertainty principle). The filter function $\tilde{F}(\mathbf{u})$ of a kernel with N coefficients is given by its Fourier transform, Equation (5.2).

$$\tilde{F}(\mathbf{u}) = \sum_{n=1}^{N} f_n e^{-i\mathbf{x}_n \cdot \mathbf{u}} \tag{5.2}$$

where $\mathbf{x}_n$ is the coordinate of coefficient f_n.

5.2 DISTANCE MEASURE

The task now becomes one of finding kernels having a filter function $\tilde{F}(\mathbf{u})$ which is as close to the reference function $F(\mathbf{u})$ as possible. The distance measure to use can of course be discussed but a weighted mean square distance would seem appropriate; the definition of the chosen measure is given by Equation (5.3).

$$D = \sum_{\Omega} W^2(\mathbf{u}) \left[F(\mathbf{u}) - \tilde{F}(\mathbf{u})\right]^2 \tag{5.3}$$

If the sample distances in both dimensions of the kernel are normalized to unity, then

$$\Omega = \{\mathbf{u} : |u_i| < \pi\} \tag{5.4}$$

where u_i is the coefficient for the i-th frequency dimension. The required number of samples summed over Ω is determined by the spatial size of the kernel, an effect of the smoothness restriction. The samples should be evenly distributed over Ω. The effect of the weighting function $W(\mathbf{u})$ is discussed in the next section.

5.3 MINIMIZING THE DISTANCE

As the distance measure is of quadratic type, finding the minimum of D can be done by computing the partial derivatives of D with respect to the kernel coefficients f_n and solving Equation (5.5).

$$\frac{\partial D}{\partial f_n} = 0 \qquad\qquad n = 1, 2, \ldots N \tag{5.5}$$

Expressing the partial derivatives of D as functions of the kernel coefficients is straightforward but cumbersome and will for that reason be omitted from this presentation. The requirement that all partial derivatives be zero simultaneously results in a set of $2N$ coupled equations and the solution will be N complex coefficients. These coefficients constitute the optimized kernel $f(\mathbf{x})$. As before, the real part of the analytic kernel will be even and the imaginary part odd. The two parts can be thought of as generalized edge and line detectors respectively.

5.4 THE WEIGHTING FUNCTION

The weighting function $W(\mathbf{u})$ determines the importance of a close fit for different spatial frequencies. In general, this weighting function should be determined from all a-priori information available about the situations in which the optimized filters are to be used. No such attempt has been made - only some general arguments about energy spectra of typical images have been taken into account; note that optimizing without use of a weighting function is equivalent to setting $W(\mathbf{u}) = 1$. The assumption made when choosing the weighting function has been that the importance of a close fit is roughly proportional to the expected energy contribution.

There is no a-priori information available for the local orientation in an average image. There is, however, some a-priori information available about the spectral energy as a function of the radius, $\rho = \sqrt{\sum u_i^2}$, for an average image. It is known that most of the energy is usually concentrated around the center and that it decreases monotonically as ρ increases. Also, as there does not seem to be a large difference in terms of energy spectrum when imaging the real world at very different scales (say a microscope image vs a satellite image), one can argue that the average energy distribution should satisfy the following equation.

$$E(\rho) = aE(b\rho) \tag{5.6}$$

where a and b are constants.

This means that a scale change is reflected as only a change of magnitude. A class of functions that exhibit both of the above properties is given by the following equation.

$$E(\rho) = \rho^{\alpha} \qquad\qquad \alpha < 0 \tag{5.7}$$

Two examples of images having this type of energy spectrum are given by:

1. A large number of lines placed at random $E(\rho) = \rho^{-1}$
2. A large number of edges placed at random $E(\rho) = \rho^{-3}$

Another fact about images that are representable in computers is that they have been digitized. Two side effects are usually caused by this procedure: the first is a bandwidth limitation of the image (corresponding to the "cosine-terms" in Equation (5.8)); the second is an addition of noise (corresponding to a constant, ϵ_n, since the added noise generally is of broadband type). In most cases it will be impossible to define an optimality criterion for the weighting function but the above discussion provides a useful background for making an apropriate choice. The weighting function used in the following examples is given by:

$$W(\mathbf{u}) \;=\; \rho^{-1} \prod_i \cos^2\left(\frac{u_i}{2}\right) \;+\; \epsilon_n \tag{5.8}$$

where ϵ_n is determined by the expected ratio between noise and signal strength. Note that $W(\mathbf{u}) \to \infty$ when $\rho \to 0$, this is equivalent to the constraint that the sum of the kernel coefficients should be exactly equal to the value given by the ideal function at $\rho = 0$.

5.5 OPTIMIZATION RESULTS

In this section a few examples of optimization results are given. To simplify visualization of the results all optimized kernels are two-dimensional. The reference functions used for the examples all are of the form:

$$F(\rho,\varphi) = \begin{cases} 2^{-\left(\frac{2}{B\ln 2}\right)^2 \ln^2(\rho/\rho_i)} \cos^{2A}(\varphi-\varphi_k) & \text{if} \quad |\varphi-\varphi_k| < \frac{\pi}{2} \\ 0 & \text{else} \end{cases} \tag{5.9}$$

where:

ρ_i is the radial center frequency.

B is the radial 6 dB sensitivity bandwidth in octaves.

φ_k is the filter orientation angle.

A determines the angular selectivity of the filter.

The advantages of this particular class of filters are discussed in the following chapters, here it serves only as an example. The parameters determining bandwidth, angle selectivity and angle orientation are, for all optimizations in this section, kept constant: $B = 2$, $A = 2$ and $\varphi_k = 22.5°$.

This filter is a quadrature filter and the optimized kernel will have complex coefficients. The real part of the coefficients corresponds to the even part of the filter function, $\tilde{F}_e(\mathbf{u})$. The imaginary part of the coefficients corresponds to the odd part of the filter function, $\tilde{F}_o(\mathbf{u})$. To facilitate studying the errors for the odd and even parts separately, they were kept separate in the optimization. An overall measure of how much an optimized kernel deviates from its ideal is given by the distortion, Δ, calculated according to Equation (5.10).

$$\Delta = \left(\|W(\mathbf{u})\|_\Omega \ \|F(\mathbf{u})\|_\Omega \right)^{-1} \sqrt{\sum_\Omega W^2(\mathbf{u})[F(\mathbf{u}) - \tilde{F}(\mathbf{u})]^2} \tag{5.10}$$

where the norm of a function sampled on Ω and N_Ω is the number of samples is defined by:

$$\|F(\mathbf{u})\|_\Omega = \sqrt{\frac{1}{N_\Omega} \sum_\Omega F^2(\mathbf{u})} \tag{5.11}$$

The algorithm used for angle estimation in the experimental section uses four kernel pairs. Using $\varphi_k = 22.5°$, it is possible to construct these kernel pairs having orientations 22.5°, 67.5°, 112, 5° and 157, 5° by simply reflecting the kernel coefficients in the x_2-axis and the $x_1 = x_2$ diagonal. Thus only one kernel instead of four needs to be optimized.

Figure 5.1 (a-f) shows an optimization example having a kernel size of 15×15 elements where $\rho_i = \pi \cdot 2^{-2.5}$ and $\epsilon_n = 1$. π is the Nyquist frequency when the distance between kernel elements is normalized to unity. Figure 5.1 (a) and Figure 5.1 (b) are the even and odd reference functions respectively, Figure 5.1 (c) and Figure 5.1 (d) show the optimized filter results and Figure 5.1 (e) and Figure 5.1 (f) depict the difference between the optimized filters and the corresponding reference function. The distortion figures for this case are $\Delta_e =$ 13.0% and $\Delta_o = 19.3\%$ for the even and odd filters respectively.

As can be seen both the even and the odd optimized filters exhibit some 'ringing' errors. This is due to the fact that a 15×15 element kernel is slightly too small to cope with the peaked frequency response of the reference functions.

Figure 5.2 (a,b) is a plot of the even and odd 15×15 kernels having the frequency response shown in Figure 5.1 (c,d). The kernel sample values are found in the "net" intersections. The envelope corresponding to the kernel pair in Figure 5.2 (a,b) is plotted in Figure 5.2 (c).

It can be seen that the envelope is roughly circularly symmetric and of approximately Gaussian type.

Figure 5.3 (a-f) shows an optimization example using the same reference functions as in Figure 5.2 but with a different weighting function and using $\epsilon_n =$ 0.01, Equation (5.8). This weighting function puts much less emphasis on the high frequencies.

It is clear from Figure 5.3 (c-f) that the change of weighting function dramatically affects the distribution of the error. The error in the low frequency parts is slightly reduced at the expense of an increased error in the high frequency areas. The former is hard to see and as the filter behaviour around the peaks is in this case fundamentally restricted by the uncertainty principle, nothing much can be done about the errors in these parts.

Figure 5.4 (a-f) is an example of optimization of a 15×15 element kernel where $\rho_i = \pi \cdot 2^{-1.5}$ and $\epsilon_n = 1$.

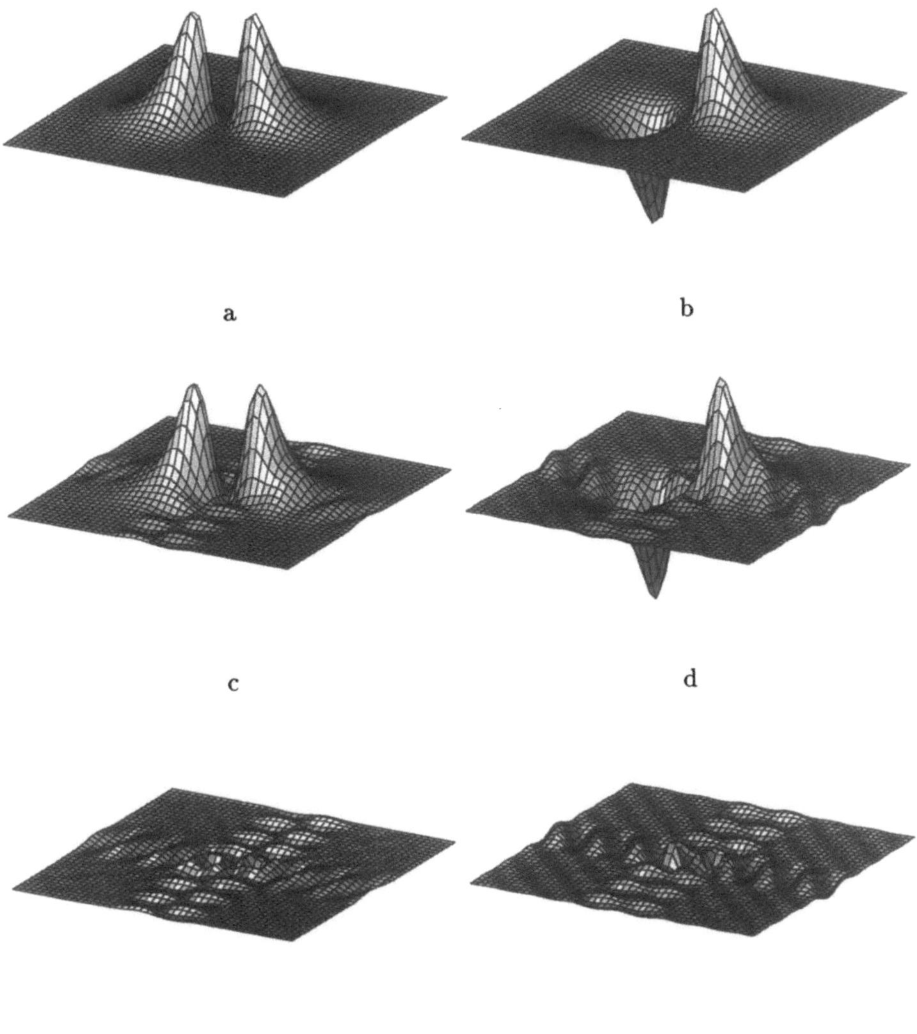

Figure 5.1 a, b: Even and odd reference functions. $A = 2, B = 2, \varphi_k = 22.5^\circ, \rho_i = \pi \cdot 2^{-2.5}$. c, d: Optimized filters corresponding to a, b. Kernel size 15 $\times$ 15, $\epsilon_n = 1$, $\Delta_e = 13.0\%$, $\Delta_o = 19.3\%$. e, f: Difference between optimized filters and the corresponding reference function.

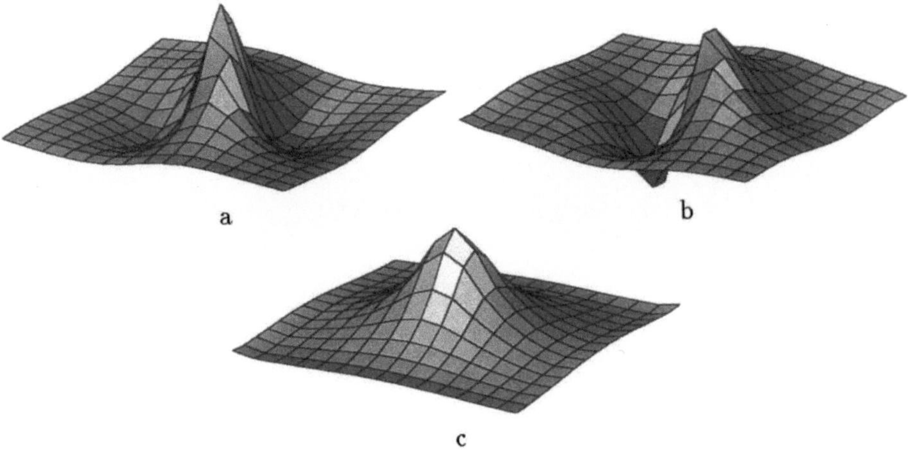

Figure 5.2 a, b: Optimized 15 $\times$ 15 even and odd kernel respectively. $A = 2, B = 2, \varphi_k = 22.5°, \rho_i = \pi \cdot 2^{-2.5}$ and $\epsilon_n = 1$. c: Envelope of the kernel pair in a, b.

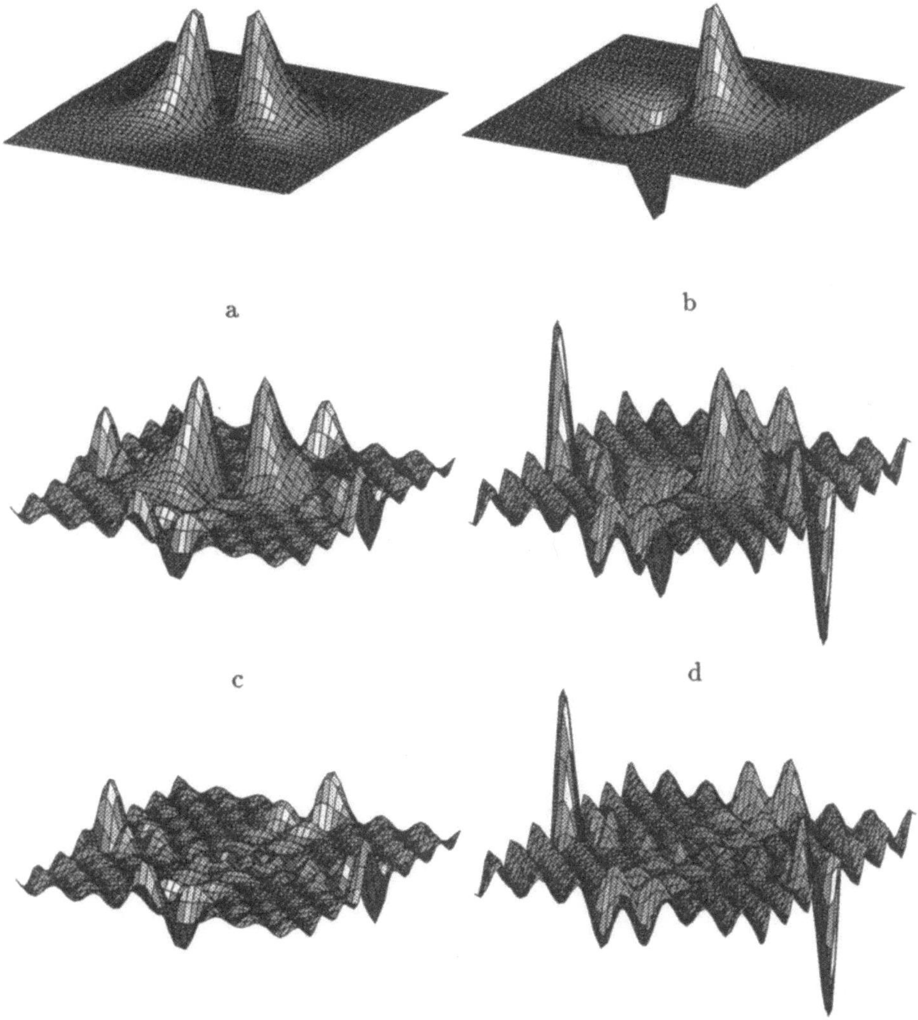

a b

c d

e f

Figure 5.3 a, b: Even and odd reference functions $A = 2, B = 2, \varphi_k = 22.5°, \rho_i = \pi \cdot 2^{-2.5}$. c, d: Optimized filters corresponding to a, b. Kernel size 15×15, $\epsilon_n = 0.01$, $\Delta_e = 28.9\%$, $\Delta_o = 51.6\%$. e, f: Difference between the optimized filters and the corresponding reference function.

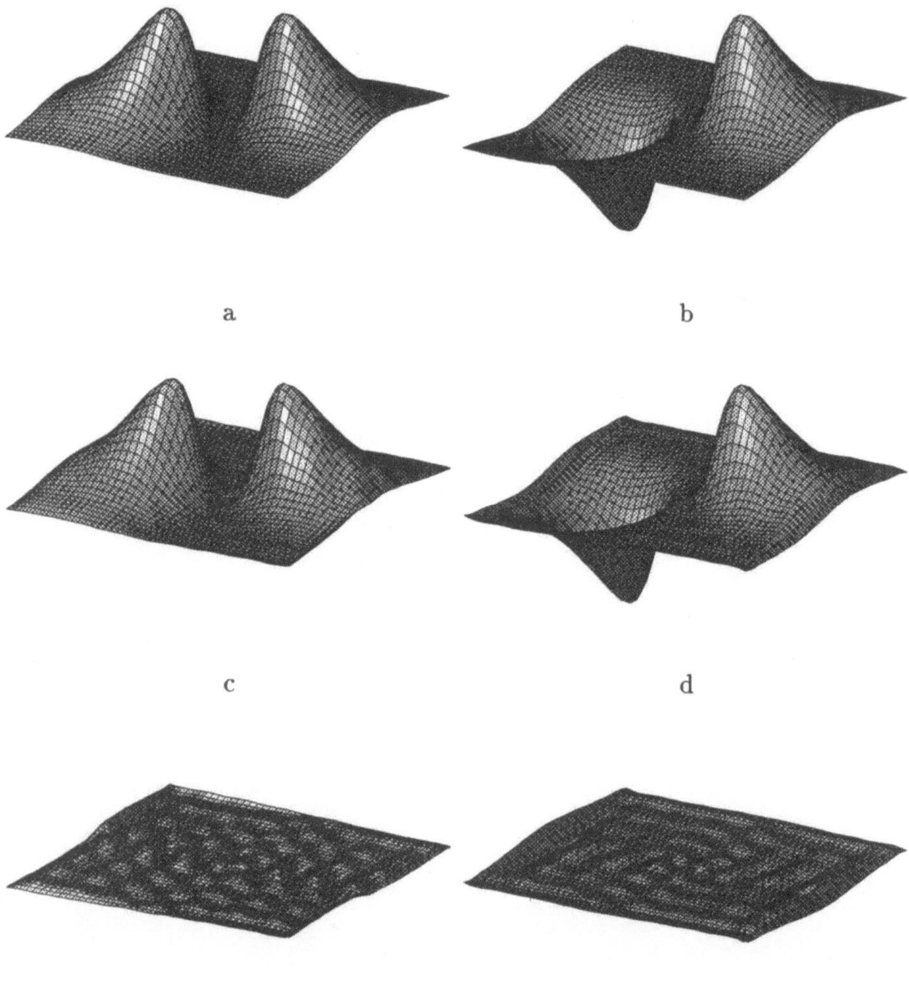

Figure 5.4 a, b: Even and odd reference functions $A = 2, B = 2, \varphi_k = 22.5^\circ, \rho_i = \pi \cdot 2^{-1.5}$. c, d: Optimized filters corresponding to a, b. Kernel size 15×15, $\epsilon_n = 1$, $\Delta_e = 3.6\%$, $\Delta_o = 5.5\%$. e, f: Difference between the optimized filters and the corresponding reference function.

Here it can be seen that the errors are very small, proving the optimized filters to be good approximations of their reference functions. A good fit is also indicated by the distortion figures which for this case are $\Delta_e = 3.6\%$ and $\Delta_o = 5.5\%$. It is the smoothness of the reference functions that makes this good fit possible.

Figure 5.5 (a-f) shows an example of optimization using the same parameters as for Figure 5.4 except that the kernel size has been reduced to 5×5. Even though the reference functions are smooth, they are not smooth enough for a kernel as small as 5×5. A large error is also indicated by the distortion figures for this case, $\Delta_e = 30.6\%$ and $\Delta_o = 27.4\%$.

Figure 5.6 (a-f) together with Figure 5.7 (a-f) provides an interesting example of optimization using different weighting functions when the reference functions are significantly non-zero at the Nyquist frequency limits. The center frequency here was $\rho_i = \pi \cdot 2^{-0.5}$ and $\epsilon_n = 1$, $\epsilon_n = 0.01$ was used in Figure 5.6 and Figure 5.7 respectively.

This is a case where the periodicity of the kernel frequency response caused by the sampling, together with the smoothness restriction caused by the limited kernel size and the uncertainty principle, is in sharp contrast with the behaviour of the reference function at the Nyquist limits. As $\epsilon_n = 1$ (Figure 5.6) means that a relatively large weight is given to high frequencies, this accentuates the incompatibility mentioned above. The effect is that a 'ringing' type of error is introduced.

When $\epsilon_n = 0.01$ is used (Figure 5.7) substantially less weight is given to the high frequencies. The effect is a marked reduction of the errors in the low frequency parts at the expense of a slightly increased error for the high frequencies. It is also interesting to note that the error in Figure 5.7 is of a smoother kind than that in Figure 5.6. The distortion figures for these two cases support the visual impression, using $\epsilon_n = 1$ yields $\Delta_e = 8.5\%$ and $\Delta_o = 15.1\%$ while using $\epsilon_n = 0.01$ yields $\Delta_e = 2.2\%$ and $\Delta_o = 4.0\%$.

5.6 KERNEL EVALUATION

To evaluate the performance of the optimized kernels in an image processing task, a number of experiments have been carried out using the GOP Image Processor [44].

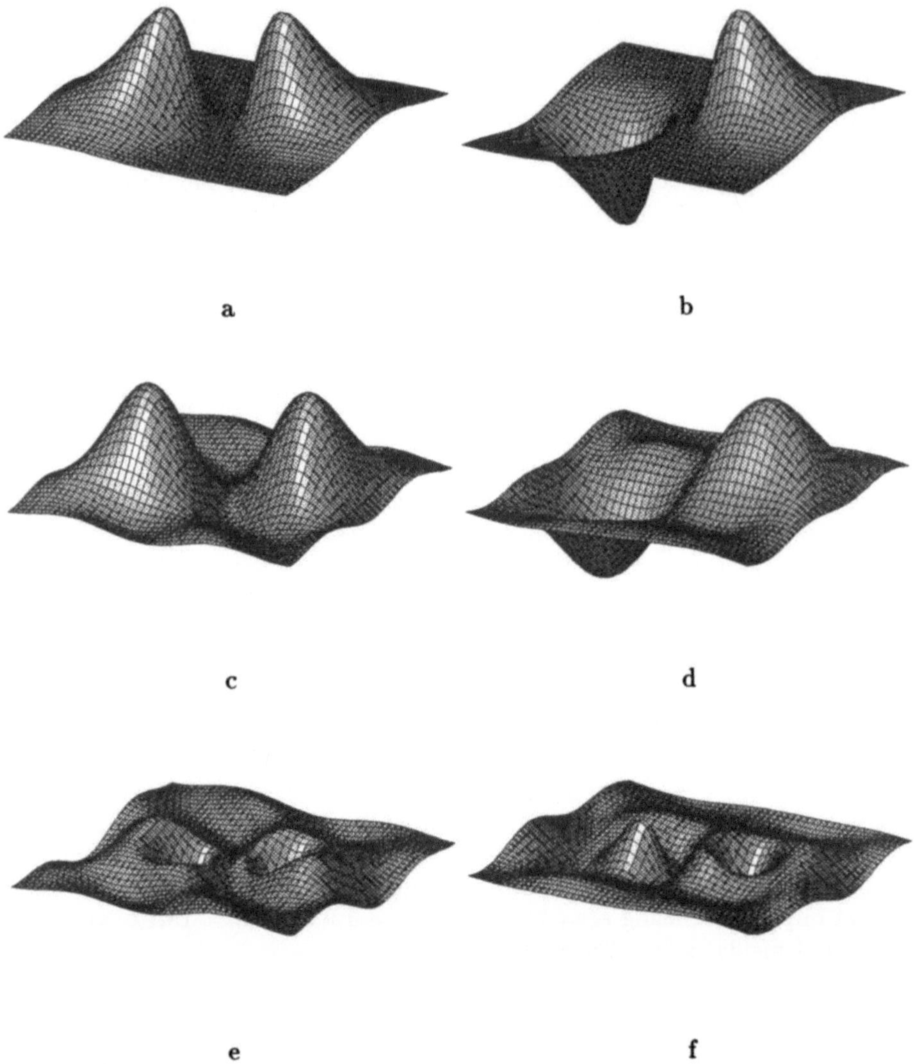

Figure 5.5 a, b: Even and odd reference function $A = 2, B = 2, \varphi_k = 22.5°, \rho_i = \pi \cdot 2^{-1.5}$. c, d: Optimized filters corresponding to a, b. Kernel size 5×5, $\epsilon_n = 1$, $\Delta_e = 30.6\%$, $\Delta_o = 27.4\%$. e, f: Difference between the optimized filters and the corresponding reference function.

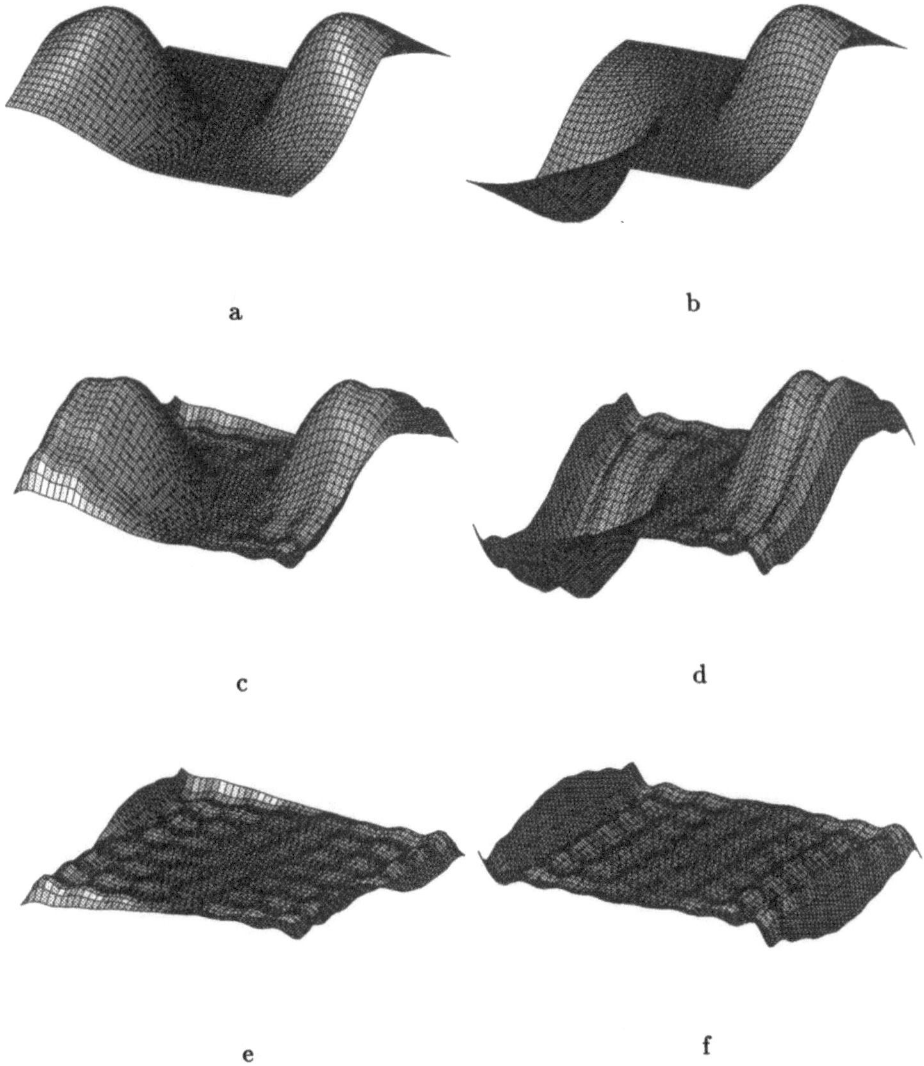

Figure 5.6 a, b: Even and odd reference functions $A = 2, B = 2, \varphi_k = 22.5°, \rho_i = \pi \cdot 2^{-0.5}$. c, d: Optimized filters corresponding to a, b. Kernel size 15×15, $\epsilon_n = 1$, $\Delta_e = 8.5\%$, $\Delta_o = 15.1\%$. e, f: Difference between the optimized filters and the corresponding reference function.

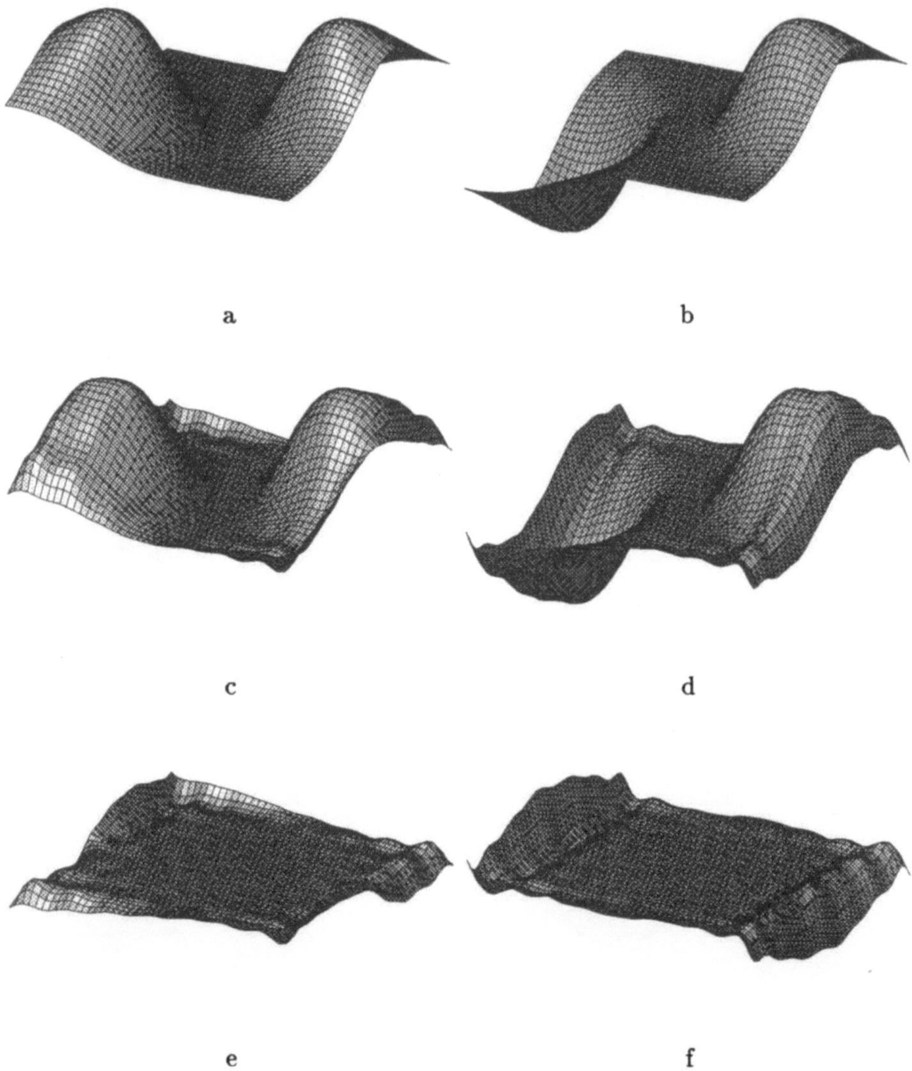

Figure 5.7 a, b: Even and odd reference function $A = 2, B = 2, \varphi_k = 22.5^\circ, \rho_i = \pi \cdot 2^{-0.5}$. c, d: Optimized filters corresponding to a, b. Kernel size 15×15, $\epsilon_n = 0.01$, $\Delta_e = 2.2\%$, $\Delta_o = 4.0\%$. e, f: Difference between the optimized filters and the corresponding reference function.

5.6.1 Test patterns

All experiments used the same testpattern, with different amounts of noise added.

Figure 5.8 Test pattern used in the experiments.

The test pattern, shown in Figure 5.8, contains local frequencies ranging from π (which is the Nyquist frequency if the distance between samples is normalized to unity) to approximately $\pi/10$. Starting at the outer edge the frequency increases exponentially toward the center so that frequencies having a constant ratio are separated by the same distance. As the pattern is circularly symmetric, all frequencies are represented in every orientation. To avoid edge effects the amplitude is smoothly decreased in the periphery and toward the center. A mathematical description of the test pattern is given by:

$$f_{test}(r) = \begin{cases} 127 & r < 56 \\ 127 \cdot [1 + g(r) \cdot \cos^2(\frac{\pi \cdot r}{16} - 4\pi)] & 56 < r < 64 \\ 127 \cdot [1 + g(r)] & 64 < r < 224 \\ 127 \cdot [1 + g(r) \cdot \sin^2(\frac{\pi \cdot r}{64} - 4\pi)] & 224 < r < 256 \\ 127 & 256 < r \end{cases}$$

where:

$$g(r) = \sin\left[\frac{112\pi}{\log 2}(2^{-\frac{r}{56}} - 2^{-\frac{256}{56}})\right]$$

and r is the distance from the center of the image in pixel distance units.

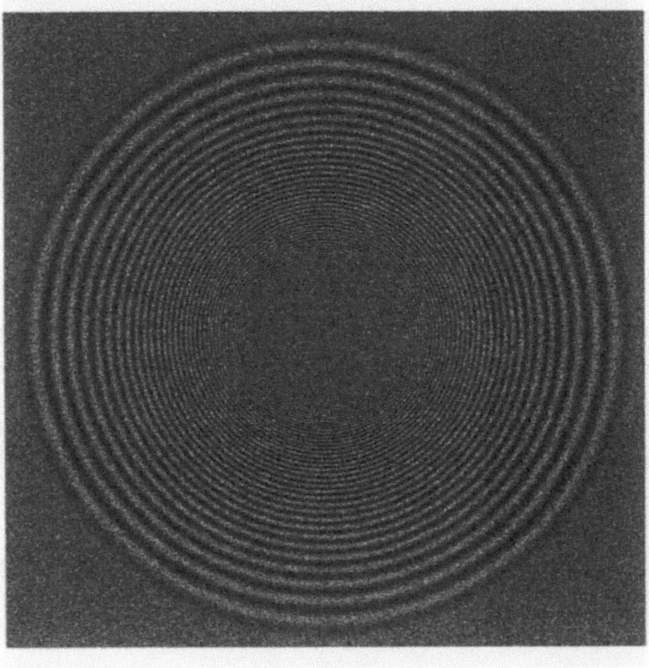

Figure 5.9 Test pattern with white Gaussian noise added, SNR = 0 dB.

White Gaussian noise was added to this test pattern to produce images with different signal-to-noise ratio (SNR). Figure 5.9 shows the test patterns with 0 dB SNR.

5.6.2 Quadrature filter pair outputs

All kernel pairs used in the test pattern experiments in this section have $A = 2, B = 2$, and $\varphi_k = 22.5°$. In Figure 5.10 (a-d) a kernel pair having $\rho_i = \pi \cdot 2^{-1.5}$ was used to produce a local orientational RMS value for test patterns having no noise (clean), 20 dB SNR, 10 dB SNR and 0 dB SNR as indicated in the figure. (Each small square corresponds to a full test pattern.)

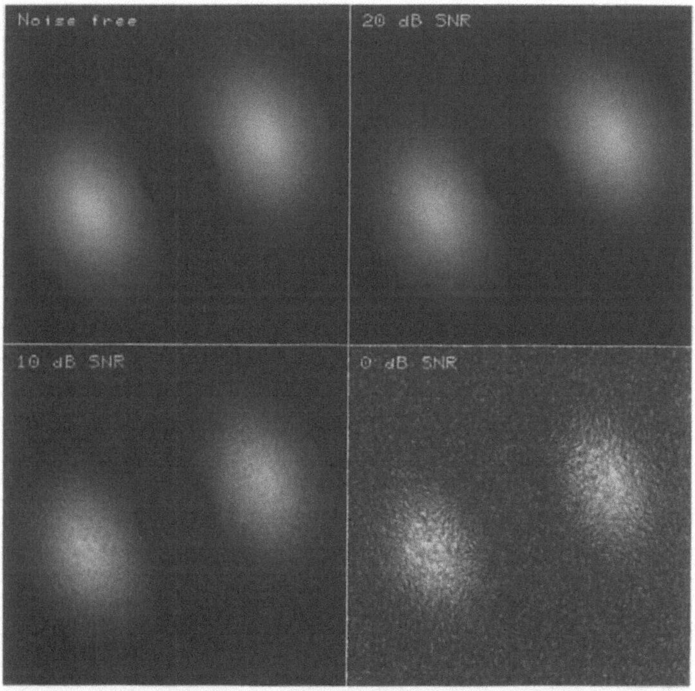

Figure 5.10 Outputs from a quadrature filter pair with $A = 2, B = 2, \varphi_k = 22.5°$ and $\rho_i = \pi \cdot 2^{-1.5}$ used to estimate the local orientational RMS value for test patterns with different SNR.

The output for the "clean" test pattern can be seen to be very smooth and unimodal in orientation and frequency, with a maximum at approximately 22.5°

and at the center of the frequency range. The lack of phase dependence indicates that the kernel pair provides a good quadrature filter.

The output when noise is added to the test pattern is rather predictable: the output becomes noisier as the SNR ratio decreases. For the 0 dB case there is a substantial amount of noise covering the whole image although the filter peak is still clearly visible. Note that the noise is no longer white but has low-pass character.

Figure 5.11 shows the orientational RMS outputs from quadrature kernel pairs having different center frequencies ρ_i. HF, MF and LF corresponds to ρ_i being $\pi \cdot 2^{-0.5}, \pi \cdot 2^{-1.5}$ and $\pi \cdot 2^{-2.5}$ respectively. (0) indicates $\epsilon_n = 1$ and (40) indicates $\epsilon_n = 0.01$ (i e ϵ_n using a dB scale). For this display a staircase type of attenuation curve was deliberately introduced to mark the 3 dB, 6 dB, 12 dB and 24 dB lines of the output.

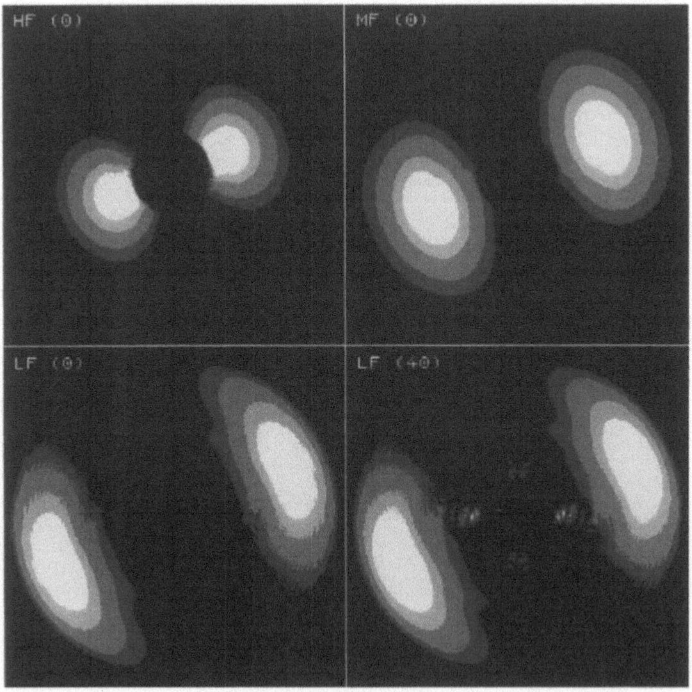

Figure 5.11 Orientational RMS outputs showing 3 dB, 6 dB, 12 dB and 24 dB lines using kernels with different center frequencies ρ_i.

Again the MF filter pair appears to be of smooth variation. The output from the HF filter pair has a somewhat "ragged" behaviour toward the highest frequencies. This is, however, partly obscured by the display resolution. The two outputs using LF kernel pairs with different ϵ_n both exhibit some irregularities. The 24 dB line towards higher frequencies is not so smooth and a slight phase dependence can be seen in the low frequency parts. The increased error in the high frequency area is clearly visible for $\epsilon_n = 0.01$ and close scrutiny reveals a minor reduction in phase dependence at low frequencies for $\epsilon_n = 1$.

Comparing Figure 5.11 (a-d) with Figure 5.6, Figure 5.4, Figure 5.1 and Figure 5.3 respectively is helpful in trying to gain some insight into the correspondence between filter shapes and outputs from test pattern processing.

To demonstrate the types of errors that would occur when using a kernel with too few elements to properly approximate the given reference function, two 5×5 kernel pairs optimized for the medium frequency range, $\rho_i = \pi \cdot 2^{-1.5}$, were tested. At the top of Figure 5.12 the outputs from kernel pairs optimized using $\epsilon_n = 1$, and $\epsilon_n = 0.01$, are shown. The lower half of Figure 5.12 depicts the absolute value of the 'error' in these outputs multiplied by two.

In order to estimate the error in the output, a 15×15 element kernel pair having the same parameters ($A = 2, B = 2, \varphi_k = 22.5°$ and $\rho_i = \pi \cdot 2^{-1.5}$) was used to produce a reference output. The error was then taken to be the difference between the output from the 5×5 kernel pair and the reference output. This procedure should be valid as the 15×15 kernel pair is very close to the ideal when compared with the 5×5 kernel pairs.

The suppression of low frequencies is unsatisfactory for both $\epsilon_n = 1$ and $\epsilon_n = 0.01$ and the banded outer edges indicate a strong phase dependence. The effect of using different ϵ_n in the optimization is that $\epsilon_n = 1$ gives relatively small errors for high frequencies, while the largest error occurs at the filter peak. Using $\epsilon_n = 0.01$ reduces the error at the filter peak, but increases the error at high frequencies. To be able to make a comparison between the distortion figures for the filters and the errors produced when using the filter for estimating the orientational RMS value, a relative error figure was calculated, Equation (5.12).

$$err = \sqrt{\frac{E_{err}}{E_{ref}}} \tag{5.12}$$

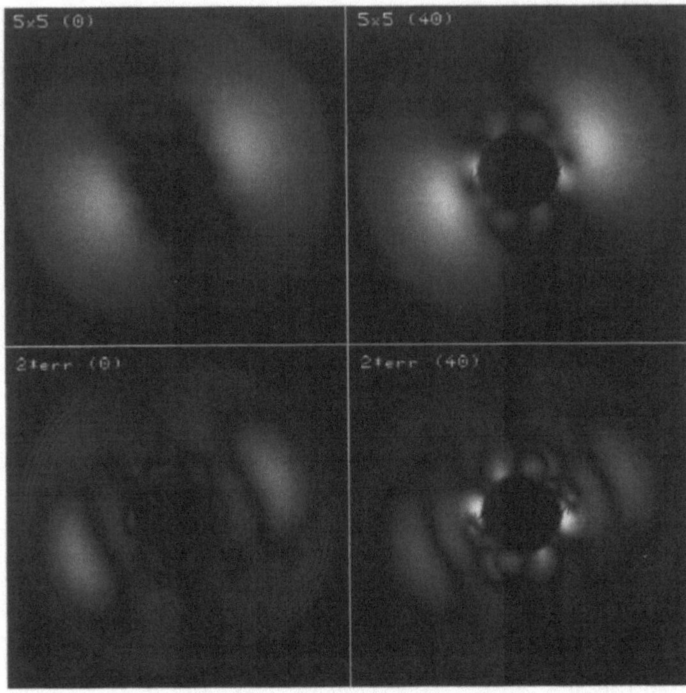

Figure 5.12 Outputs and errors using 5 × 5 kernel pairs.

Here E_{err} is the error energy and E_{ref} the energy of the reference output.

For $\epsilon_n = 1$, the relative error was $err = 0.30$ which should be compared with $\Delta_e = 30.5\%$ and $\Delta_o = 27.4\%$. For $\epsilon_n = 0.01$, the relative error was $err = 0.26$, compared with $\Delta_e = 47.8\%$ and $\Delta_o = 58.3\%$. Although the distortion values using $\epsilon_n = 0.01$ are higher the value of err is slightly lower. This indicates that the energy distribution for the test pattern is closer to the weighting function having $\epsilon_n = 0.01$ than $\epsilon_n = 1$.

6

ORIENTATION AND VELOCITY

This chapter presents a unified approach and a theory for estimation of local orientation and velocity in multi-dimensional spatial or spatio-temporal signals. The presentation is intended to serve as a basis for the design of efficient multi-dimensional signal analysis algorithms.

Multi-dimensional signals can be seen as functions of more than one variable and there are in principle no restrictions on what these variables may represent. However, signals as functions of space or space-time are by far the most frequently used. New techniques to produce and process multi-dimensional data are constantly being developed in a number of different fields.

The techniques presented in this chapter are based on spatial or spatio-temporal filtering and tensor representations. For each neighborhood a number of quadrature filters are employed. The filters are sensitive to signals in different orientations in the signal space. Each quadrature filter produces a magnitude and a phase. Phase can be used to describe local symmetry properties of the signal. The magnitudes of the different filters are combined to give a tensor description of local orientation. In addition to orientation the resulting tensors contain information about how *plane-like* or *line-like* the neighborhoods are.

Velocity estimates can be directly obtained from the orientation tensor. In line-like, i.e. "moving point" type regions, a true velocity estimate is produced. In plane-like, i.e. "moving line" type regions, the velocity component perpendicular to the local structure is produced. In addition, discontinuous motion can be detected by operating on the tensor field produced by the orientation algorithm.

A few examples are given below of commonly available multi-dimensional data sets to which the techniques presented in this chapter could be applied.

- **Images (2D)**

  Photograph
  X-ray
  Satellite
  Infrared

- **Image sequences (2D + time)**

  Video
  Ultrasound
  Satellite

- **Volumes (3D)**

  Magnetic Resonance (MR)
  Computed Tomography (CT)
  Positron Emission Tomography (PET)

- **Volume sequences (3D + time)**

  Magnetic Resonance (MR)

Local orientation

A basis for the approach taken in this chapter is the fact that local gradient directions contain important information. Equally important is the local simplicity hypothesis [42, 74, 65]. The basis for this hypothesis is that the spatial variation of the gradient directions is in general much slower than the spatial variation of the image itself. In summary, local orientation is an important and, for most neighborhoods, well defined local feature of multi-dimensional data. That local orientation plays an important role in the mammalian visual system has been demonstrated by the physiological findings of Hubel and Wiesel and others [61], as discussed in Chapter 2.

Orientation and local Fourier transforms

To see how local orientation analysis can be attained it is useful to study the behavior of local Fourier transforms (see Chapter 4). If the image is locally simple, in the above sense, this will be apparent in the Fourier domain by the fact that most of the energy is concentrated in a narrow sector oriented at the same angle as the local image gradients. The less the variation of local orientation the narrower the sector will be.

The distribution of energy along the sector in the radial direction will reflect the frequency properties of the neighborhood changes in the gradient direction. The behavior thus becomes analogous to that of one-dimensional Fourier transforms.

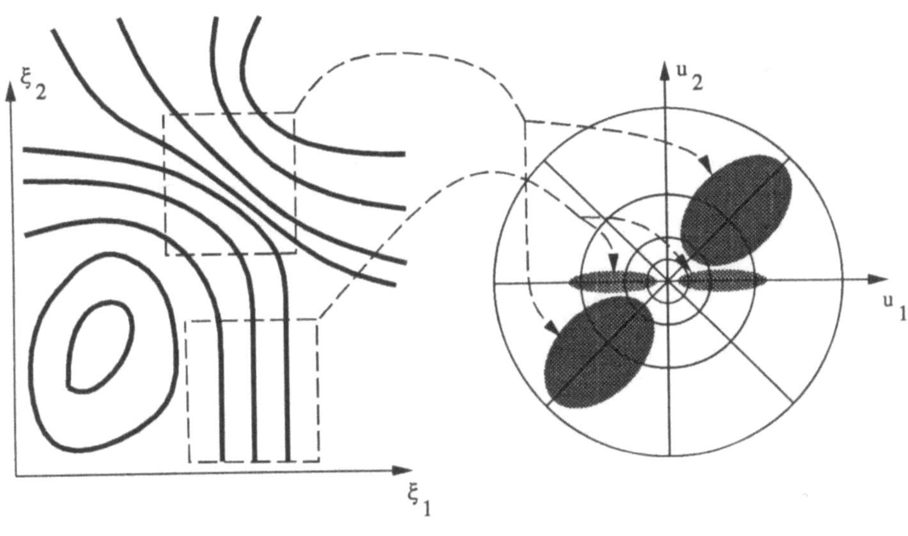

Figure 6.1 Two neighborhoods and the corresponding energy contributions in the Fourier domain.

From the above it can be concluded that a description of a neighborhood in terms of local orientation and local frequency content can be obtained by partitioning the local Fourier transform and studying the energy contribution in the different parts (Figure 6.1).

6.1 TENSORS - A SHORT INTRODUCTION

It is obvious that processing of higher-dimensional data sets puts high demands on computer power and storage capacity. Perhaps less obvious is that increasing the dimensionality of the data also has profound implications for the analysis of the data. In the analysis of data with a dimensionality higher than two it turns out that using scalars and vectors is no longer always convenient [67]. An important extension in this chapter is the introduction of a generalization of the vector concept - the *tensor*.

Tensor analysis is a generalization of the notions from vector analysis. The need for such a theory is motivated by the fact that there are many physical quantities of complicated nature that cannot naturally be described or represented by scalars or vectors. Examples are the stress at a point in a solid body due to internal forces, the deformation of an arbitrary infinitesimal element of volume of an elastic body, and the moments of inertia and conductivity in anisotropic materials. These quantities can be described and represented adequately only by the more sophisticated mathematical entities called tensors. Scalars and vectors actually also belong to this family of elements. Thus, scalars and vectors are special cases of tensors.

Associated with a tensor is the order of that tensor. The order of a tensor can be thought of as the complexity of the entity it represents, e.g. a zeroth order tensor is a scalar, a first order tensor is a vector and a second order tensor can be thought of as a matrix. Tensors will be denoted by bold capital letters, e.g. , $\mathbf{T}$.

The scalar product

Tensors are similar to vectors in that they exist in a vector space and the scalar product between tensors is defined:

$$\mathbf{T} \bullet \mathbf{U} \equiv \sum_{ij} t_{ij} u_{ij} \tag{6.1}$$

where:

t_{ij} and u_{ij} are the components of $\mathbf{T}$ and $\mathbf{U}$ respectively.

The norm

The definition of the scalar product automatically generates a definition of the norm of a tensor. Again, just as for vectors, the square of the norm is the sum of the squares of the components, i.e. the *norm* of a second order tensor $\mathbf{T}$ is defined by:

$$\|\mathbf{T}\|^2 \equiv \mathbf{T} \bullet \mathbf{T} = \sum_{ij} t_{ij}^2 = \sum_{n} \lambda_n^2 \tag{6.2}$$

where:

λ_n are the eigenvalues of $\mathbf{T}$.

This norm is known as the *Frobenius norm.*

Basis tensors

Another important concept from vector analysis that directly applies to tensors is the concept of a basis. Any tensor $\mathbf{T}$ can be expressed as a weighted sum of a set of basis tensors $\mathbf{B}_i$

$$\mathbf{T} = \sum_{i} c_i \, \mathbf{B}_i \tag{6.3}$$

where:

c_i is the coefficient for basis tensor $\mathbf{B}_i$.

In general the number of basis tensors required is the same as the number of elements in $\mathbf{T}$. Orthonormality can be thought of in the same manner as for vectors and a tensor basis set $\mathbf{B}_i$ is orthonormal if:

$$\begin{cases} \mathbf{B}_i \bullet \mathbf{B}_j = 1 & if \quad i = j \\ \mathbf{B}_i \bullet \mathbf{B}_j = 0 & if \quad i \neq j \end{cases} \tag{6.4}$$

Outer product tensors

A special case of tensor, which will be extensively used in this chapter, is constructed as an outer product of two vectors **a** and **b**.

$$\mathbf{T} = \mathbf{a}\mathbf{b}^T \tag{6.5}$$

It is not hard to show that the scalar product between two tensors of this type is given by:

$$\mathbf{T}_1 \bullet \mathbf{T}_2 = \mathbf{a}_1\mathbf{b}_1^T \bullet \mathbf{a}_2\mathbf{b}_2^T = (\mathbf{a}_1 \cdot \mathbf{a}_2)(\mathbf{b}_1 \cdot \mathbf{b}_2) \tag{6.6}$$

From the above it follows that the norm of an outer product tensor is given by:

$$||\mathbf{T}||^2 = \mathbf{T} \bullet \mathbf{T} = (\mathbf{a} \cdot \mathbf{a})(\mathbf{b} \cdot \mathbf{b}) = ||\mathbf{a}||^2 \; ||\mathbf{b}||^2 \tag{6.7}$$

Although a great deal more could be said about this subject, using tensors of orders 0, 1 and 2 together with the short introduction above will suffice for the following presentation.

6.2 REPRESENTING ORIENTATION

To begin, note that, in the mathematical analysis of local orientation, time and space can be treated equally. Correct interpretation of local orientation, however, requires knowledge of which dimension corresponds to time.

The analysis in this section is general, in the sense that it is valid for signals of any dimension, but will deal only with signals that locally can vary in only one direction. These signals will be denoted *simple* signals.

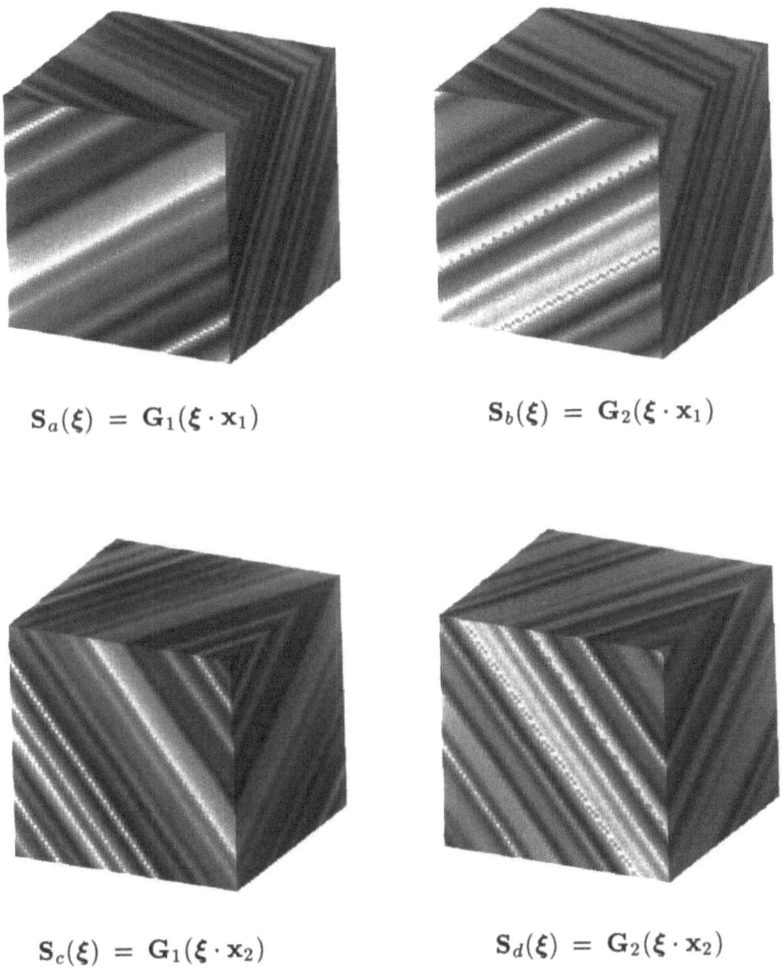

Figure 6.2 Four simple neighborhoods in 3D. The neighborhoods are constructed using two different signal functions ($\mathbf{G}_1$ and $\mathbf{G}_2$) and two different signal orienting vectors ($\mathbf{x}_1$ and $\mathbf{x}_2$).

A *simple* neighborhood of any dimension can be expressed as:

$$\mathbf{S}(\boldsymbol{\xi}) = \mathbf{G}(\boldsymbol{\xi} \cdot \mathbf{x}) \tag{6.8}$$

where:

$\mathbf{S}$ and $\mathbf{G}$ are non-constant tensors of any order.

$\boldsymbol{\xi}$ is the spatial coordinate and

$\mathbf{x}$ is a constant vector oriented along the axis of maximal signal variation.

For this type of neighborhood it is possible to define the local orientation as a tensor $\mathbf{T}$ of order two. The size of the tensor will depend on the dimensionality of the signal, e.g. 3-dimensional signals are described by a 3×3 tensor.

6.2.1 Definition of orientation

The entity representing orientation should meet two basic requirements:

The *Invariance* requirement:

It is evident that the *orientation* of the neighborhood will be the same for all possible $\mathbf{G}$ in Equation (6.8). In other words the entity representing orientation must be invariant to $\mathbf{G}$.

$$\frac{\delta \mathbf{T}}{\delta \mathbf{G}} = 0 \tag{6.9}$$

where $\mathbf{T}$ represents orientation.

The *Equivariance* requirement:

The orientation tensor should locally preserve the angle metric of the original space, i.e

$$\|\delta \hat{\mathbf{T}}\| \propto \|\delta \hat{\mathbf{x}}\| \tag{6.10}$$

where the "^" indicates normalization, i.e. :

$\hat{\mathbf{T}} = \frac{\mathbf{T}}{\|\mathbf{T}\|}$ and

$\hat{\mathbf{x}} = \frac{\mathbf{x}}{\|\mathbf{x}\|}$.

The implications of the invariance and equivariance requirements can easily be explained by studying Figure 6.2. Invariance to **G** implies that **T** should be the same for neighborhoods $\mathbf{S}_a$ and $\mathbf{S}_b$ even though **G** is different. Equivariance implies that in moving from $\mathbf{S}_a$ to $\mathbf{S}_c$, **T** should change proportionally to the change in local orientation.

6.2.2 The orientation tensor

A representation **T** for orientation that meets the above criteria is given by:

$$\mathbf{T} \equiv A\hat{\mathbf{x}}\hat{\mathbf{x}}^T \tag{6.11}$$

where A can be any constant greater than zero.

The norm

Calculating the norm of **T** gives:

$$\|\mathbf{T}\|^2 = A^2(\sum_i x_i^2)^{-2}\sum_{ij} x_i^2 x_j^2 = A^2 \tag{6.12}$$

where $x_{i(j)}$ are the components of $\mathbf{x}$.

Thus the norm of **T** is equal to A. Since the orientation represented by **T** is independent of the norm of **T** A can be used to represent another property. In Section 6.3 it will be shown that A can be made to represent the local amplitude of the signal in a natural way.

Invariance

By its construction **T** is trivially invariant to the signal function **G**. However, as can be expected, making the actual orientation estimates invariant to **G** is by no means trivial. A discussion of this topic appears in Section 6.3.

Equivariance

To show that the *Equivariance* requirement is met by the mapping is fairly straightforward. Rotate $\hat{\mathbf{x}}$ by adding a small perpendicular vector $\epsilon\hat{\mathbf{v}}$ to $\hat{\mathbf{x}}$ and calculate the difference in the norm of $\mathbf{T}$. To start, define $\mathbf{B}$ to be:

$$\mathbf{B} = \lim_{\epsilon \to 0} \frac{\mathbf{T}(\hat{\mathbf{x}} + \epsilon\hat{\mathbf{v}}) - \mathbf{T}(\hat{\mathbf{x}})}{\epsilon} \tag{6.13}$$

where: $\hat{\mathbf{v}} \cdot \hat{\mathbf{x}} = 0$

then

$$||\delta\mathbf{T}|| = ||\mathbf{B}||\,||\delta\hat{\mathbf{x}}|| \tag{6.14}$$

Carrying out the limit calculation is simple and yields:

$$\mathbf{B} = A(\hat{\mathbf{x}}\hat{\mathbf{v}}^T + \hat{\mathbf{v}}\hat{\mathbf{x}}^T) \tag{6.15}$$

The eigenvectors and eigenvalues of $\mathbf{B}$ are easily found by letting $\mathbf{B}$ operate on the combined vectors $(\hat{\mathbf{x}} + \hat{\mathbf{v}})$ and $(\hat{\mathbf{x}} - \hat{\mathbf{v}})$.

$$\begin{cases} \mathbf{B}(\hat{\mathbf{x}} + \hat{\mathbf{v}}) = A(\hat{\mathbf{x}} + \hat{\mathbf{v}}) \\ \mathbf{B}(\hat{\mathbf{x}} - \hat{\mathbf{v}}) = A(\hat{\mathbf{v}} - \hat{\mathbf{x}}) \end{cases} \tag{6.16}$$

This shows that $(\hat{\mathbf{x}}+\hat{\mathbf{v}})$ and $(\hat{\mathbf{x}}-\hat{\mathbf{v}})$ are eigenvectors of $\mathbf{B}$, the eigenvalues being A and $-A$ respectively. Since all other eigenvectors of $\mathbf{B}$ are orthogonal to $\hat{\mathbf{x}}$ and $\hat{\mathbf{v}}$ it follows from Equation (6.15) that all other eigenvalues are zero. Then, according to Equation (6.2), the norm of $\mathbf{B}$ is given by:

$$||\mathbf{B}|| = \sqrt{\lambda_1^2 + \lambda_2^2} = \sqrt{2}\,A \tag{6.17}$$

and since A is constant independent of the orientation it is shown that the *Equivariance* requirement is met.

6.2.3 The 2D orientation vector

If the neighborhood is simple and two-dimensional, a two-component vector, or a complex number, **z**, will suffice to meet the two requirements of orientation representation. The relation between **z** and **T** is given by:

$$\begin{aligned} \mathbf{T} &= A\begin{pmatrix} x_1^2 & x_1x_2 \\ x_1x_2 & x_2^2 \end{pmatrix} \\ \mathbf{z} &= A\begin{pmatrix} x_1^2 - x_2^2 \\ 2\,x_1x_2 \end{pmatrix} \end{aligned} \tag{6.18}$$

The norm

The norm of **z** is given by:

$$\|\mathbf{z}\|^2 = A^2[(x_1^2 - x_2^2)^2 + (2x_1x_2)^2] = A^2(x_1^2 + x_2^2)^2 = A^2 \tag{6.19}$$

i.e. , the norm of **z** is the same as the norm of **T**.

Invariance

In the same manner as **T**, **z** is trivially invariant to **G**.

Equivariance

Let

$$\hat{\mathbf{x}} = \begin{pmatrix} x_1 \\ x_2 \end{pmatrix} = \begin{pmatrix} \cos\varphi \\ \sin\varphi \end{pmatrix} \tag{6.20}$$

Then

$$\mathbf{z} = A\begin{pmatrix} \cos^2\varphi - \sin^2\varphi \\ 2\cos\varphi\sin\varphi \end{pmatrix} = A\begin{pmatrix} \cos(2\varphi) \\ \sin(2\varphi) \end{pmatrix} \tag{6.21}$$

Thus, the representation of orientation is a vector having an argument that is twice that of the signal orientation. Differentiating $\mathbf{z}$ with respect to φ and calculating the norm yields:

$$\|\frac{\partial \mathbf{z}}{\partial \varphi}\| = A\sqrt{(-2\sin(2\varphi))^2 + (2\cos(2\varphi))^2} = 2A \tag{6.22}$$

showing that $\mathbf{z}$ is equivariant to the signal orienting vector $\hat{\mathbf{x}}$.

6.3 ORIENTATION ESTIMATION

Having found a suitable orientation representation, we ask the question: *Can the representation be implemented using measurements on actual image data, where lines or other structures are represented as local gray scale correlations?* It is shown that by combining the outputs from polar separable quadrature filters, it is possible to produce a representation corresponding exactly to Equation (6.11). The exactness relies on the image data being locally *simple*, Equation (6.8), i.e. on the existence of a locally well defined orientation. The case where the simplicity assumption does not hold is discussed in Section 6.5.

In the following, the above procedure is discussed in detail. The analysis is restricted to *real* valued *simple* neighborhoods, i.e. neighborhoods that can be expressed as:

$$s(\boldsymbol{\xi}) = g(\boldsymbol{\xi} \cdot \hat{\mathbf{x}}) \tag{6.23}$$

where f and g are real functions.

The quadrature filter concept

In order to realize the *invariance* requirement, the estimation procedure is designed using quadrature filters [65]. The quadrature filter concept forms a basis for minimizing the sensitivity to phase changes in the signal.

A quadrature filter can, independently of the dimensionality of the signal space, be defined as a filter that is zero over one half of the Fourier space, or more precisely defined by:

$$F_k(\mathbf{u}) = 0 \quad if \quad \mathbf{u} \cdot \hat{\mathbf{n}}_k \leq 0 \tag{6.24}$$

where:

$\hat{\mathbf{n}}_k$ is the filter directing vector.

$\mathbf{u}$ is the frequency.

The output $\mathbf{q}_k$ of the corresponding quadrature filter will be a complex number. As an example of the phase insensitive property of quadrature filters it can be mentioned that, if g is a sinusoidal function, the magnitude $q_k = \|\mathbf{q}_k\|$ of $\mathbf{q}_k$ will be completely phase invariant. The argument $\arg(\mathbf{q}_k)$ will represent the local phase.

Spherically separable filters

For *simple* signals the Fourier transform is non-zero only on the line defined by:

$$\mathbf{u} \propto \hat{\mathbf{x}} \tag{6.25}$$

More precisely the Fourier transform of $s(\boldsymbol{\xi})$ in Equation 6.23 can be expressed as:

$$S(\mathbf{u}) = G(\mathbf{u} \cdot \hat{\mathbf{x}})\delta_{\hat{\mathbf{x}}}^{n-1}(\mathbf{u}) \tag{6.26}$$

Let the filter functions be real and spherically separable [65], ie separable into one function of radius, R, and one function of direction D.

$$F(\mathbf{u}) = R(\rho)\, D(\hat{\mathbf{u}}) \tag{6.27}$$

where: $\rho = \|\mathbf{u}\|$

The result $\mathbf{q}$ of filtering S by F then becomes:

$$\mathbf{q} = \int_{-\infty}^{\infty} F(\mathbf{u})S(\mathbf{u})\, d\mathbf{u} = \int_{-\infty}^{\infty} R(\rho)\, D(\hat{\mathbf{u}})G(\mathbf{u}\cdot\hat{\mathbf{x}})\delta_{\hat{\mathbf{x}}}^{n-1}(\mathbf{u})\, d\mathbf{u} \qquad (6.28)$$

Setting $\mathbf{u} = \rho\hat{\mathbf{x}}$ yields:

$$\mathbf{q} = D(\hat{\mathbf{x}}) \int_{0}^{\infty} R(\rho)G(\rho)d\rho + D(-\hat{\mathbf{x}}) \int_{0}^{\infty} R(\rho)G(-\rho)d\rho \qquad (6.29)$$

Note that the value of the integrals does not depend on the signal orientation. Let $\mathbf{a}$ be the value of the first integral in Equation (6.29).

$$\mathbf{a} = \int_{0}^{\infty} R(\rho)G(\rho)d\rho \qquad (6.30)$$

Then, since the signal $s(\boldsymbol{\xi})$ is real, $G(\rho)$ will be Hermitian and the value of the second integral becomes $\mathbf{a}^*$. Thus Equation (6.29) simplifies to:

$$\mathbf{q} = \mathbf{a}\, D(\hat{\mathbf{x}}) + \mathbf{a}^*\, D(-\hat{\mathbf{x}}) \qquad (6.31)$$

Taking the quadrature requirement into account it is clear that either $D(\hat{\mathbf{x}})$ or $D(-\hat{\mathbf{x}})$ will be zero, implying that the two components will not interfere and the magnitude of the quadrature filter output can be written:

$$q = A\,[D(\hat{\mathbf{x}}) + D(-\hat{\mathbf{x}})] \qquad (6.32)$$

showing that the *invariance* requirement is met and that the magnitude of the quadrature filter output is separable into two components, one orientation invariant component, $A = \|\mathbf{a}\|$, which can be thought of as the local signal amplitude, and one component (in brackets) invariant to the signal function G.

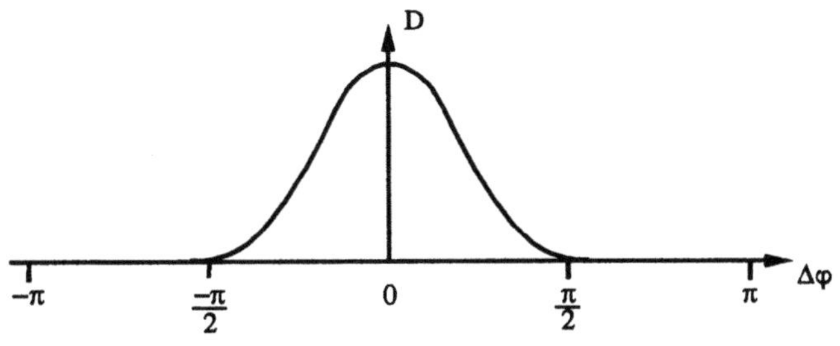

Figure 6.3 One-dimensional plot of the directional function with $\Delta\phi$, i.e. the angle between $\mathbf{u}$ and $\hat{\mathbf{n}}_k$, as parameter.

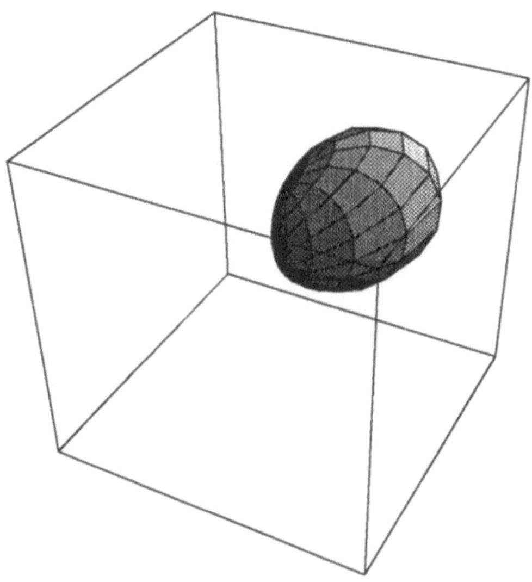

Figure 6.4 Angular plot, i.e. *radius* $= D(\hat{\mathbf{u}})$, of the directional function in 3-dimensional signal space.

The directional function

To meet the *equivariance* requirement, Equation (6.24), it is required that the frequency response of the filters have particular interpolation properties. Directional functions having the necessary properties were first suggested by

Knutsson in [65] for the 2D case and in [66] for the 3D case. Regardless of dimension these functions can be written:

$$\begin{cases} D_k(\hat{\mathbf{u}}) = (\hat{\mathbf{u}} \cdot \hat{\mathbf{n}}_k)^2 & if \ \mathbf{u} \cdot \hat{\mathbf{n}}_k > 0 \\ D_k(\hat{\mathbf{u}}) = 0 & otherwise \end{cases} \tag{6.33}$$

where $\hat{\mathbf{n}}_k$ is the filter directing vector. Thus $D(\hat{\mathbf{u}})$ varies as $\cos^2(\varphi)$, where φ is the difference in angle between $\mathbf{u}$ and the filter direction $\hat{\mathbf{n}}_k$. See Figure 6.4 for a visualization of the directional function.

The filter outputs.

Finally, combining Equations (6.32) and (6.33), the output magnitude from a quadrature filter in direction k is found to be:

$$q_k = A(\hat{\mathbf{x}} \cdot \hat{\mathbf{n}}_k)^2 \tag{6.34}$$

where A is independent of the filter orientation and depends only on radial distribution of the signal spectrum $G(\rho)$ and the radial filter function $R(\rho)$. See Figure 6.5 for a visualization of the output magnitude q as a function of $\hat{\mathbf{x}}$.

The radial function

It is clear from the preceding analysis that the radial function $R(\rho)$ can be chosen arbitrarily without violating the basic requirements. This makes the choice of $R(\rho)$ subject to considerations similar to those traditionally found in *one – dimensional* filter design. Typically $R(\rho)$ is a band-pass function having design parameters such as *center frequency* and *bandwidth.* Perhaps even more important than in traditional one-dimensional filter design are the concepts of locality and scale. Good $R(\rho)$s are therefore found by studying the resulting filter simultaneously in the space-time and the Fourier domains. A radial filter function with useful properties, first suggested in [74], is given by:

$$R(\rho) = e^{-\frac{4}{B^2 \ln 2}\ln^2(\rho/\rho_i)} \tag{6.35}$$

This class of functions are termed *lognormal* functions. B is the relative bandwidth and ρ_i is the center frequency.

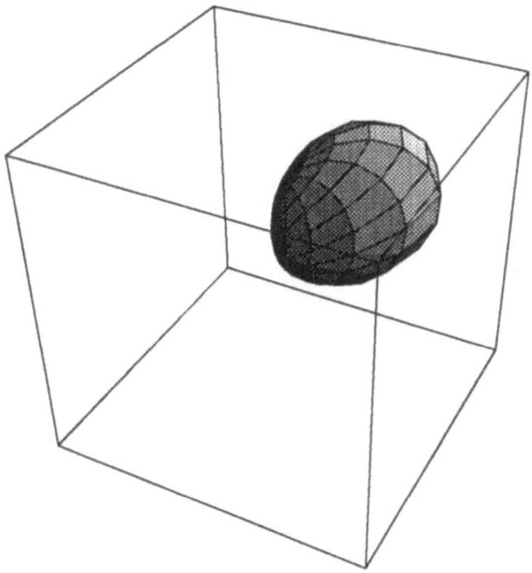

Figure 6.5 Angular plot of the quadrature output magnitude, q_k, as a function of the signal orienting vector $\hat{\mathbf{x}}$ for 3-dimensional simple signals.

6.3.1 Filter sets

In the following the quadrature filter sets, F_k, suitable for orientation estimation will be discussed. The discussion shows that a natural choice is to use 3 quadrature filters for 2D, 6 for 3D and 12 for 4D, [65, 66, 67, 68].

It will be required that the filter axes are uniformly distributed over the orientation space, i.e. the filter axes should pass through the vertices of a regular polytope. This is a reasonable requirement as the final result, $\mathbf{T}$, by definition is invariant to rotation of the filters. As before the analysis will deal only with real valued simple neighborhoods. It is helpful in the following discussion to remember that:

1. The Fourier transform is invariant to rotation of the coordinate system.
2. The Fourier transform of a simple signal is a line through the origin parallel to the signal orienting vector, Equations (6.8) and (6.25).

3. The quadrature filter output is invariant to rotation around its axis (given by $\hat{\mathbf{n}}_k$) and also diametrically symmetric so that $q(\mathbf{x}) = q(-\mathbf{x})$, Equation (6.34).

Minimum number of filters

Consider the case of 2^{N-1} quadrature filters, having symmetry axes passing through the corners of a cube in N dimensions, giving a fully symmetric distribution of filters. Consider the contribution to the filters from frequencies on a line through the center of two opposing cube faces. Since the angle between the line and any filter axis will be the same it is clear that all the filters will give the same output. Consequently the filter set is incapable of giving information sufficient to determine which pair of cube faces the line passes through and thus, the orientation of the signal is undecidable. It can be concluded, therefore, that more than 2^{N-1} quadrature filters must be used.

The minimum number of filters and their directions in 2D, 3D and 4D are given below. (For $N > 4$ no regular polyhedron having more vertices than a cube exists; it is, however, still possible to estimate orientation tensors for $N > 4$ although calculations will be more complex.)

2D filter set

It is shown in [65] that the minimum number of filters required when $N = 2$ is 3 and the orientations of the filters are given by vectors pointing to three adjacent vertices of a regular hexagon, see Figure 6.6.

The filter orienting vectors are given by:

$$\begin{aligned} \hat{\mathbf{n}}_1 &= (\ 1\ ,\ 0\ )^T \\ \hat{\mathbf{n}}_2 &= (\ a\ ,\ b\ )^T \\ \hat{\mathbf{n}}_3 &= (-a\ ,\ b\ )^T \end{aligned} \tag{6.36}$$

where:

$$\begin{aligned} a &= 0.5 \\ b &= \sqrt{3}/2 \end{aligned}$$

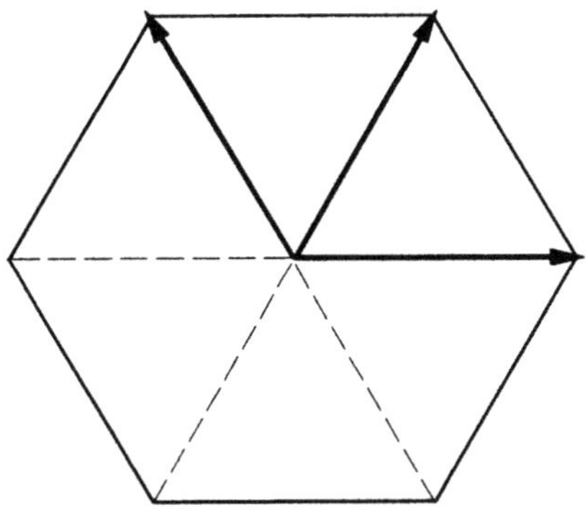

Figure 6.6 A hexagon and the corresponding filter orienting vectors.

3D filter set

For N = 3 the number of filters must be greater than 4 but, since there does not exist a way of distributing 5 filters in 3-D in a fully symmetrical fashion, the next possible number is 6. In fact the only possible numbers are those given by half the number of vertices (or faces) of a diametrically symmetric regular polyhedron, leaving only the numbers 3, 4, 6 and 10. Note that this is in contrast to the 2-D case where the only symmetry restriction is $K > 2$. It turns out that the minimum required number of quadrature filters K *is* 6.

The orientations of the filters are given by vectors pointing to the vertices of a *hemi-icosahedron*, see Figure 6.7.

The 6 normal vectors are given in Cartesian coordinates by:

$$\begin{array}{llll} \hat{\mathbf{n}}_1 = c(\;a\;,\;0\;,\;b\;)^T & \hat{\mathbf{n}}_2 = c(-a\;,\;0\;,\;b\;)^T \\ \hat{\mathbf{n}}_3 = c(\;b\;,\;a\;,\;0\;)^T & \hat{\mathbf{n}}_4 = c(\;b\;,-a\;,\;0\;)^T \\ \hat{\mathbf{n}}_5 = c(\;0\;,\;b\;,\;a\;)^T & \hat{\mathbf{n}}_6 = c(\;0\;,\;b\;,-a\;)^T \end{array} \tag{6.37}$$

where:

$$a = 2$$
$$b = (1+\sqrt{5})$$
$$c = (10+2\sqrt{5})^{-1/2}$$

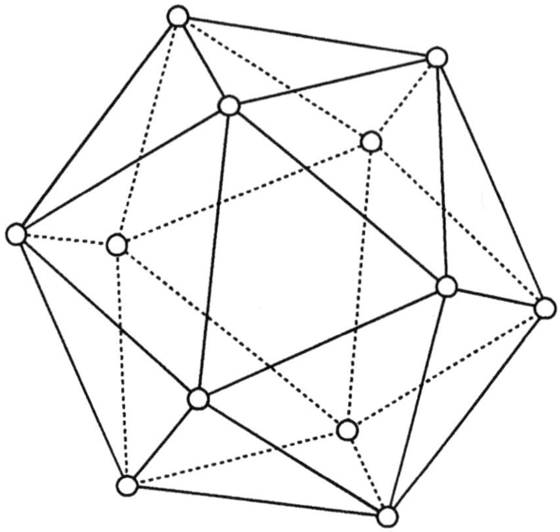

Figure 6.7 An icosahedron, one of the 5 Platonic polyhedra.

4D filter set

As before, the filters should be distributed symmetrically over half of the Fourier space. This implies that the filters should be distributed in accordance with the vertices of a regular polytope. The choice is further limited by the restriction that K, the number of filters, should be greater than 8. This leaves the 24–cell (Coxeter [32]) as the only alternative, see Figure 6.8. Note that computational complexity makes the 120–cell and the 600–cell unrealistic alternatives.

The 12 filter directions are given in Cartesian coordinates by:

$$\begin{array}{llll} \hat{\mathbf{n}}_1 = c(1, & 1, & 0, & 0)^T \\ \hat{\mathbf{n}}_3 = c(1, & 0, & 1, & 0)^T \\ \hat{\mathbf{n}}_5 = c(1, & 0, & 0, & 1)^T \\ \hat{\mathbf{n}}_7 = c(0, & 1, & 1, & 0)^T \\ \hat{\mathbf{n}}_9 = c(0, & 1, & 0, & 1)^T \\ \hat{\mathbf{n}}_{11} = c(0, & 0, & 1, & 1)^T \end{array} \quad \begin{array}{llll} \hat{\mathbf{n}}_2 = c(1, & -1, & 0, & 0)^T \\ \hat{\mathbf{n}}_4 = c(1, & 0, & -1, & 0)^T \\ \hat{\mathbf{n}}_6 = c(1, & 0, & 0, & -1)^T \\ \hat{\mathbf{n}}_8 = c(0, & 1, & -1, & 0)^T \\ \hat{\mathbf{n}}_{10} = c(0, & 1, & 0, & -1)^T \\ \hat{\mathbf{n}}_{12} = c(0, & 0, & 1, & -1)^T \end{array} \tag{6.38}$$

where $c = \frac{1}{\sqrt{2}}$.

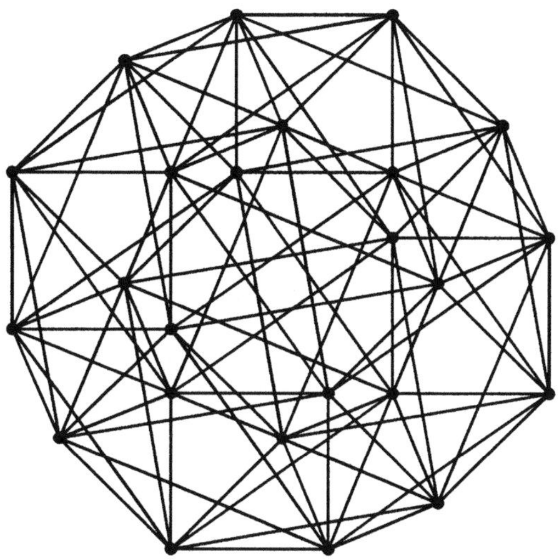

Figure 6.8 A projection of the 24–cell

6.4 TENSOR CONSTRUCTION

In this section it is shown that the final result **T** can, independent of signal dimensionality, be obtained by linear summation of the quadrature filter output magnitudes as expressed by Equations (6.39) and (6.40).

$$\mathbf{T} = \sum_k q_k \, \mathbf{M}_k \tag{6.39}$$

$$\mathbf{M}_k = \alpha \, \hat{\mathbf{n}}_k \hat{\mathbf{n}}_k^T - \beta \mathbf{I} \tag{6.40}$$

where: q_k is the output from quadrature filter k.

$\mathbf{M}_k$ is a tensor associated with quadrature filter k.

$\hat{\mathbf{n}}_k$ is the orientation of quadrature filter k.

$\mathbf{I}$ is the identity tensor.

α is $\frac{4}{3}$ in 2D, $\frac{5}{4}$ in 3D and 1 in 4D.

β is $\frac{1}{3}$ in 2D, $\frac{1}{4}$ in 3D and $\frac{1}{6}$ in 4D.

6.4.1 Some concepts from tensor theory

It may be appropriate here to briefly touch upon a few concepts from tensor theory. The tensors $\mathbf{M}_k$ are termed the *dual* elements of the tensor set $\mathbf{N}_k$. The basic relation between a tensor set and the set of dual tensors is given by:

$$\sum_{k=1}^{K} \mathbf{M}_k \left(\mathbf{T} \bullet \mathbf{N}_k \right) = \mathbf{T} \qquad \forall\, \mathbf{T} \tag{6.41}$$

Each dual tensor can be obtained from a linear combination of the tensors in the original set.

$$\mathbf{M}_k = \sum_{l=1}^{K} g'_{kl} \mathbf{N}_l \tag{6.42}$$

where g'_{kl} are the weighting coefficients.

In the case where the tensors $\mathbf{N}_k$ constitute a *basis*, the constraint given by Equation (6.41) can be simplified to:

$$\mathbf{M}_k \bullet \mathbf{N}_l = \delta_{kl} \tag{6.43}$$

where $\delta_{kl} = 1$ if $k = l$, and 0 otherwise. If $\mathbf{N}_k$ constitutes a basis, the collection of coefficients g'_{kl} can be obtained as the components of the inverse of a matrix having elements:

$$g_{kl} = \mathbf{N}_k \bullet \mathbf{N}_l \tag{6.44}$$

The coefficients g_{kl} are the coordinates of the *metric* for the space spanned by the basis tensor set $\mathbf{N}_k$.

Given what is stated above, it is clear that the estimation procedures presented in this chapter are not unique. Many different sets of basis elements $\mathbf{N}_k$ with corresponding filters $F_k(\mathbf{u})$ could be used to obtain the results. However, the present choice leads to a particularly simple relation between the basis elements and the dual elements, Equation (6.40).

2D tensor construction

It is convenient in the following to define tensors $\hat{\mathbf{N}}_k$ being the outer product of the filter orienting vectors $\hat{\mathbf{n}}_k$, i.e.

$$\hat{\mathbf{N}}_k = \hat{\mathbf{n}}_k \hat{\mathbf{n}}_k^T \tag{6.45}$$

Combining Equations (6.36) and (6.45) gives the elements of the filter associated tensors $\hat{\mathbf{N}}_k$:

$$\begin{aligned}
\hat{\mathbf{N}}_1 &= \begin{pmatrix} 1 & 0 \\ 0 & 0 \end{pmatrix} \\
\hat{\mathbf{N}}_2 &= \begin{pmatrix} a^2 & ab \\ ab & b^2 \end{pmatrix} \\
\hat{\mathbf{N}}_3 &= \begin{pmatrix} a^2 & -ab \\ -ab & b^2 \end{pmatrix}
\end{aligned} \tag{6.46}$$

where:

$$\begin{aligned}
a &= 0.5 \\
b &= \sqrt{3}/2
\end{aligned}$$

Let the signal orienting vector be given by:

$$\mathbf{x} = (x_1, x_2)^T \tag{6.47}$$

then the magnitude of the outputs from the three quadrature filters are, according to Equation (6.34), given by:

$$\begin{aligned}
q_1 &= Ax^{-2}x_1^2 \\
q_2 &= Ax^{-2}(a^2x_1^2 + 2abx_1x_2 + b^2x_2^2) \\
q_3 &= Ax^{-2}(a^2x_1^2 - 2abx_1x_2 + b^2x_2^2)
\end{aligned} \tag{6.48}$$

Next, calculating the sum

$$\mathbf{T}' = \sum_k q_k \hat{\mathbf{N}}_k \tag{6.49}$$

yields the components of $\mathbf{T}'$:

$$\begin{aligned} t_{11} &= \tfrac{3}{4}A(x_1^2 x^{-2} + \tfrac{1}{2}) \\ t_{22} &= \tfrac{3}{4}A(x_2^2 x^{-2} + \tfrac{1}{2}) \\ t_{12} = t_{21} &= \tfrac{3}{4}A x_1 x_2 x^{-2} \end{aligned} \tag{6.50}$$

It is evident that if the quantity $\frac{3}{8}A$ is subtracted from the diagonal elements of $\mathbf{T}'$ the result will be of the desired form.

$$\mathbf{T}' - \tfrac{3}{8}A\,\mathbf{I} = \tfrac{3}{4}A\,\hat{\mathbf{x}}\hat{\mathbf{x}}^T \tag{6.51}$$

Finally calculate the sum of all quadrature filter output magnitudes.

$$\sum_k q_k = \frac{3}{2}A \tag{6.52}$$

Combining Equations (6.49), (6.51) and (6.52) yields the desired result:

$$\mathbf{T} = A\,\mathbf{x}\mathbf{x}^T = \sum_k q_k(\frac{4}{3}\hat{\mathbf{N}}_k - \frac{1}{3}\mathbf{I}) = \sum_k q_k\,\mathbf{M}_k \tag{6.53}$$

Note that the dual elements, $\mathbf{M}_k$, are constant tensors and can be precalculated. Thus the orientation tensor can be estimated by weighted summation of fixed tensors, the weights being the quadrature filter output magnitudes q_k. In the following it will be shown that this statement is also true for 3D and 4D signal spaces. Although the numbers and number of numbers differ in higher dimensions, the method for arriving at the final result is identical to the one used in 2D.

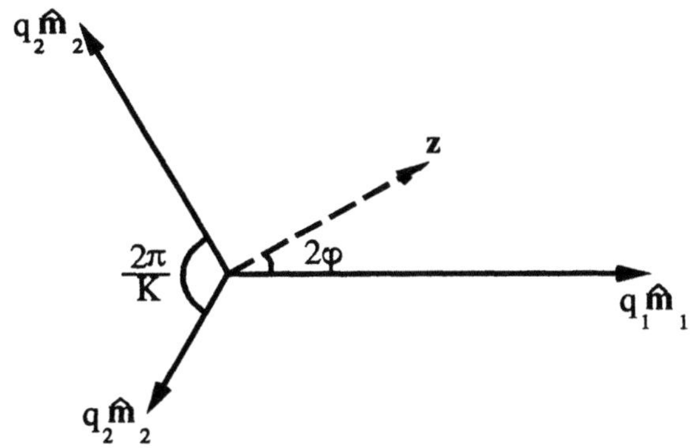

Figure 6.9 In 2 dimensions a vector representing the dominant orientation can be constructed using three quadrature filters.

2D vector construction

For two-dimensional signals it is possible to construct a 2-component vector representing the dominant orientation in a neighborhood. This is done using the same principles as for the tensor construction, i.e. by multiplying the quadrature outputs with fixed elements and summing the results.

$$\mathbf{z} = \sum_k q_k \hat{\mathbf{m}}_k \tag{6.54}$$

where:

$$\hat{\mathbf{m}}_k = \begin{pmatrix} \cos(2\varphi_k) \\ \sin(2\varphi_k) \end{pmatrix} \tag{6.55}$$

and

$$\varphi_k = \arg(\hat{\mathbf{n}}_k) \tag{6.56}$$

The minimum number of quadrature filters needed is 3, the same as for the tensor case, see Figure 6.9. It can be shown, [65], that the vector construction

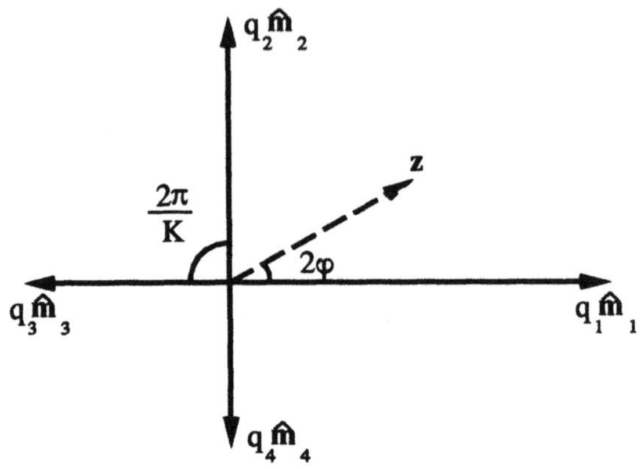

Figure 6.10 Using four filters leads to a particularly simple vector construction algorithm.

can be done using any number, greater or equal to 3, of evenly distributed filters. A particularly simple case occurs if 4 filters are used, see Figure 6.10.

$$\varphi_k = (k-1)\frac{\pi}{4} \qquad k = 1 \ldots 4 \tag{6.57}$$

$$\mathbf{z} = \sum_{k=1}^{4} q_k \begin{pmatrix} \cos(2\varphi_k) \\ \sin(2\varphi_k) \end{pmatrix} = \begin{pmatrix} q_1 - q_3 \\ q_2 - q_4 \end{pmatrix} \tag{6.58}$$

Expressing Equation (6.34) in trigonometric functions yields:

$$q_k = d\cos^2(\arg(\hat{\mathbf{x}}) - \varphi_k) \tag{6.59}$$

Carrying out the sums leads to the final result:

$$\mathbf{z} = A \begin{pmatrix} \cos(2\arg(\hat{\mathbf{x}})) \\ \sin(2\arg(\hat{\mathbf{x}})) \end{pmatrix} \tag{6.60}$$

showing that the vector construction procedure leads to the desired result, Equation (6.21).

3D tensor construction

Combining Equations (6.37) and (6.45) gives the elements of the filter associated tensors $\hat{\mathbf{N}}_k$:

$$\hat{\mathbf{N}}_1 = c^2 \begin{pmatrix} a^2 & 0 & ab \\ 0 & 0 & 0 \\ ab & 0 & b^2 \end{pmatrix} \qquad \hat{\mathbf{N}}_2 = c^2 \begin{pmatrix} a^2 & 0 & -ab \\ 0 & 0 & 0 \\ -ab & 0 & b^2 \end{pmatrix}$$

$$\hat{\mathbf{N}}_3 = c^2 \begin{pmatrix} b^2 & ab & 0 \\ ab & a^2 & 0 \\ 0 & 0 & 0 \end{pmatrix} \qquad \hat{\mathbf{N}}_4 = c^2 \begin{pmatrix} b^2 & -ab & 0 \\ -ab & a^2 & 0 \\ 0 & 0 & 0 \end{pmatrix} \qquad (6.61)$$

$$\hat{\mathbf{N}}_5 = c^2 \begin{pmatrix} 0 & 0 & 0 \\ 0 & b^2 & ab \\ 0 & ab & a^2 \end{pmatrix} \qquad \hat{\mathbf{N}}_6 = c^2 \begin{pmatrix} 0 & 0 & 0 \\ 0 & b^2 & -ab \\ 0 & -ab & a^2 \end{pmatrix}$$

where:

$$\begin{aligned} a &= 2 \\ b &= (1+\sqrt{5}) \\ c &= (10+2\sqrt{5})^{-1/2} \end{aligned}$$

Let the signal orienting vector be given by:

$$\mathbf{x} = (x_1, x_2, x_3)^T \qquad (6.62)$$

then the magnitude of the outputs from the 6 quadrature filters are, according to Equation 6.34, given by:

$$\begin{aligned} q_1 &= Ac^2x^{-2}(a^2x_1^2 + 2abx_1x_3 + b^2x_3^2) \\ q_2 &= Ac^2x^{-2}(a^2x_1^2 - 2abx_1x_3 + b^2x_3^2) \\ q_3 &= Ac^2x^{-2}(b^2x_1^2 + 2abx_1x_2 + a^2x_2^2) \\ q_4 &= Ac^2x^{-2}(b^2x_1^2 - 2abx_1x_2 + a^2x_2^2) \\ q_5 &= Ac^2x^{-2}(b^2x_2^2 + 2abx_2x_3 + a^2x_3^2) \\ q_6 &= Ac^2x^{-2}(b^2x_2^2 - 2abx_2x_3 + a^2x_3^2) \end{aligned} \qquad (6.63)$$

Next, calculating the sum

$$\mathbf{T}' = \sum_k q_k \hat{\mathbf{N}}_k \tag{6.64}$$

yields the components of $\mathbf{T}'$:

$$\begin{aligned} t_{11} &= \tfrac{4}{5}A(x_1^2 x^{-2} + \tfrac{1}{2}) & t_{12} = t_{21} &= \tfrac{4}{5}Ax_1x_2x^{-2} \\ t_{22} &= \tfrac{4}{5}A(x_2^2 x^{-2} + \tfrac{1}{2}) & t_{13} = t_{31} &= \tfrac{4}{5}Ax_1x_3x^{-2} \\ t_{33} &= \tfrac{4}{5}A(x_3^2 x^{-2} + \tfrac{1}{2}) & t_{23} = t_{32} &= \tfrac{4}{5}Ax_2x_3x^{-2} \end{aligned} \tag{6.65}$$

If the quantity $\frac{2}{5}A$ is subtracted from the diagonal elements of $\mathbf{T}'$ the result will be of the desired form.

$$\mathbf{T}' - \tfrac{2}{5}A\,\mathbf{I} = \tfrac{4}{5}A\,\hat{\mathbf{x}}\hat{\mathbf{x}}^T \tag{6.66}$$

Finally calculate the sum of all quadrature filter output magnitudes.

$$\sum_k q_k = 2A \tag{6.67}$$

Combining Equations (6.64), (6.66) and (6.67) yields the desired result:

$$\mathbf{T} = A\,\mathbf{x}\mathbf{x}^T = \sum_k q_k(\frac{5}{4}\hat{\mathbf{N}}_k - \frac{1}{4}\mathbf{I}) = \sum_k q_k\,\mathbf{M}_k \tag{6.68}$$

4D tensor construction

Combining Equations (6.38) and (6.45) gives the elements of the filter associated tensors $\hat{\mathbf{N}}_k$:

$$
\begin{array}{ll}
\hat{\mathbf{N}}_1 = c^2 \begin{pmatrix} 1 & 1 & 0 & 0 \\ 1 & 1 & 0 & 0 \\ 0 & 0 & 0 & 0 \\ 0 & 0 & 0 & 0 \end{pmatrix} &
\hat{\mathbf{N}}_2 = c^2 \begin{pmatrix} 1 & -1 & 0 & 0 \\ -1 & 1 & 0 & 0 \\ 0 & 0 & 0 & 0 \\ 0 & 0 & 0 & 0 \end{pmatrix} \\
\hat{\mathbf{N}}_3 = c^2 \begin{pmatrix} 1 & 0 & 1 & 0 \\ 0 & 0 & 0 & 0 \\ 1 & 0 & 1 & 0 \\ 0 & 0 & 0 & 0 \end{pmatrix} &
\hat{\mathbf{N}}_4 = c^2 \begin{pmatrix} 1 & 0 & -1 & 0 \\ 0 & 0 & 0 & 0 \\ -1 & 0 & 1 & 0 \\ 0 & 0 & 0 & 0 \end{pmatrix} \\
\hat{\mathbf{N}}_5 = c^2 \begin{pmatrix} 1 & 0 & 0 & 1 \\ 0 & 0 & 0 & 0 \\ 0 & 0 & 0 & 0 \\ 1 & 0 & 0 & 1 \end{pmatrix} &
\hat{\mathbf{N}}_6 = c^2 \begin{pmatrix} 1 & 0 & 0 & -1 \\ 0 & 0 & 0 & 0 \\ 0 & 0 & 0 & 0 \\ -1 & 0 & 0 & 1 \end{pmatrix} \\
\hat{\mathbf{N}}_7 = c^2 \begin{pmatrix} 0 & 0 & 0 & 0 \\ 0 & 1 & 1 & 0 \\ 0 & 1 & 1 & 0 \\ 0 & 0 & 0 & 0 \end{pmatrix} &
\hat{\mathbf{N}}_8 = c^2 \begin{pmatrix} 0 & 0 & 0 & 0 \\ 0 & 1 & -1 & 0 \\ 0 & -1 & 1 & 0 \\ 0 & 0 & 0 & 0 \end{pmatrix} \\
\hat{\mathbf{N}}_9 = c^2 \begin{pmatrix} 0 & 0 & 0 & 0 \\ 0 & 1 & 0 & 1 \\ 0 & 0 & 0 & 0 \\ 0 & 1 & 0 & 1 \end{pmatrix} &
\hat{\mathbf{N}}_{10} = c^2 \begin{pmatrix} 0 & 0 & 0 & 0 \\ 0 & 1 & 0 & -1 \\ 0 & 0 & 0 & 0 \\ 0 & -1 & 0 & 1 \end{pmatrix} \\
\hat{\mathbf{N}}_{11} = c^2 \begin{pmatrix} 0 & 0 & 0 & 0 \\ 0 & 0 & 0 & 0 \\ 0 & 0 & 1 & 1 \\ 0 & 0 & 1 & 1 \end{pmatrix} &
\hat{\mathbf{N}}_{12} = c^2 \begin{pmatrix} 0 & 0 & 0 & 0 \\ 0 & 0 & 0 & 0 \\ 0 & 0 & 1 & -1 \\ 0 & 0 & -1 & 1 \end{pmatrix}
\end{array}
\qquad (6.69)
$$

where $c = \frac{1}{\sqrt{2}}$.
Let the signal orienting vector be given by:

$$\mathbf{x} = (x_1, x_2, x_3, x_4)^T \qquad (6.70)$$

Then the magnitude of the outputs from the 12 quadrature filters are given by

$$\begin{aligned}
q_1 &= Ac^2x^{-2}\ (x_1^2 + 2x_1x_2 + x_2^2) \\
q_2 &= Ac^2x^{-2}\ (x_1^2 - 2x_1x_2 + x_2^2) \\
q_3 &= Ac^2x^{-2}\ (x_1^2 + 2x_1x_3 + x_3^2) \\
q_4 &= Ac^2x^{-2}\ (x_1^2 - 2x_1x_3 + x_3^2) \\
q_5 &= Ac^2x^{-2}\ (x_1^2 + 2x_1x_4 + x_4^2) \\
q_6 &= Ac^2x^{-2}\ (x_1^2 - 2x_1x_4 + x_4^2) \\
q_7 &= Ac^2x^{-2}\ (x_2^2 + 2x_2x_3 + x_3^2) \\
q_8 &= Ac^2x^{-2}\ (x_2^2 - 2x_2x_3 + x_3^2) \\
q_9 &= Ac^2x^{-2}\ (x_2^2 + 2x_2x_4 + x_4^2) \\
q_{10} &= Ac^2x^{-2}\ (x_2^2 - 2x_2x_4 + x_4^2) \\
q_{11} &= Ac^2x^{-2}\ (x_3^2 + 2x_3x_4 + x_4^2) \\
q_{12} &= Ac^2x^{-2}\ (x_3^2 - 2x_3x_4 + x_4^2)
\end{aligned} \tag{6.71}$$

Next, calculating the sum

$$\mathbf{T}' = \sum_k q_k \hat{\mathbf{N}}_k \tag{6.72}$$

yields the components of $\mathbf{T}'$

$$\begin{aligned}
t_{11} &= A(x_1^2x^{-2} + \tfrac{1}{2}) & t_{22} &= A(x_2^2x^{-2} + \tfrac{1}{2}) \\
t_{33} &= A(x_3^2x^{-2} + \tfrac{1}{2}) & t_{44} &= A(x_4^2x^{-2} + \tfrac{1}{2}) \\
t_{12} &= t_{21} = Ax_1x_2x^{-2} & t_{13} &= t_{31} = Ax_1x_3x^{-2} \\
t_{14} &= t_{41} = Ax_1x_4x^{-2} & t_{23} &= t_{32} = Ax_2x_3x^{-2} \\
t_{24} &= t_{42} = Ax_2x_4x^{-2} & t_{34} &= t_{43} = Ax_3x_4x^{-2}
\end{aligned} \tag{6.73}$$

If the quantity $\frac{1}{2}A$ is subtracted from the diagonal elements of $\mathbf{T}'$ the result will be of the desired form.

$$\mathbf{T}' - \tfrac{1}{2}A\mathbf{I} = A\hat{\mathbf{x}}\hat{\mathbf{x}}^T \tag{6.74}$$

Finally, calculate the sum of all quadrature filter output magnitudes.

$$\sum_k q_k = 3A \tag{6.75}$$

Combining Equations (6.72), (6.74) and (6.75) yields the desired result:

$$\mathbf{T} \;=\; A\,\mathbf{x}\mathbf{x}^T \;=\; \sum_k q_k(\hat{\mathbf{N}}_k - \frac{1}{6}\mathbf{I}) \;=\; \sum_k q_k\,\mathbf{M}_k \tag{6.76}$$

6.5 INTERPRETATION OF THE ORIENTATION TENSOR

In the physical world, acquired data are seldom exactly *simple*. It is, however, still possible to find a best approximation to $\mathbf{T}$ corresponding to a *simple* neighborhood, Equation 6.8. This is done by finding the $\mathbf{T_S}$ that minimizes:

$$\Delta \;=\; ||\mathbf{T} - \mathbf{T_S}|| \tag{6.77}$$

where:

$$\mathbf{T_S} = A\hat{\mathbf{x}}\hat{\mathbf{x}}^T \tag{6.78}$$

The norm of a tensor is invariant under rotation of the coordinate system and Equation (6.77) can be rewritten as:

$$\Delta \;=\; ||\mathbf{C}^{-1}(\mathbf{T} - A\hat{\mathbf{x}}\hat{\mathbf{x}}^T)\mathbf{C}|| \tag{6.79}$$

giving:

$$\Delta = \|\mathbf{C}^{-1}\mathbf{T}\mathbf{C} - \mathbf{C}^{-1}A\hat{\mathbf{x}}\hat{\mathbf{x}}^T\mathbf{C}\| \tag{6.80}$$

where $\mathbf{C}$ is an orthogonal matrix.

Let $\mathbf{C}$ be such that $\mathbf{C}^{-1}\mathbf{T}\mathbf{C}$ is diagonal and note that only one eigenvalue of $\hat{\mathbf{x}}\hat{\mathbf{x}}^T$ is nonzero. Then, since the norm of $\mathbf{C}^{-1}\mathbf{T}\mathbf{C}$ is the sum of the squares of its elements, it is clear that Δ is minimized if $\mathbf{C}^{-1}A\hat{\mathbf{x}}\hat{\mathbf{x}}^T\mathbf{C}$ removes the largest of these values, i.e. the largest eigenvalue of $\mathbf{C}^{-1}\mathbf{T}\mathbf{C}$. Thus, if the eigenvalues are numbered in decreasing order, Δ is given by:

$$\Delta = \sqrt{\sum_{n \neq 1} \lambda_n^2} \tag{6.81}$$

Then, since $\mathbf{T}$ and $A\hat{\mathbf{x}}\hat{\mathbf{x}}^T$ are subject to identical rotation, it is clear that the A and $\hat{\mathbf{x}}$ which minimize Δ are given by:

$$\begin{aligned} \hat{\mathbf{x}} &= \hat{\mathbf{e}}_1 \\ A &= \lambda_1 \end{aligned} \tag{6.82}$$

where $\mathbf{e}_1$ is the eigenvector corresponding to the largest eigenvalue of $\mathbf{T}$.

Thus, the tensor most similar to $\mathbf{T}$ corresponding to a *simple* neighborhood is given by:

$$\mathbf{T}_\mathbf{S} = \lambda_1 \hat{\mathbf{e}}_1 \hat{\mathbf{e}}_1^T \tag{6.83}$$

where:

λ_1 is the largest eigenvalue of $\mathbf{T}$ and

$\hat{\mathbf{e}}_1$ is the corresponding eigenvector.

The value of Δ indicates how well the one-dimensionality hypothesis fits the neighborhood: the smaller the value the better the fit.

Spatial domain **Fourier domain**

Figure 6.11a **Plane case:** A planar autocorrelation function in the spatial domain corresponds to energy being distributed on a line in the Fourier domain. (Iso-surface plots, $\lambda_1 \gg \lambda_2 \simeq \lambda_3$.)

Figure 6.11b **Line case:** An autocorrelation function concentrated on a line in the spatial domain corresponds to a planar energy distribution in the Fourier domain. (Iso-surface plots, $\lambda_1 \simeq \lambda_2 \gg \lambda_3$.)

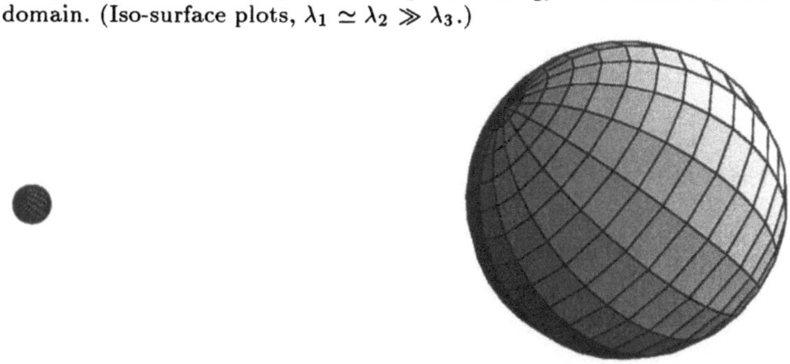

Figure 6.11c **Isotropic case:** A spherical autocorrelation function in the spatial domain corresponds to a spherical energy distribution in the Fourier domain. (Iso-surface plots, $\lambda_1 \simeq \lambda_2 \simeq \lambda_3$.)

Higher rank neighborhoods

Simple neighborhoods are represented by tensors, $\mathbf{T_S}$, having rank 1. In higher dimensional data there exist highly structured neighborhoods that are not simple. The rank of the representing tensor will then reflect the complexity of the neighborhood. The eigenvalue distributions and the corresponding tensor representations are given below for three particular cases of $\mathbf{T}$ in 3D. The relations between the local spatial autocorrelation function and the corresponding energy distributions in the Fourier domain can be studied in Figure 6.11. $\lambda_1 \geq \lambda_2 \geq \lambda_3 \geq 0$ are the eigenvalues of $\mathbf{T}$ in decreasing order, and $\hat{\mathbf{e}}_\mathbf{i}$ is the eigenvector corresponding to λ_i.

1. **Plane case** (*simple neighborhood:* $\lambda_1 \gg \lambda_2 \simeq \lambda_3$)

   $$\mathbf{T} \simeq \lambda_1 \mathbf{T_1} = \lambda_1 \hat{\mathbf{e}}_1 \hat{\mathbf{e}}_1^T \tag{6.84}$$

   This case corresponds to a neighborhood that is approximately *planar*, i.e. is constant on planes in a given orientation. The orientation of the *normal vectors* to the planes is given by $\hat{\mathbf{e}}_1$.

2. **Line case** (*rank 2 neighborhood:* $\lambda_1 \simeq \lambda_2 \gg \lambda_3$)

   $$\mathbf{T} \simeq \lambda_1 \mathbf{T_2} = \lambda_1 \left(\hat{\mathbf{e}}_1 \hat{\mathbf{e}}_1^T + \hat{\mathbf{e}}_2 \hat{\mathbf{e}}_2^T\right) \tag{6.85}$$

   This case corresponds to a neighborhood that is approximately constant on *lines*. The orientation of the lines is given by the eigenvector corresponding to the smallest eigenvalue, $\hat{\mathbf{e}}_3$.

3. **Isotropic case** (*rank 3 neighborhood:* $\lambda_1 \simeq \lambda_2 \simeq \lambda_3$)

   $$\mathbf{T} \simeq \lambda_1 \mathbf{T_3} = \lambda_1 \left(\hat{\mathbf{e}}_1 \hat{\mathbf{e}}_1^T + \hat{\mathbf{e}}_2 \hat{\mathbf{e}}_2^T + \hat{\mathbf{e}}_3 \hat{\mathbf{e}}_3^T\right) \tag{6.86}$$

   This case corresponds to an approximately *isotropic* neighborhood, meaning that there exists energy in the neighborhood but no typical orientation, e.g. in the case of noise.

The eigenvalues and eigenvectors are easily computed with standard methods such as the Jacobi method, e.g. [113]. In general **T** will be somewhere in between these cases, but note that the spectrum theorem states that all 3D tensors can be expressed as a linear combination of these three cases, i.e. **T** can always be expressed as:

$$\mathbf{T} = (\lambda_1 - \lambda_2)\,\mathbf{T_1} + (\lambda_2 - \lambda_3)\,\mathbf{T_2} + \lambda_3\,\mathbf{T_3} \tag{6.87}$$

6.6 TIME SEQUENCES - VELOCITY

A natural way of estimating velocity in an image sequence is to estimate 3D-orientation in the sequence, as described above. The orientation estimate in three dimensions –two spatial and one time-dimension– contains information of both the local spatial orientation and the local velocity. Note that for time sequences, a 3D plane means a line in the 2D image plane and a 3D line means a point in the image plane.

The velocity can be obtained by an eigenvalue analysis of the estimated representation tensor. The projection of the eigenvector corresponding to the largest eigenvalue onto the image plane will give the line velocity field. For moving lines or linear structures only the velocity component perpendicular to the structure can be determined. The velocity component along the line is indeterminable since motion in this direction induces no change in the local signal. This fact is commonly, but somewhat misleadingly, referred to as the 'aperture problem'. It is a fundamental problem for all velocity algorithms if the entire neighborhood is constant on parallel lines. The 'aperture problem' does not exist for moving non-linear structures, e.g. points, and in this case the correct velocity can be estimated.

By examining the relations between the eigenvalues in the orientation tensor it is possible to estimate which of the above categories the neighborhood belongs to. Depending on the category different strategies can be chosen, see Section 6.5. Case number one in Section 6.5, the plane case, corresponds to a moving line and gives a correct estimation only for the velocity component perpendicular to the line. Case number two in Section 6.5, the line case, corresponds to a moving point and gives a correct estimation of the velocity in the image plane.

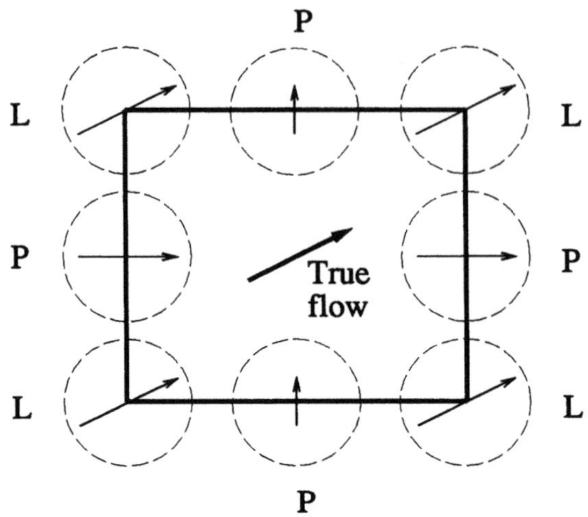

Figure 6.12 Illustration of the 'aperture problem'.

Haglund [47] tested the algorithm using a synthetic test sequence consisting of a rotating and translating star together with a fair amount of Gaussian noise, Figure 6.13 (top left). The star is rotated 1.8° counter-clockwise around its center, and translates 0.5 pixel up and 1 pixel to the right between each two successive frames. In Figure 6.13 (top right) the correct velocity field is given with arrows: white arrows correspond to the moving line case and black arrows to the moving point case.

To categorize the tensor the following functions were chosen:

$$\begin{aligned} P_1 &= \frac{\lambda_1 - \lambda_2}{\lambda_1} \\ P_2 &= \frac{\lambda_2 - \lambda_3}{\lambda_1} \\ P_3 &= \frac{\lambda_3}{\lambda_1} \end{aligned} \tag{6.88}$$

Using these coefficients a general 3D tensor can be written:

$$\mathbf{T} = \lambda_1 \sum P_i \mathbf{T_i} \tag{6.89}$$

The coefficients P_i can be thought of as the probability for each case and the velocity calculation is done according to wich case has the highest probability, (see also Equation (6.87)). In Figure 6.13 (bottom left) the probability for the moving line case, P_1, is shown.

The calculation of the velocity is done using Equation 6.90 for the moving line case and Equation 6.91 for the moving point case, $(P_2 > P_1)$. In neighborhoods classified as isotropic no velocity is computed. In neighborhoods classified as belonging to the moving line case the velocity is computed by:

$$\begin{cases} \mathbf{v}_{line} &= -x_3\,(x_1\hat{\boldsymbol{\xi}}_1 + x_2\hat{\boldsymbol{\xi}}_2)/(x_1^2 + x_2^2) \\ x_1 &= \hat{\mathbf{e}}_1 \cdot \hat{\boldsymbol{\xi}}_1 \\ x_2 &= \hat{\mathbf{e}}_1 \cdot \hat{\boldsymbol{\xi}}_2 \\ x_3 &= \hat{\mathbf{e}}_1 \cdot \hat{\mathbf{t}} \end{cases} \tag{6.90}$$

where $\hat{\boldsymbol{\xi}}_1$ and $\hat{\boldsymbol{\xi}}_2$ are the orthogonal unit vectors defining the image plane and $\hat{\mathbf{t}}$ is a unit vector in the time direction.

In neighborhoods classified as belonging to the moving point case the velocity is computed by:

$$\begin{cases} \mathbf{v}_{point} &= (x_1\hat{\boldsymbol{\xi}}_1 + x_2\hat{\boldsymbol{\xi}}_2)/x_3 \\ x_1 &= \hat{\mathbf{e}}_3 \cdot \hat{\boldsymbol{\xi}}_1 \\ x_2 &= \hat{\mathbf{e}}_3 \cdot \hat{\boldsymbol{\xi}}_2 \\ x_3 &= \hat{\mathbf{e}}_3 \cdot \hat{\mathbf{t}} \end{cases} \tag{6.91}$$

In Figure 6.13 the results from Equation 6.90 (white arrows) and Equation 6.91 (black arrows) are given.

Discontinuous motion

Objects suddenly appearing, disappearing or abruptly changing direction or speed often signify important events. Such events will be reflected by corresponding changes in the orientation tensor and can be detected robustly using

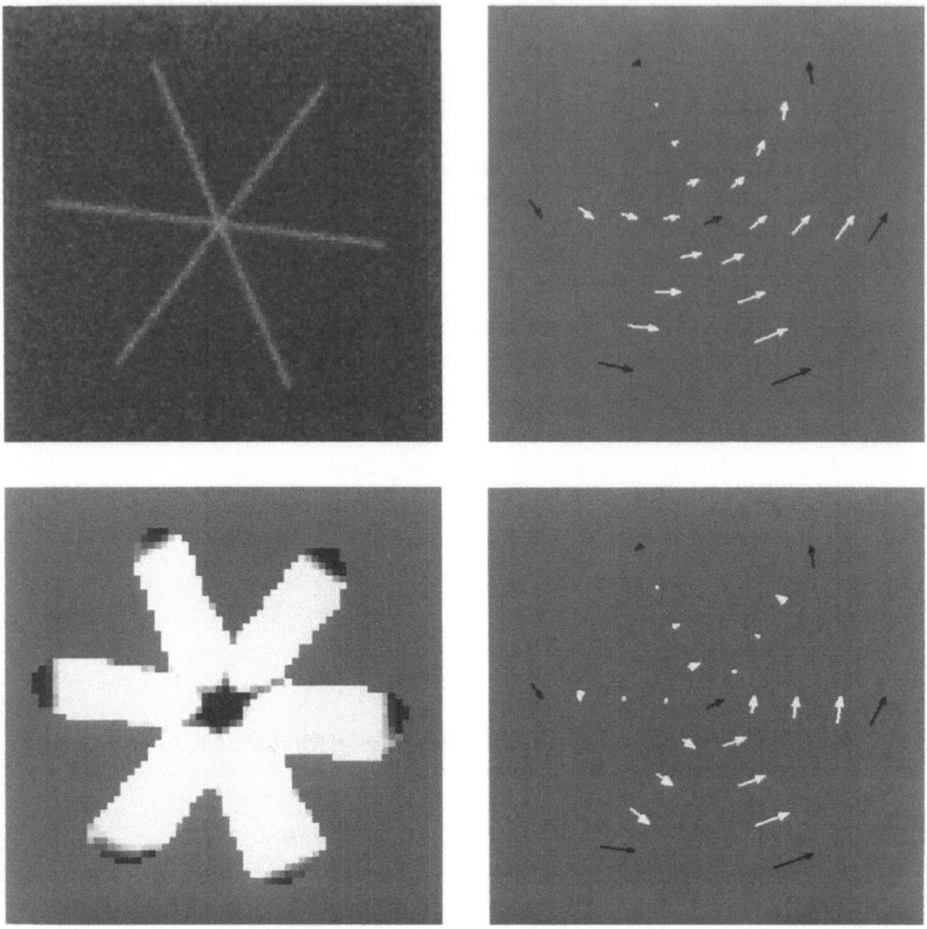

Figure 6.13 **Top left:** One frame from the original sequence of the translating and rotating star, with white Gaussian noise added. **Top right:** The correct velocity vectors from the star sequence. Black vectors correspond to the moving point case and white ones to the moving line case. **Bottom left:** The probability for a neighborhood belonging to the moving line case according to Equation 6.88 for the test sequence. **Bottom right:** The result of the velocity algorithm. Black vectors correspond to the moving point case and white ones to the moving line case.

a tensor field filtering technique termed *normalized convolution* [75]. Gradual changes such as curvature and acceleration can also be estimated using local operators [8].

6.7 PERFORMANCE MEASURES

The performance of the algorithm was evaluated in 2D and 3D using a synthetic test patterns. The test patterns were generated to be locally simple with all directions equally represented, i.e. the test patterns reflect case 1 in section 6.5.

The 2D test pattern is circularly symmetric containing all orientations over a wide frequency range. The pattern is realized as a 512×512 image, see Figure 5.8. The 3D test pattern is spherically symmetric containing all 3D plane orientations over a wide frequency range. The pattern is realized as a $64 \times 64 \times 64$ data volume, see Figure 6.14.

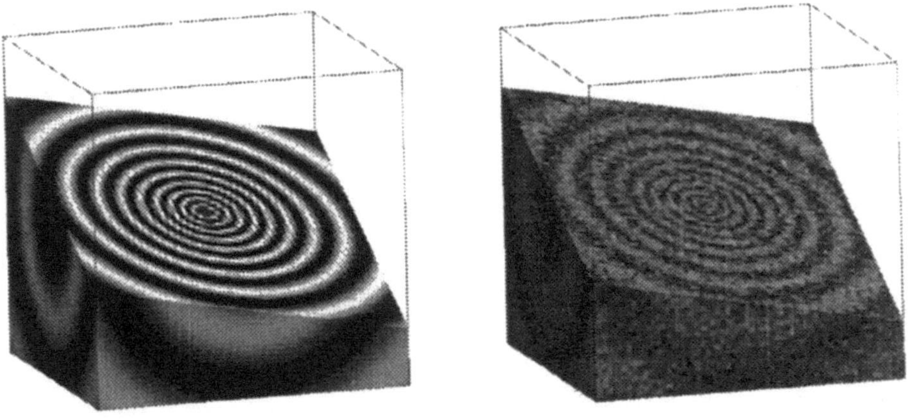

Figure 6.14 Test volumes for 3D orientation estimation. The volumes have been 'cut' along a slanted plane to show the interior signal. Left - without noise, Right - SNR = 0 dB.

Three instances of the test patterns were used for the evaluation, one without noise and two with added white Gaussian noise. The noisy test patterns have a signal-to-noise ratio (SNR) of 10 dB and 0 dB respectively, with SNR defined as:

$$\text{SNR} = 20 \log[\frac{\text{SDEV(pattern)}}{\text{SDEV(noise)}}] \tag{6.92}$$

The pattern is a one-variable function of the radius, $r = \|\xi\|$, the signal orienting vector for the test patterns given by $\mathbf{x} = \xi$ and the correct normalized tensor field is given by:

$$\hat{\mathbf{T}}_1 = \hat{\xi}\hat{\xi}^T \tag{6.93}$$

The estimated orientation tensor $\mathbf{T}$ was compared with the ideal, rank one, tensor using the error estimate given by:

$$\Delta\varphi = \arcsin\left(\sqrt{\frac{1}{2m}\sum_{i=1}^{m}\|\hat{\mathbf{e}}_{1i}\hat{\mathbf{e}}_{1i}^T - \hat{\xi}_i\hat{\xi}_i^T\|^2}\right) \tag{6.94}$$

where $\Delta\varphi$ is the RMS angular error and m is the number of measurement points. The results are given in Table 6.1.

SNR	$\Delta\varphi_{2D}$	$\Delta\varphi_{3D}$
∞ dB	0.8°	0.7°
10 dB	8.0°	3.1°
0 dB	21.2°	10.7°

Table 6.1 Performance figures in 2D and 3D for the orientation estimation algorithm.

The quadrature filters for the estimation were optimized using the principles described in [65] and Chapter 5. The filter size was 15×15 in 2D and $7 \times 7 \times 7$ in 3D. The radial function was a lognormal function having a relative bandwidth of two octaves and a center frequency of $\frac{\pi}{2\sqrt{2}}$ in both 2D and 3D. Note that averaging the orientation tensor will stabilize the estimates and reduce the errors, see Chapter 9.

7

LOCAL PHASE ESTIMATION

7.1 WHAT IS LOCAL PHASE?

Most people are familiar with the global Fourier phase. The shift theorem, describing how the Fourier phase is affected by moving the signal, is common knowledge. The phase in signal representations based on local operations, e.g. lognormal filters, is on the other hand not so well known.

The *local phase* has a number of interesting invariance and equivariance properties that makes it an important feature in image processing.

Local phase estimates are invariant to signal energy. The phase varies in the same manner regardless if there are small or large signal variations.

Local phase estimates and spatial position are equivariant. The local phase generally varies smoothly and monotonically with the position of the signal except for the modulo 2π wrap-around. Section 7.2 discusses cases where the local phase behaves differently.

The spatial derivative of local phase estimates is equivariant with spatial frequency. In high frequency areas the phase changes faster than in low frequency areas. The slope of the phase curve is therefore steep for high frequencies. The phase derivative is called *local or instantaneous frequency.*

There are many ways to approach the concept of local phase. One way is to start from the analytic function of a signal and design filters that locally estimate the instantaneous phase described in Chapter 4. An alternative approach used in this chapter is to relate local phase to the detection of lines and edges in images.

This chapter starts with discussing one-dimensional signals. The extension of the concept of phase in two or more dimensions is discussed in Section 7.3.

Figure 7.1 shows the intensity profile over a number of lines and edges. The lines and edges will be called events in the rest of this chapter. For illustration purposes the ideal step and Dirac functions have been blurred more than what corresponds to the normal fuzziness of a naturalistic image. The low pass filter used is a Gaussian with $\sigma = 1.8$ pixels.

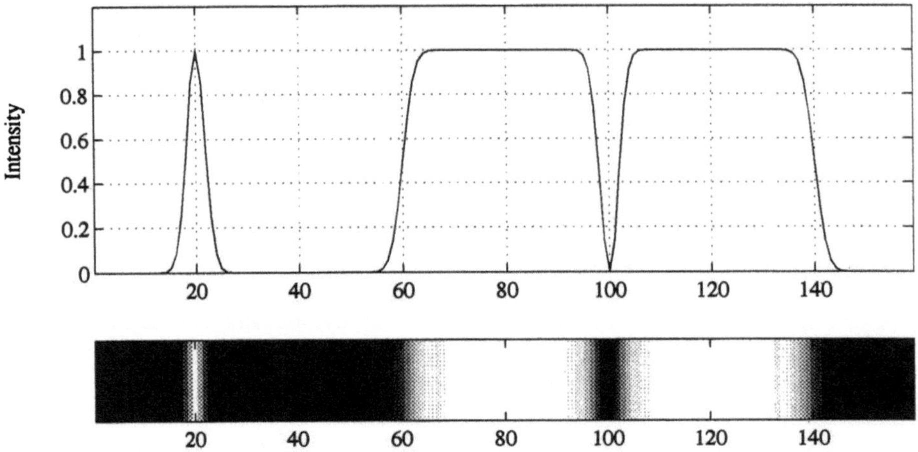

Figure 7.1 Intensity profiles for a bright line on dark background at position $\xi = 20$, an edge from dark to bright at position $\xi = 60$, a dark line on bright background at position $\xi = 100$, and an edge from bright to dark at position $\xi = 140$. All lines and edges are ideal functions blurred with a Gaussian, ($\sigma = 1.8$).

When designing the filters for line and edge detection it is important that they are insensitive to the DC component in the image since flat surfaces are not of any interest for edge and line detection. The simplest line detector is:

$$h_{line}(\xi) \quad = \quad -\delta(\xi + 1) + 2\delta(\xi) - \delta(\xi - 1) \tag{7.2}$$

However, this filter is of too high frequency to fit the frequency spectrum of the signal in Figure 7.1. Convolving the filter with a Gaussian, $\sigma = 2.8$, tunes the filter to the appropriate frequency band (see Figure 7.2 left). The problem is now to design an edge filter that "matches" the line filter. There are two requirements on an edge/line filter pair:

1. Detection of both lines and edges with equal localization acuity.

2. Discrimination between the type of events.

Is there a formal way to define a line/edge filter pair such that these requirements are met?

The answer is *yes*. In order to see how to generate such a filter pair, study the properties of lines and edges centered in a window. Setting the origin to the center of the window reveals that lines are even functions, $f(-\xi) = f(\xi)$. Thus, lines have an even real Fourier transform (see Chapter 4). Edges are odd functions plus a DC term. The DC term can be neglected without loss of generality since neither of the line and edge filters should be sensitive to it. Thus, consider edges simply as odd functions, $f(-\xi) = -f(\xi)$, having an odd imaginary transform.

Now, take a line, $f_{line}(\xi)$, and an edge, $f_{edge}(\xi)$, with exactly the same magnitude function in the Fourier domain,

$$\|F_{edge}(u)\| = \|F_{line}(u)\|. \tag{7.2}$$

For such signals the line and edge filter should give identical outputs when applied to their respective target events,

$$H_{edge}(u)F_{edge}(u) = H_{line}(u)F_{line}(u). \tag{7.3}$$

Combining Equations (7.2) and (7.3) gives:

$$\|H_{edge}(u)\| = \|H_{line}(u)\|. \tag{7.4}$$

Equation (7.4) in combination with the fact that the line filter is an even function with an even real Fourier transform, while the edge filter is an odd function having an odd imaginary Fourier transform, shows that an edge filter can be generated from a line filter using the Hilbert transform (see Chapter 4).

$$H_{edge}(u) = \begin{cases} -iH_{line}(u), & \text{if } u < 0 \\ iH_{line}(u), & \text{if } u \geq 0 \end{cases} \tag{7.5}$$

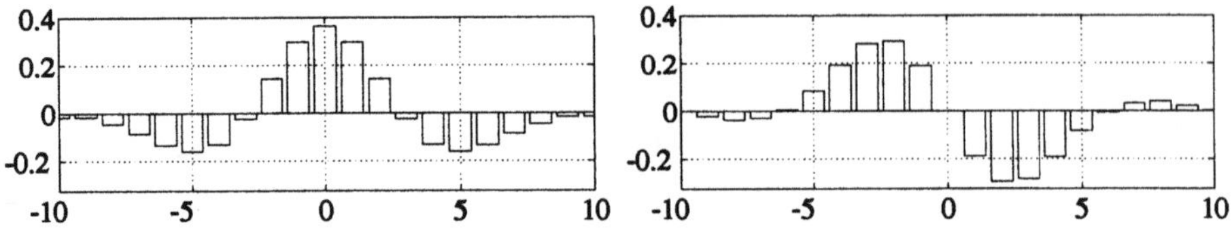

Figure 7.2 The line detector (left) and its Hilbert transform as edge detector (right).

The line and edge detector in Equation (7.5) are both real-valued which makes it possible to combine them into a complex filter with the line filter as the real part and the edge filter as the imaginary part:

$$h(\xi) = h_{line}(\xi) - ih_{edge}(\xi). \tag{7.6}$$

The phase is then represented as a complex value where the magnitude reflects the signal energy and the argument reflects the relationship between the evenness and oddness of the signal (see Figure 7.3).

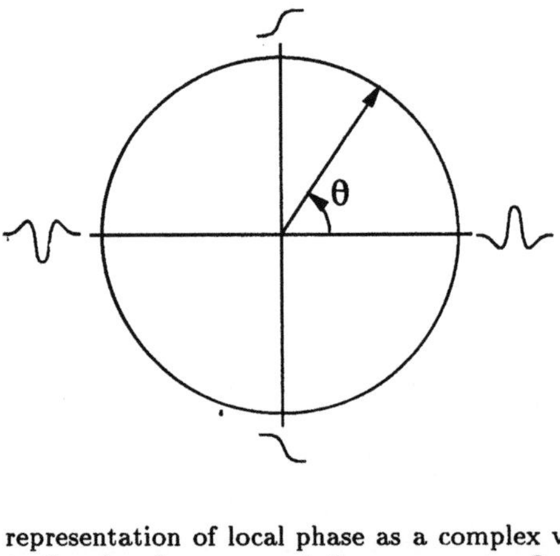

Figure 7.3 A representation of local phase as a complex vector where the magnitude reflects the signal energy and the argument reflects the evenness and oddness relationship.

A filter fulfilling Equations (7.5) and (7.6) is called a *quadrature filter* and is in fact the analytic signal of the line filter. Figure 7.4 shows that the output magnitude from a quadrature filter depends only on the signal energy and on how well the signal matches the filter pass band, and *not* on whether the signal is even, odd or any mixture thereof. The filter phase, on the other hand, depends on the relation between evenness and oddness of the signal relative to the filter center. The phase is not affected by signal energy. The polar plots at the bottom show the trajectory of the phase vector in Figure 7.3 when traversing the neighborhood around each event. Note how the phase value points out the type of event when the magnitude has its peak value.

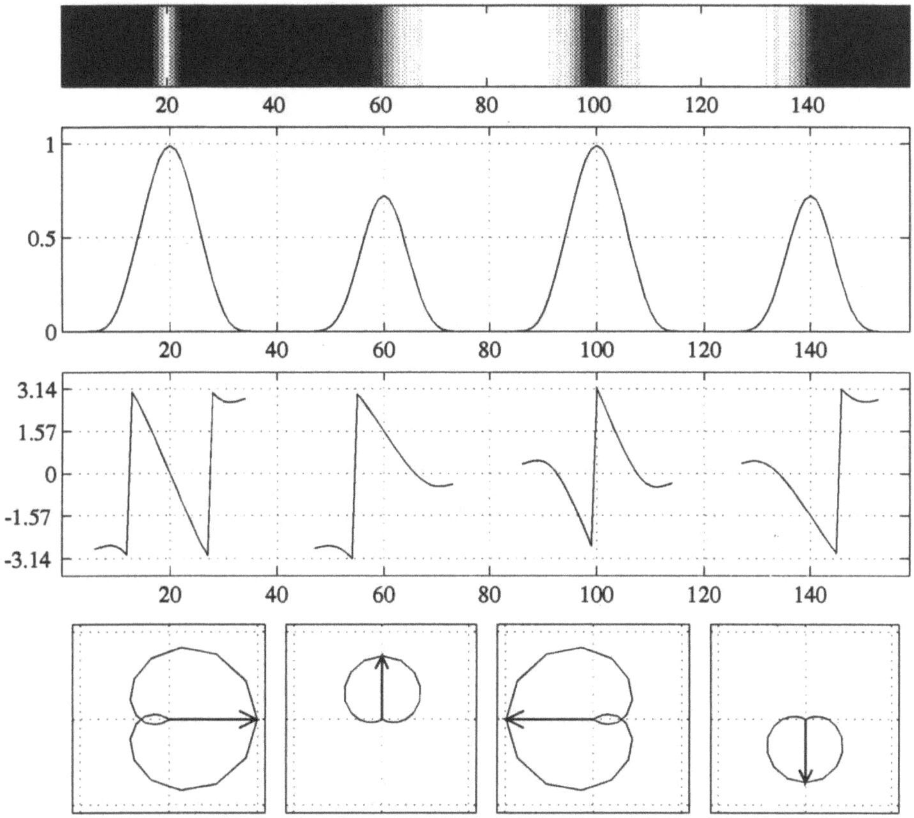

Figure 7.4 Line and edge detection using the quadrature filter in Figure 7.2. Top: The input image. Second: The quadrature filter output magnitude has one peak for each event and the peak value depends only on the signal energy and how well the signal fits the filter pass band. Third: The quadrature filter output phase indicates the kind of event. Bright lines have $\theta = 0$, dark lines have $\theta = \pi$, dark to bright edges have $\theta = \pi/2$, and bright to dark edges have $\theta = -\pi/2$. Bottom: Polar plots showing the phase vector in a neighborhood around the lines and edges.

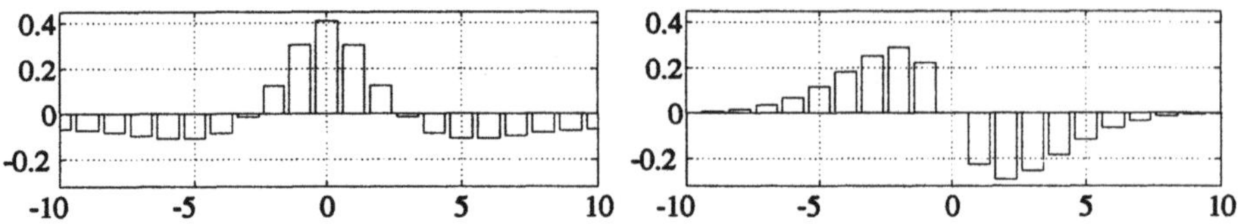

Figure 7.5 A lognormal quadrature filter with $\rho_0 = \pi/8$ and $\beta = 2.0$.

Both the center frequency and the bandwidth of the filter are affected when tuning the quadrature filter in Figure 7.2 with a Gaussian. A filter with a high center frequency and a narrow bandwidth can therefore not be created using that method.

The lognormal filters presented in Chapter 8 have the center frequency, ρ_0, and the relative bandwidth, β, as separate design parameters. This makes it easier to adjust the filter to a desired spectrum. Figure 7.5 shows a lognormal quadrature filter with $\rho_0 = \pi/8$ and $\beta = 2.0$. The result from filtering the test signal with this filter is shown in Figure 7.6. The lognormal filter fits the signal better and gives more equal output magnitude for lines and edges.

How is the phase from the quadrature filter related to the instantaneous phase of the analytic function of the signal? It is easy to show that convolving a signal with a quadrature filter is the same as convolving the analytic signal with the real part of the filter:

$$\begin{aligned} & h(\xi) * f(\xi) = (h_{line}(\xi) - i h_{edge}(\xi)) * f(\xi) = \\ & h_{line}(\xi) * f(\xi) - i\mathcal{H}i\{h_{line}(\xi)\} * f(\xi) = \\ & h_{line}(\xi) * f(\xi) - i h_{line}(\xi) * \mathcal{H}i\{f(\xi)\} = \\ & h_{line}(\xi) * (f(\xi) - i\mathcal{H}i\{f(\xi)\}) = h_{line}(\xi) * f_A(\xi). \end{aligned} \tag{7.7}$$

Since h_{line} is sensitive to changes in a signal and f_A is a signal with continuously changing phase the filter output phase will be an estimate of the instantaneous phase. Narrow band filters generally estimate phase better than broadband filters. If the signal is a sine function the instantaneous phase will be estimated exactly.

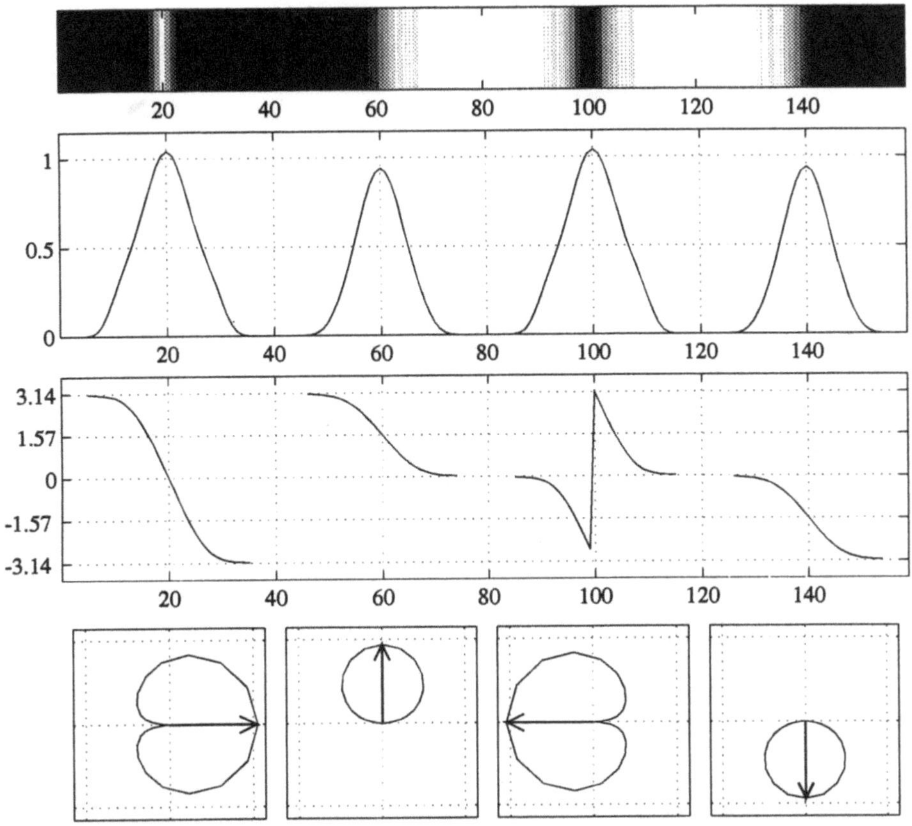

Figure 7.6 Using a lognormal quadrature filter with $\rho_0 = \pi/8$ and $\beta = 2.0$ instead. Top: The input image. Second: The peaks of filter output magnitude are more similar since the lognormal filter fits the signal spectrum better than the filter in Figure 7.4. Third: The quadrature filter output phase indicates the kind of event. Bright lines have $\theta = 0$, dark lines have $\theta = \pi$, dark to bright edges have $\theta = \pi/2$, and bright to dark edges have $\theta = -\pi/2$. Bottom: Polar plots showing the phase vector in a neighborhood around the lines and edges.

7.2 LOCAL PHASE IN SCALE–SPACE

7.2.1 Resolution pyramids and scale–spaces

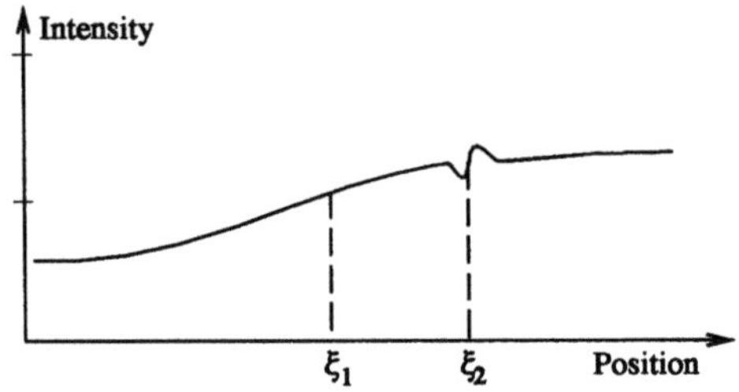

Figure 7.7 Two edges on different scales. A phase estimator tuned for low frequencies will detect the edge at $\xi = \xi_1$ while the edge at $\xi = \xi_2$ is too high in frequency. A phase estimator tuned for high frequencies, on the other hand, will detect the edge at $\xi = \xi_2$ and miss the edge at $\xi = \xi_1$.

Figure 7.7 shows an example of two edges in two scales. At $\xi = \xi_1$ there is an edge in a large scale and at $\xi = \xi_2$ there is an edge in a small scale. A phase estimator tuned for low frequencies will detect the edge at $\xi = \xi_1$ while the edge at $\xi = \xi_2$ is too high in frequency. A high frequency phase estimator, on the other hand, will detect the edge at $\xi = \xi_2$ and miss the edge at $\xi = \xi_1$. A broad band phase estimator will most likely also fail due to phase interference from the two edges, and generate erroneous results.

The answer to the problem of finding image features, not only phase, in all different scales is to use either a set of differently tuned filters or the same filter on differently subsampled versions of the image. The second method is the most widespread since it is much less computationally demanding than the first method. These methods are collected under the names scale-space or resolution pyramids.

7.2.2 Singular points in phase scale–space

The phase is generally very stable in scale–space. There are, however, points around which the phase has an unwanted behavior. At these points the signal goes through the origin of the complex plane, i.e. they are singular points of the analytic function. The invariances and equivariances described in Section 7.1 are generally not valid if the analytic signal is close to a singular point.

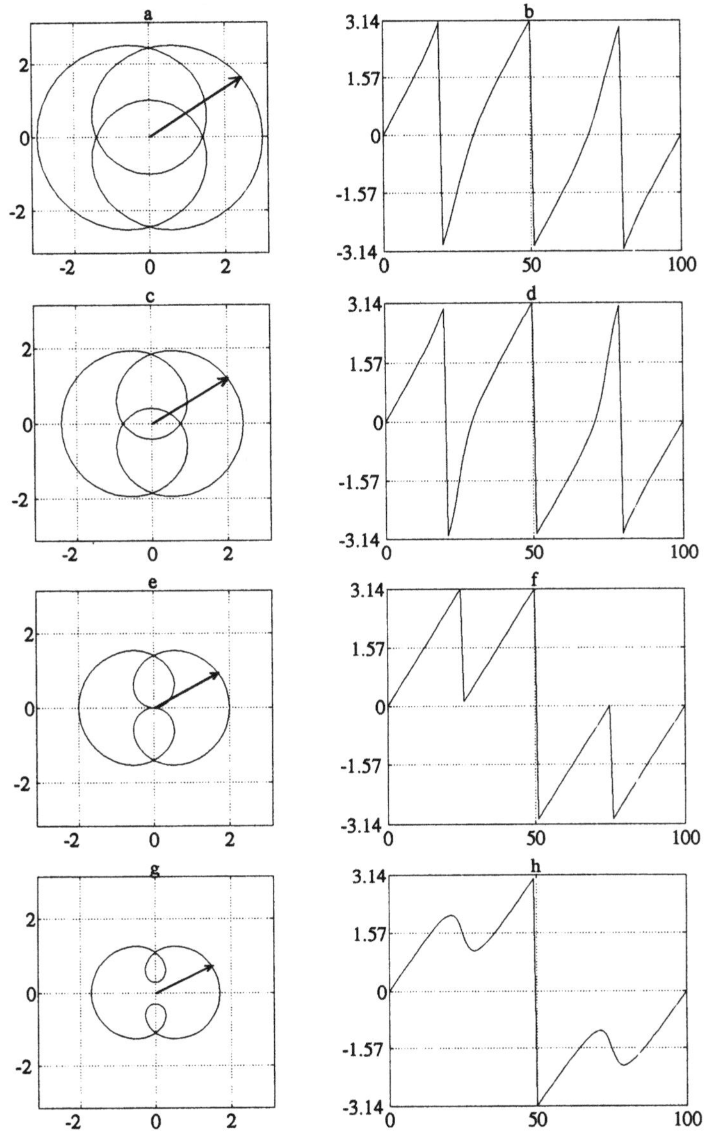

Figure 7.8 The periodic analytic signal, $f_A(\xi) = e^{iu_0\xi} + 2e^{i3u_0\xi}$, is low pass filtered in four steps, attenuating the high frequency part of the signal. On the left are polar plots of the signal vector; on the right are plots of the phase of the same signals. a) The signal circles the origin at a distance and always in the same direction. b) The slope of the phase plot, i.e. the local frequency, is positive and almost constant. c) The small loops circle the origin closely. d) The phase curve bends, i.e. the frequency is locally very high but continuous. e) The small loops pass through the origin, i.e. singular points. f) The local frequency is impulsive. g) The small loops do not circle the origin. h) The the phase curve bends downward, i.e. the frequency is locally negative but continuous.

Figure 7.8 shows a stylized example of a phase resolution pyramid. Suppose we have an analytic signal consisting of two positive frequencies, u_0 and $3u_0$,

$$\begin{aligned} F_A(u) &= \delta(u-u_0) + 2\delta(u-3u_0) && (7.9) \\ f_A(\xi) &= e^{iu_0\xi} + 2e^{i3u_0\xi}. && (7.10) \end{aligned}$$

Suppose now that we create the resolution pyramid using a filter that attenuates the high frequency part of the signal, a factor $1/\sqrt{2}$, but leaves the low frequency part unaffected. The behavior of the signal can be studied using a polar plot of the signal vector, since the signal is periodic.

Figure 7.8a shows a polar plot of the original signal vector. The signal vector runs counter-clockwise all the time and its magnitude is fairly large. The phase is therefore monotonic and increasing, except for the wrap-around caused by the modulo 2π representation (see Figure 7.8b). The local frequency, i.e. the slope, is positive and almost constant. In figure Figure 7.8c the amplitude of the high frequency signal is reduced and the signal vector now circles the origin closely. The local frequency is much higher when the signal vector passes close to the origin causing the phase curve to bend, however it is still monotonic and increasing. Further low pass filtering causes the signal vector to pass through the origin (Figure 7.8e). In these points the phase jumps discontinuously and the local frequency becomes impulsive. In Figure 7.8g, the signal vector moves clockwise when going through the small loops. This means that the phase decreases and that the local frequency is negative (see Figure 7.8h).

This behavior of the phase in scale–space is due to the fact that the high-frequency part of the signal disappears at lower resolution. In Figure 7.8b the phase has three cycles for each period of the signal since the high-frequency part of the signal dominates. In Figure 7.8h, on the other hand, there is only one cycle per signal period since the high-frequency part is attenuated.

7.2.3 Phase Diagrams

Phase diagrams are used to study the scale–space behavior of phase from different kinds of filters. The diagram is created by convolving a representative test signal, Figure 7.9, with the filter to be tested; the convolution is done on different scales. The highest-frequency filter output is plotted at the bottom of the diagram and the lowest-frequency filter at the top. In Figure 7.10 the

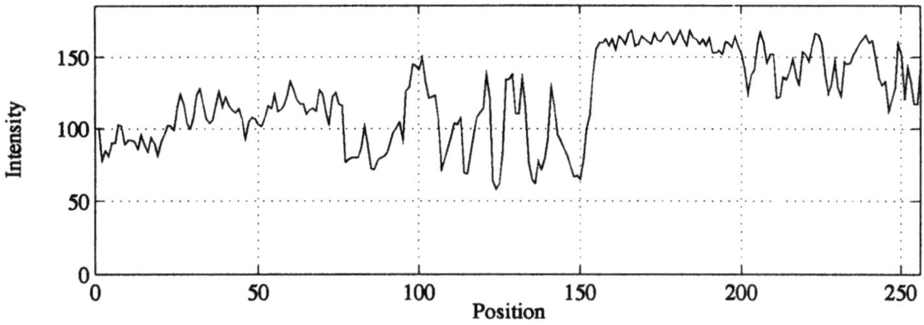

Figure 7.9 The signal used to show the scale–space behavior of a set of lognormal filters.

lines are isovalue lines connecting points with equal phase. The filter center frequency is a function of the scale parameter λ:

$$\rho_0 = \frac{\pi}{4}2^{-\lambda} \quad \text{where } 0 \leq \lambda \leq 4. \tag{7.10}$$

A feature that is stable in scale–space maintains its spatial position in all scales. If the phase were completely stable in scale, the isophase pattern would then consist of vertical lines only. The singular points described above can easily be observed in the phase diagram. The positions in Figure 7.10 where the isophase curves converges are singular points. Just above them the isophase curves turn downward indicating areas with decreasing phase, i.e. negative local frequency (Cf. Figure 7.8g). The high density of isophase curves just below the singular points shows that the local frequency is very high (Cf. Figure 7.8d).

All filters estimating local phase must have singular points since this is a consequence of fine structure disappearing on coarser resolution. However, different types of filters require different distances to the singular points, both spatially and in scale, to return to normal phase behavior again. The size and shape of the areas disturbed by the singular points can therefore be used as a quality measure of the filters. The phase diagram in Figure 7.10 is generated using a lognormal filter, $\rho_0 = \pi/4$ and $\beta = 2.0$, and the test signal in Figure 7.9.

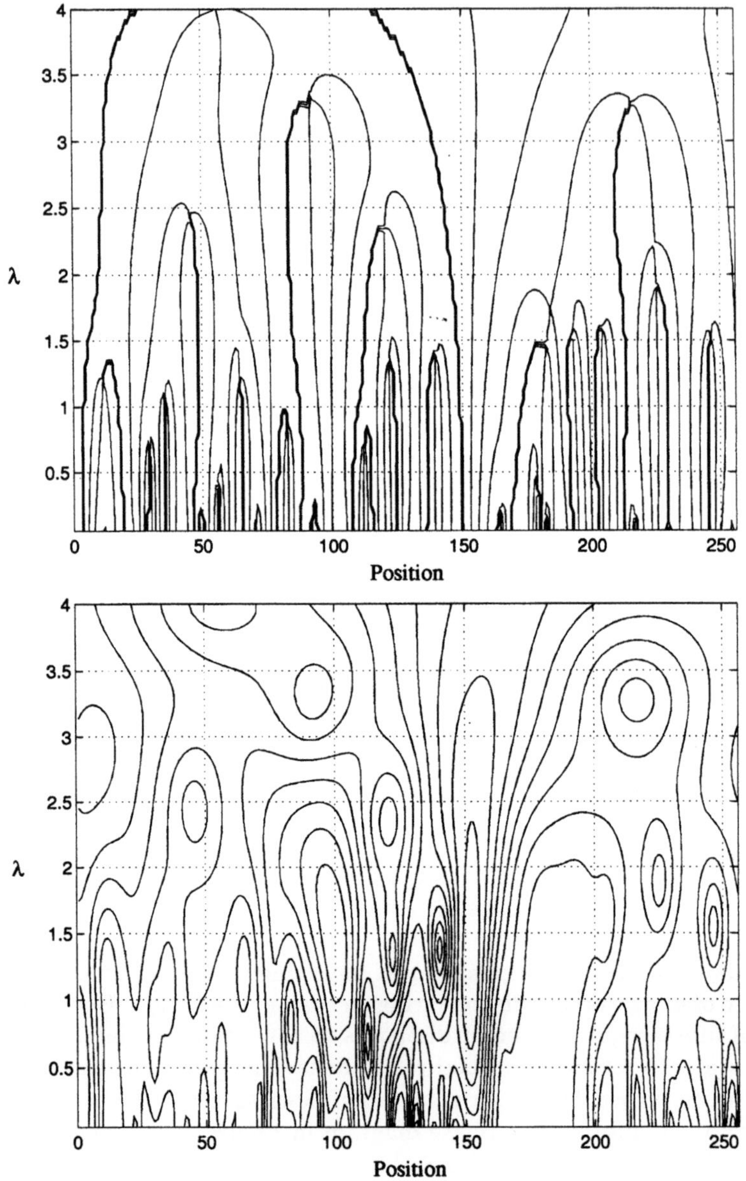

Figure 7.10 Above: A phase diagram of Lognorm phase scale–space. The dark broad lines are due to phase wrap-around. The positions where the isophase curves converge are singular points. Below: Isomagnitude plot of Lognorm phase scale–space. $\rho_0 = \frac{\pi}{4}2^{-\lambda}$ and $\beta = 2.0$

7.3 PHASE IN HIGHER DIMENSIONS

So far, only one-dimensional signals have been discussed. The extension of the phase concept into two or more dimensions is not trivial [47]. The phase is uniquely defined only if the signal is simple. In Section 7.1 it was shown that the local phase is connected to the analytic function and hence to the Hilbert transform. In Chapter 4 a direction of reference was introduced in order to make a multi–dimensional definition of the Hilbert transform possible. Local phase needs a direction of reference as well.

An algorithm for phase-invariant orientation estimation, in two or more dimensions, is presented in Chapter 3 and Chapter 6. Phase-invariant means that the orientation of a simple signal can be estimated regardless if it is an edge or a line, i.e. regardless of the phase. For simple signals the eigenvector corresponding to the largest eigenvalue of the orientation tensor can then be used to measure the phase across the one-dimensional structure.

7.3.1 Representation of phase in higher dimensions

If a continuous representation is desired, the phase cannot be represented with only a single value, although it is a scalar. This is due to the fact that an eigenvector can be defined in either of two opposite directions. Since the phase is measured along the eigenvector it will change sign if the eigenvector changes to the opposite direction. Two neighboring points may have eigenvectors in opposite directions, and thus, phase values with opposite signs, although they belong to the same image structure. The phase values along the horizontal center line in the bottom left Color image 3 display this behavior. Averaging in such a neighborhood would be meaningless. In order to overcome this problem, the phase representation has to include *both* the phase value *and* the reference direction. This means that if the dimensionality of the signal space is N, the dimensionality of the phase representation is N+1.

The phase representation for a simple signal in an N-dimensional space is:

$$\begin{cases} x_1 & = \lambda_1(\boldsymbol{\xi}_1 \cdot \hat{\mathbf{e}}_1)\sin(\theta) \\ x_2 & = \lambda_1(\boldsymbol{\xi}_2 \cdot \hat{\mathbf{e}}_1)\sin(\theta) \\ \ldots\ldots & \\ x_N & = \lambda_1(\boldsymbol{\xi}_n \cdot \hat{\mathbf{e}}_1)\sin(\theta) \\ x_{N+1} & = \lambda_1 \cos(\theta) \end{cases} \tag{7.11}$$

where

λ_1	is the largest eigenvalue,
$\hat{\mathbf{e}}_1$	is the corresponding eigenvector,
$\boldsymbol{\xi}_i$	are the base vectors, and
θ	is the phase value.

A 2D example

Figure 7.11 shows an example in 2D where the neighboring regions A and F, and regions C and D have phase estimates, θ, with opposite signs. This makes a meaningful averaging impossible. For instance, if f_C and f_D denote the phase filter outputs, the vector average between region C and D is:

$$\begin{aligned} f_{aver} &= \tfrac{1}{2}(\|f_C\|e^{i\theta_C} + \|f_D\|e^{i\theta_D}) \\ &= \tfrac{1}{2}(\|f_C\|e^{i\pi/2} + \|f_D\|e^{i(-\pi/2)}) \\ &= \tfrac{i}{2}(\|f_C\| - \|f_D\|) \end{aligned} \tag{7.12}$$

The average phase can be $\pi/2$, $-\pi/2$, or even undefined depending on the relationship between the filter magnitudes in the regions.

The reason for the shifting sign on the phase value is the definition of the reference direction, marked with arrows in Figure 7.11. It could be argued that choosing the phase reference directions such that they all point out from the object would solve the problem, but it is impossible to locally determine what is inside or outside of an object. With only local information available, region A could for instance be a region F on a white disc on a dark background.

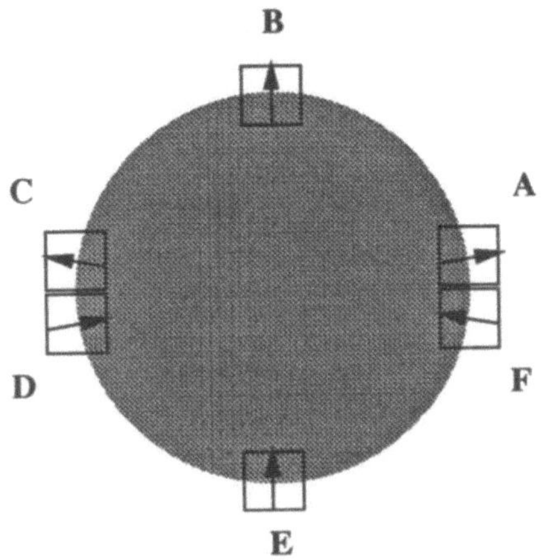

Region	$\hat{\mathbf{e}}_1$ φ	phase θ
A	$\gtrsim 0$	$\pi/2$
B	$\pi/2$	$\pi/2$
C	$\lesssim \pi$	$\pi/2$
D	$\gtrsim 0$	$-\pi/2$
E	$\pi/2$	$-\pi/2$
F	$\lesssim \pi$	$-\pi/2$

Figure 7.11 A dark disc on a bright background. The eigenvectors corresponding to the largest eigenvalues, marked with arrows, are used as the phase reference direction. Note the opposite signs on the phase in regions A and F, and in regions C and D. The table contains the the phase reference direction as an angle to the horizontal axis, φ, and the phase value, θ (c.f. Figure 7.4.)

The phase representation in Equation (7.11) handles the change of sign on the reference direction and produces an averageble result. In 2D the scalar product between the eigenvectors and the basis axes can be written as:

$$\begin{cases} \boldsymbol{\xi}_1 \cdot \hat{\mathbf{e}}_1 &= \cos(\varphi) \\ \boldsymbol{\xi}_2 \cdot \hat{\mathbf{e}}_1 &= \sin(\varphi) \end{cases} \tag{7.13}$$

Combining Equation (7.13) with Equation (7.11) gives:

$$\begin{cases} x_1 &= \lambda_1 \cos(\varphi)\sin(\theta) \\ x_2 &= \lambda_1 \sin(\varphi)\sin(\theta) \\ x_3 &= \lambda_1 \cos(\theta) \end{cases} \tag{7.14}$$

where

$$\begin{aligned} \lambda_1 &\in [0,\infty] \\ \varphi &\in [0,\pi[\\ \theta &\in [-\pi,\pi[. \end{aligned}$$

Figure 7.12 shows the representation which can be interpreted as a vector of length λ_1 rotated an angle θ in a plane defined by the eigenvector $\hat{\mathbf{e}}_1$. The shaded circle corresponds to the phase representation in 1D shown in Figure 7.3.

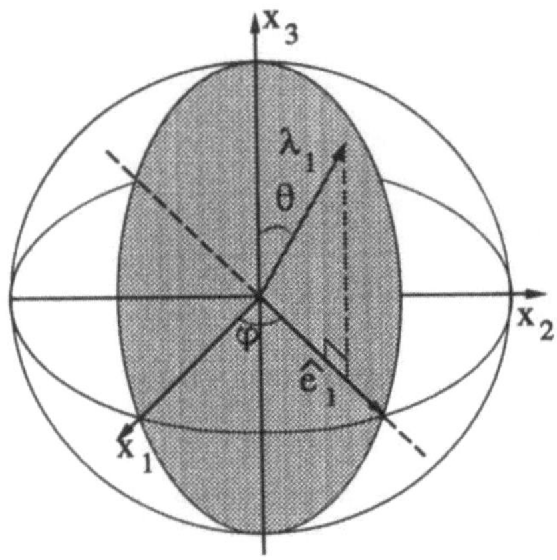

Figure 7.12 A 3D representation of phase in 2D. The shadowed circle corresponds to the phase representation in Figure 7.3.

The phase estimate in region C in Figure 7.11 is then:

$$\begin{cases} x_{1_C} &= \lambda_{1_C}\cos(\pi)\sin(\pi/2) &= -\lambda_{1_C} \\ x_{2_C} &= \lambda_{1_C}\sin(\pi)\sin(\pi/2) &= 0.0 \\ x_{3_C} &= \lambda_{1_C}\cos(\pi/2) &= 0.0 \end{cases} \tag{7.15}$$

and in region D it is:

$$\begin{cases} x_{1_D} &= \lambda_{1_D}\cos(0)\sin(-\pi/2) &= -\lambda_{1_D} \\ x_{2_D} &= \lambda_{1_D}\sin(0)\sin(-\pi/2) &= 0.0 \\ x_{3_D} &= \lambda_{1_D}\cos(-\pi/2) &= 0.0 \end{cases} \tag{7.16}$$

The average phase in a neighborhood is simply the average of the components, x_i, respectively:

$$\begin{cases} x_{1_{aver}} &= \frac{1}{2}(x_{1_C} + x_{1_D}) &= \frac{1}{2}((-\lambda_{1_C}) + (-\lambda_{1_D})) \\ x_{2_{aver}} &= \frac{1}{2}(x_{2_C} + x_{2_D}) &= 0.0 \\ x_{3_{aver}} &= \frac{1}{2}(x_{3_C} + x_{3_D}) &= 0.0 \end{cases} \tag{7.17}$$

Note that the direction of the average phase vector is now independent of the signal energy in the two filter as opposed to the case in Equation (7.12).

7.4 APPLICATIONS USING LOCAL PHASE ESTIMATES

7.4.1 Disparity estimation

The fact that phase is locally equivariant with position can be used to estimate local displacement between two images, [120, 145, 38, 134]. In a stereo image pair the local displacement is a measure of depth and in an image sequence the local displacement is an estimate of velocity.

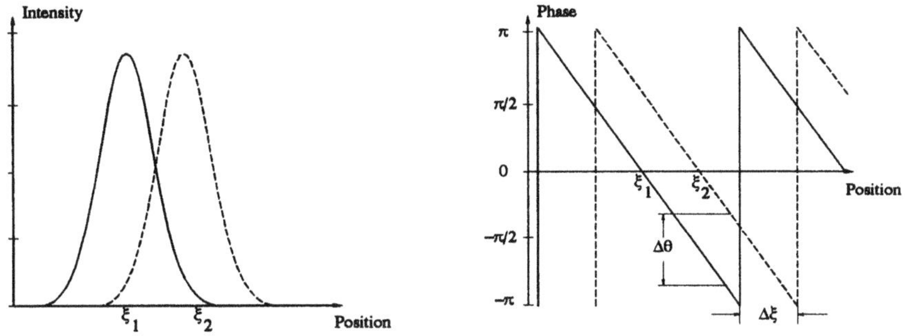

Figure 7.13 Left: A superimposed stereo image pair of a line. In the left image the line is located at ξ_1 (solid) and in the right image it is located at ξ_2. Right: The phase curves corresponding to the line in the two images. The displacement $\Delta\xi$ can be estimated by calculating the phase difference, $\Delta\theta$, and the slope of the phase curve, i.e. the local frequency $d\theta/d\xi$.

One of the advantages of using phase for displacement estimation is that subpixel accuracy can be obtained without having to change the sampling density. Figure 7.13 shows an example where the displacement of a line in a stereo image pair is estimated using phase differences. Traditional displacement estimation would calculate the position of a significant feature, e.g. the local maximum of the intensity, and then calculate the difference. If subpixel accuracy were needed the feature locations would have to be stored using some sort of subpixel representation.

The local phase, on the other hand, is a continuous variable sensitive to changes much smaller than the spatial quantization. That the phase function is sampled with a certain density does not restrict the phase differences to the same accuracy. Thus, a subpixel displacement generates a phase shift giving phase differences with subpixel accuracy without a subpixel representation of image features. In Figure 7.13 the displacement estimate is:

$$\Delta\xi = \frac{\Delta\theta}{\theta'} \tag{7.18}$$

A hierarchical algorithm for disparity from phase

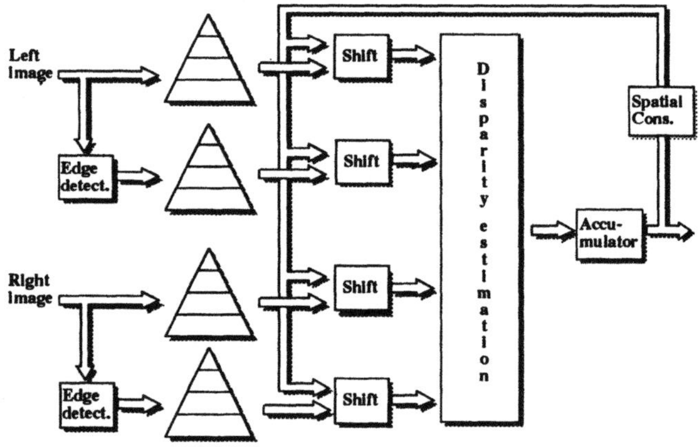

Figure 7.14 Computation structure for the hierarchical stereo algorithm.

Figure 7.14 shows a block diagram of a hierarchical stereo algorithm that uses a phase based disparity, [145, 134]. In order to optimize the computational performance, a multiresolution representation of the left and right image is used. An edge detector, tuned to vertical structures, is used to produce a pair of images containing edge information. The edge images reduces the influence of singular points since the singular points in the original images and the edge images generally do not coincide. The edge images together with the corresponding original image pair are used to build the resolution pyramids. It is one octave between the levels. The number of levels needed depends on the maximum disparity in the stereo image pair.

The algorithm starts at the top of the pyramids, i.e. at the coarsest resolution. The disparity accumulator holds and updates the disparity estimates for each pixel. The four input images are shifted according to the current disparity estimates. This means that the images are locally warped towards each other, reducing the disparity. After the shift, the new disparity estimates are calculated using the phase differences and the local frequencies. The disparity estimates from the gray level images an the edge images are weighted together to form the disparity measures between the shifted images. The new disparity measures updates the current estimate in the disparity accumulator. For each resolution level a refinement of the disparity estimate can be done by iterating these steps. It should get closer and closer to zero during the iterations.

Between each level the disparity image is resampled to the new resolution. The steps above are repeated until the finest resolution is reached. The accumulator image then contains the final disparity estimates.

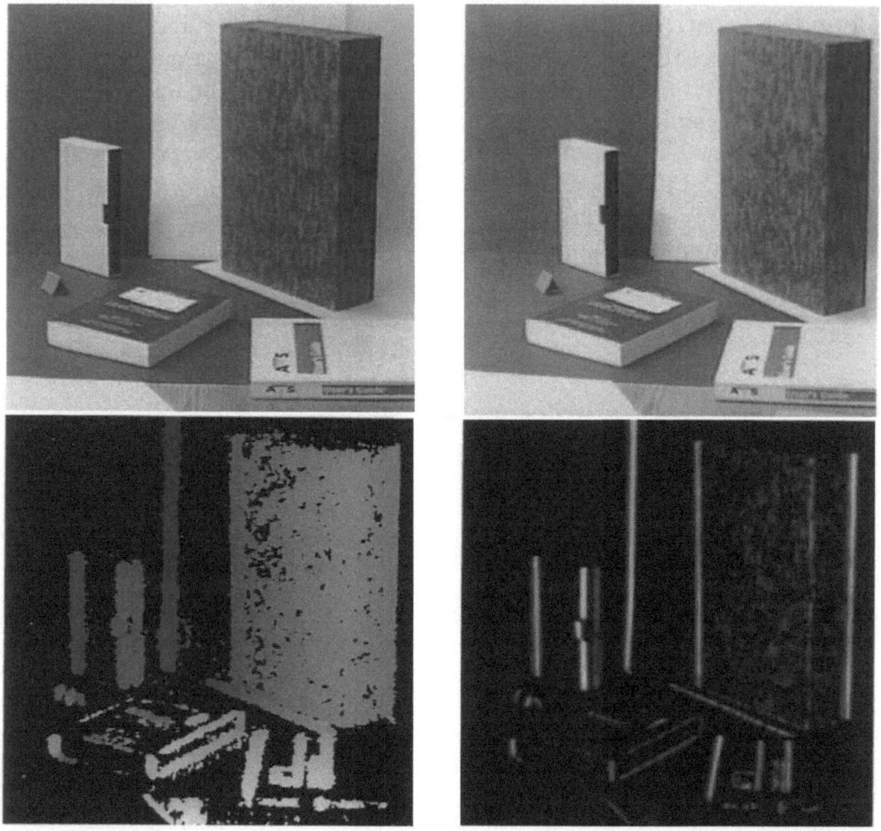

Figure 7.15 Above: Left and right image, (captured using the 'Getafix' robot head at Department of Electronic and Electrical Engineering at University of Surrey, UK). Lower left: Disparity estimates threshold using the confidence values. Lower right: Estimate confidence values. Note that the confidence values are strong on image structures where the disparity is easy to measure and weak on flat surfaces. The result is obtained with five resolution levels and two iterations on each level.

8

LOCAL FREQUENCY

This chapter describes a technique for estimation of local signal frequency and bandwidth. Local frequency is an important concept useful for local structure analysis as well as for determining the appropriate range of scales for subsequent processing. The method is based on combining local estimates of instantaneous frequency over a large number of scales. The filters used are a set of lognormal quadrature filters. The bandwidth is used to produce a measure of certainty for the estimated frequency. The algorithm is applicable to multidimensional data and examples of the performance of the method are demonstrated for one-dimensional and two-dimensional signals.

Background

The concept of frequency is a mathematically well defined entity. Any stationary signal can be represented as a weighted sum of sine and cosine functions having particular amplitudes, phases and frequencies. For non-stationary signals, however, this is not the case and a description in term of sines and cosines impossible. This fact has, since most physical signals are non-stationary, led to a number of attempts to find methods which would allow non stationary signals to be analyzed in a frequency-like manner.

One such attempt was the introduction of the notion of *instantaneous frequency*. Carson and Fry [28] and van der Pol [130] defined instantaneous frequency and applied it to frequency modulated signals. The instantaneous frequency is commonly defined as the rate of change in phase of the analytical signal.

8.0.2 Instantaneous frequency

The analytic signal, discussed in Chapter 4, is a complex signal uniquely defined by a real signal as:

$$f_A = f - i f_{\mathcal{H}i} \tag{8.1}$$

where $f_{\mathcal{H}i}$ denotes the Hilbert transform of f.

A historical review of the concept of instantaneous frequency can be found in [20, 21].

It is noteworthy that, despite it's name, instantaneous frequency is a global entity in that local alterations of the signal will effect the instantaneous frequency everywhere, (due to the infinite kernel associated with the Hilbert transform).

8.0.3 Local Fourier transforms

In 1946 Gabor [40] proposed a combined representation of time and frequency. He expanded the signal in modulated Gaussian basis functions

$$f(x) \;=\; \textstyle\sum_i \alpha_i(x)\; g_{u_i,\sigma}(x)$$

where u is the frequency variable and

$$g_{u_i,\sigma}(x) \;=\; e^{iux} e^{-\frac{x^2}{\sigma^2}}$$

These complex modulated Gaussian basis functions are today commonly referred to as *Gabor functions*. Gabor's motivation for choosing these functions was that they are maximally compact in time and frequency. The main criticism of these basis functions is that the Gabor functions are non-orthogonal and although the functions are well localized they have infinite support - forcing truncation in all practical implementations.

It should be pointed out that since the functions are non-orthogonal, the coefficients α_i in Equation (8.0.3) can *not* be obtained by convolving the signal with the Gabor basis functions. In fact, the coefficients are obtained by convolving the signal with what is known as the *dual* basis functions, and Gabor proposed an iterative method for finding the coefficients.

In many practical situations, however, orthogonality of the basis functions used is not a critical issue and the Gabor functions have found extensive use as filter functions. Such localized filters can be defined using a general window function instead of the Gaussian. This leads to the definition of a *windowed Fourier transform* , see Chapter 4, which can be expressed:

$$G(x,u) = \int_{-\infty}^{\infty} \mathrm{w}(x-x')f(x')\mathrm{e}^{-iux'}\mathrm{d}x' \tag{8.2}$$

The spectrogram provides a common example of windowed Fourier analysis resulting in a combined space-frequency representation. This representation has long been used for analyzing 1D-dimensional time varying signals, see e.g. [2].

The wavelet concept has over the last few years gained considerable attention. A wavelet basis is produced by translation and dilation of a "mother wavelet". By using dilation of the wavelet for controlling its frequency characteristic, a constant relative bandwidth is obtained. The use of such filter sets was first suggested to the computer vision community by Granlund in [42] 1978. A feature of constant relative bandwidth filters is that the spatial localization becomes proportional to the local signal wavelength. It is also appropriate to note that constant bandwidth filters are consistent with current multi-frequency channel theories in psychophysics.

A transform based on a set of scaled Gabor functions was introduced by Morlet *et al.* in 1982. A generalized wavelet transform based on multiresolution local Fourier transforms was presented by Wilson, Calway and Pearson in [142] 1992. The set of filters used for frequency estimation in this paper can be seen as a set of wavelets described by Gabor functions on a logarithmic scale and were introduced by Knutsson in [65] 1982.

8.1 FREQUENCY ESTIMATION

In the following it will be shown how an estimate of local frequency can be obtained by combining instantaneous frequency estimates over a range of scales. For each scale a narrow-band instantaneous frequency estimate is obtained using ratios of lognormal quadrature filter. A wide-range frequency estimate is produced by an energy-weighted summation of the narrow-band estimates.

8.1.1 The lognormal filter function

The lognormal filter $F(\mathbf{u})$ is a spherically separable quadrature filter with a radial frequency function that is Gaussian on a *logarithmic* scale.

$$\begin{cases} F(\mathbf{u}) &= R_i(\rho)D_k(\hat{\mathbf{u}}) \\ R_i(\rho) &= \mathrm{e}^{-C_B \ln^2(\rho/\rho_i)} \end{cases} \tag{8.3}$$

where:

ρ_i is the center frequency

$C_B = \frac{4}{B^2 \ln 2}$

$B = \frac{1}{\ln 2} \ln\left(\frac{\rho_u}{\rho_l}\right)$

$\rho = \|\mathbf{u}\|$ is the norm of the frequency vector. B is the 6 dB relative bandwidth in octaves. ρ_l and ρ_u are the values of ρ for which $R_i(\rho) = 0.5$, see Figure 8.1.

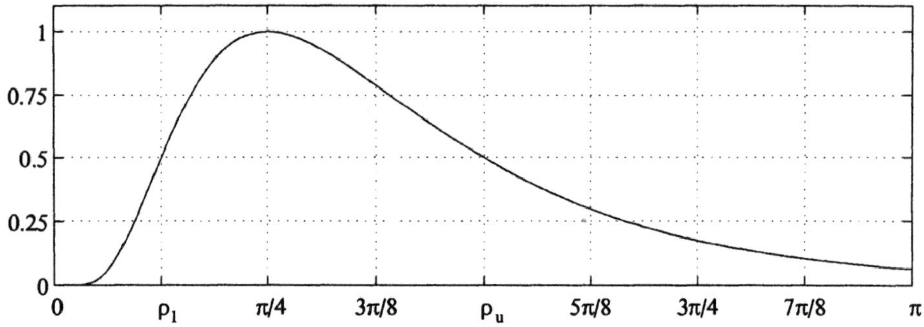

Figure 8.1 A lognormal function.

The directional function of the quadrature filters has the following form:

$$\begin{cases} D_k(\hat{\mathbf{u}}) &= (\hat{\mathbf{u}} \cdot \hat{\mathbf{n}}_k)^2 \quad \textit{if } \mathbf{u} \cdot \hat{\mathbf{n}}_k > 0 \\ D_k(\hat{\mathbf{u}}) &= 0 \quad \text{otherwise} \end{cases} \tag{8.4}$$

where $\hat{\mathbf{n}}_k$ is the filter directing vector, i.e. $D(\hat{\mathbf{u}})$ varies as $\cos^2(\varphi)$, where φ is the angular difference between $\mathbf{u}$ and the filter direction $\hat{\mathbf{n}}_k$.

For simple signals, having an orientation given by $\hat{\mathbf{x}}$, the output magnitude from a quadrature filter in direction k is given by

$$q_{ki} = d_i(\hat{\mathbf{x}} \cdot \hat{\mathbf{n}}_k)^2, \tag{8.5}$$

where d_i is independent of the filter orientation and will be determined by the radial distribution of the signal spectrum $G(\rho)$ and the radial filter function $R_i(\rho)$, (see Chapter 6).

8.1.2 Orientation invariance

An orientation invariant estimate of signal amplitude that is simultaneously local in the spatial and the frequency domain can be obtained by summing the magnitude of the outputs from a number of quadrature filters in different orientations.

$$q_i = \sum_k q_{ki} = \sum_k d_i(\hat{\mathbf{x}} \cdot \hat{\mathbf{n}}_k)^2 \tag{8.6}$$

Expanding $\hat{\mathbf{x}}$ in the filter orienting vectors $\hat{\mathbf{n}}$ yields

$$q_i = \sum_k d_i(\hat{\mathbf{n}}_k \cdot \sum_l x_l \hat{\mathbf{n}}_l)^2 = \sum_k d_i(\sum_l x_l \hat{\mathbf{n}}_l \cdot \hat{\mathbf{n}}_k)^2 \tag{8.7}$$

where x_l are the coefficients of $\hat{\mathbf{x}}$.

If the filter orienting vectors $\hat{\mathbf{n}}_k$ constitute an orthonormal basis of the same dimensionality as $\hat{\mathbf{x}}$, Equation (8.7) reduces to

$$q_i = \sum_k d_i(\hat{\mathbf{x}}_k)^2 = d_i \tag{8.8}$$

This shows that an orientation invariant estimate of the amplitude of a simple signal is obtained by summing the magnitudes of the outputs of orthogonally oriented quadrature filters. The estimate is local both spatially and in the frequency domain. The number of filters required is the same as the dimensionality of the signal.

8.1.3 The ratio of two lognormal filters

By combining the outputs from two or more sets of filters which differ only in center frequency ρ_i, it is possible to produce a local frequency estimate.

For simple single frequency neighborhoods, the contribution in the Fourier domain will be concentrated at a point at a distance ρ from the origin. The isotropic magnitude of a sum of filters is then given by Equation (8.3),

$$q_i = d_i = A e^{-C_B \ln^2(\rho/\rho_i)} \tag{8.9}$$

where A is the local signal amplitude. The ratio between the outputs from two filters differing only in their center frequencies ρ_i and ρ_j is

$$\frac{q_j}{q_i} = \frac{e^{-C_B \ln^2\left(\rho/\rho_j\right)}}{e^{-C_B \ln^2\left(\rho/\rho_i\right)}} \tag{8.10}$$

Simplification of this expression leads to a power relation [65],

$$\begin{aligned}
\frac{q_j}{q_i} &= e^{-C_B\left[\ln^2\left(\rho/\rho_i\right) - \ln^2\left(\rho/\rho_j\right)\right]} \\
&= e^{-C_B\left[\left(\ln\rho - \ln\rho_i\right)^2 - \left(\ln\rho - \ln\rho_j\right)^2\right]} \\
&= e^{-C_B\left[2\left(\ln\rho_j - \ln\rho_i\right)\ln\rho + \ln^2\rho_i - \ln^2\rho_j\right]} \\
&= e^{-C_B\left(\ln\rho_j - \ln\rho_i\right)\left[2\ln\rho - \ln\rho_j - \ln\rho_i\right]} \\
&= e^{-2C_B\left(\ln\rho_j - \ln\rho_i\right)\left[\ln\rho - \frac{1}{2}\ln(\rho_j + \rho_i)\right]} \\
&= \left(\frac{\rho}{\sqrt{\rho_i\rho_j}}\right)^{-2C_B\ln\left(\rho_j/\rho_i\right)}
\end{aligned}$$

It is convenient at this point to introduce two constants. The first, α, depends on the relative bandwidth and the ratio of the center frequencies of the filters and the second $\rho_{(ij)}$, is the geometric mean of the two center frequencies.

$$\alpha = \left[2C_B\ln\left(\frac{\rho_j}{\rho_i}\right)\right]^{-1} \tag{8.11}$$

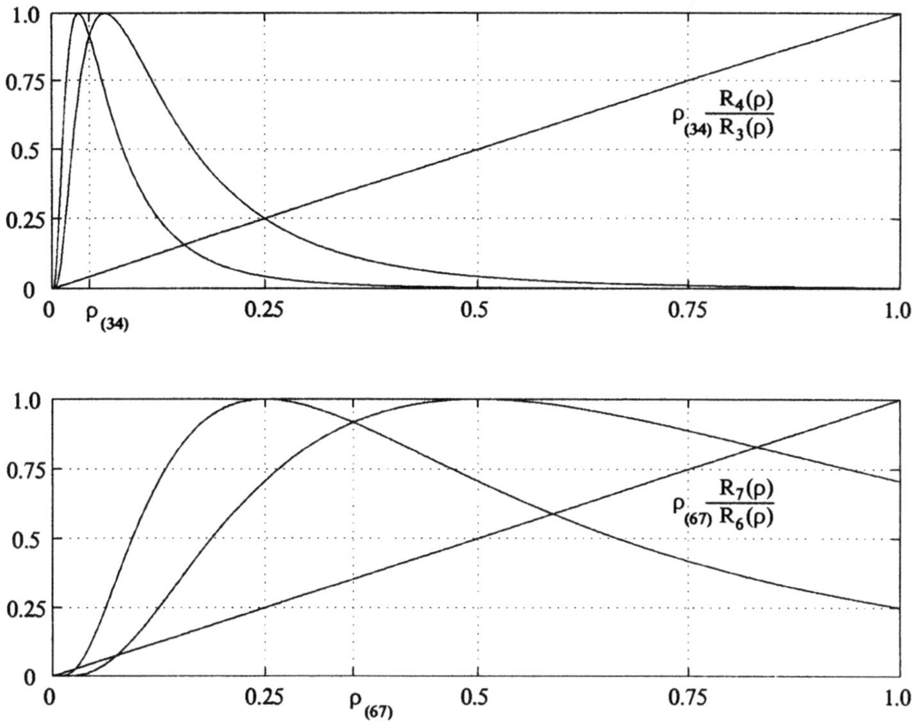

Figure 8.2 The local radial frequency ρ can be estimated as the ratio between two lognormal functions times $\rho_{i+0.5}$, the geometric mean of the two center frequencies, if $\alpha = 1$, Equation (8.14).

$$\rho_{(ij)} = \sqrt{\rho_i \rho_j} \tag{8.12}$$

The ratio between to lognormal filter outputs can then be written

$$\frac{q_j}{q_i} = \left(\frac{\rho}{\rho_{(ij)}}\right)^{\alpha^{-1}} \tag{8.13}$$

Solving for ρ yields:

$$\rho = \rho_{(ij)} \left(\frac{q_j}{q_i}\right)^{\alpha} \tag{8.14}$$

A particularly simple situation occurs if $\alpha = 1$. Using Equation (8.11) then gives the following relation between the ratio of the filter frequencies and the relative bandwidth

$$\frac{\rho_j}{\rho_i} = 2^{B^2/8} \tag{8.15}$$

A center frequency ratio of two is a natural choice and has proved to work well; for α to be one, this choice requires a relative filter bandwidth B of $2\sqrt{2}$.

8.1.4 The relation between lognormal ratios and instantaneous frequency

In this section it is shown that the magnitude of the ratio between two lognormal quadrature filters can be interpreted as the *instantaneous frequency* of a filtered simple signal. To start, note that a quadrature filter suppresses all negative frequencies and produces a filtered analytic signal s_{Ai} where i indicates filter center frequency. Also note that for simple signals the problem can be treated as one-dimensional, since $\mathbf{u} = \rho \cdot \hat{\mathbf{x}}$ for all parts in the Fourier domain where the signal contribution is non-zero. For this reason ρ can be thought of as equivalent to the one dimensional frequency variable u The instantaneous frequency ω_i for s_{Ai} can then be written

$$\omega_i = \frac{\partial}{\partial x} \arg(s_{Ai}) \tag{8.16}$$

where x is the projection of the spatial coordinate $\boldsymbol{\xi}$ on the signal directing vector $\hat{\mathbf{x}}$, i.e. $x = \boldsymbol{\xi} \cdot \hat{\mathbf{x}}$.

Equation (8.16) can be rewritten as

$$\omega_i = \mathrm{Im}[\frac{\frac{\partial}{\partial x} s_{Ai}}{s_{Ai}}] \tag{8.17}$$

The equivalent to differentiation in the spatial domain is multiplication by the frequency variable and a phase shift of $\frac{\pi}{2}$ in the Fourier domain, i.e. for the derivative in Equation (8.16):

$$\frac{\partial}{\partial x} s_{Ai} \leftrightarrow i\rho S_i \tag{8.18}$$

where S_i is the Fourier transform of s_{Ai}.

S_i is the product of the frequency response of the the quadrature filter F_i and the Fourier transform of the signal S.

$$\frac{\partial}{\partial x} f_{Ai} \leftrightarrow i\,\rho F_i\, S \tag{8.19}$$

If the lognormal quadrature filter set is designed so that $\alpha = 1$, the following relation holds, (rewriting Equation (8.14)).

$$\rho\; F_i = \rho_{(ij)} F_{i+1} \tag{8.20}$$

Combining Equation (8.19) and Equation (8.20) yields

$$\frac{\partial}{\partial x} f_{Ai} \leftrightarrow i\,\rho_{(ij)} F_{i+1}\, S \;=\; i\,\rho_{(ij)} S_{i+1} \tag{8.21}$$

Expressing the right hand side in spatial terms, Equation (8.21) gives the interesting result

$$\frac{\partial}{\partial x} s_{Ai} \;=\; i\,\rho_{(ij)}\; s_{A(i+1)} \tag{8.22}$$

This implies that, given the output from one lognormal filter, a whole set of lognormal filters outputs can be generated by repeated spatial differentiation. The filters corresponding to the outputs will have the same relative bandwidth and geometrically increasing center frequencies.

Finally, the instantaneous frequency can be written

$$\omega_i \;=\; \rho_{(ij)}\; \mathrm{Re}[\frac{s_{A(i+1)}}{s_{Ai}}] \tag{8.23}$$

The narrow-band instantaneous frequency estimates used in the present method differ somewhat from Equation (8.23) in that the *magnitudes* of the filter ratios are used, i.e.

$$\omega_i = \rho_{(ij)} \left\|\frac{s_{A(i+1)}}{s_{Ai}}\right\| = \frac{q_{i+1}}{q_i} \tag{8.24}$$

For complex exponential signals Equation (8.23) and Equation (8.24) will be identical. However, for broadband signals, the estimates produced by Equation (8.24) are less sensitive to phase interference effects and guarantee non-negativity. Negative estimates will frequently occur if Equation (8.23) is used, see Chapter 7.

8.2 WIDE RANGE FREQUENCY ESTIMATION

Using two band-pass filters to estimate local frequency will in most situations only work well if the signal spectrum falls within the range of the filters. However, a wide range local frequency estimate can be obtained by weighted summation of a number of filter pairs covering an extended frequency range. Given a set of N quadrature filters with a relative bandwidth of $2\sqrt{2}$ and the center frequencies given by,

$$\rho_i = 2^i \rho_0 \tag{8.25}$$

a wide range frequency estimate can be obtained as

$$\tilde{\rho} = \left[\sum_{i=1}^{N-1} c_i\right]^{-1} \sum_{i=1}^{N-1} c_i \frac{q_{i+1}}{q_i} \rho_{i+0.5}, \tag{8.26}$$

where

$\rho_{i+0.5} = \sqrt{\rho_i \rho_{i+1}} = \rho_0 2^{i+0.5}$

c_i are weighting coefficients.

For single frequency neighborhoods all filter pairs will estimate the same frequency, see Figure 8.2, and any choice of weighting coefficients c_i will yield the same result. For other situations, however, the choice of weighting coefficients is important. To obtain reliable estimates, the weighting coefficients should reflect the amount of signal energy that a given filter pair encounters. A choice of c_i that satisfies this requirement at the same time as it leads to a particularly simple estimation algorithm is:

$$c_i = q_i \tag{8.27}$$

With this choice Equation (8.26) becomes:

$$\tilde{\rho} = \rho_0 \left[\sum_{i=1}^{N-1} q_i \right]^{-1} \sum_{i=1}^{N-1} 2^{i+0.5} q_{i+1} \tag{8.28}$$

i.e. a wide range frequency estimate is obtained as the ratio between to sums of lognormal filter outputs. This estimation algorithm was used in the experiments below with eight filters, with a relative filter bandwidth B of $2\sqrt{2}$, one octave apart.

8.2.1 Estimation of spectrum variance

The local frequency may be considered the average of the frequencies present at a given spatial position. It is useful to have a measure defining the deviation from this value. Such a measure may be interpreted as *instantaneous bandwidth* [20], and will be directly related to the reliability of the mean frequency estimate, $\tilde{\rho}$. The broader the bandwidth, the more uncertain the local frequency estimate. An estimation of the local signal spectrum variance can be obtained by summing the squared differences between the wide range estimation of the mean frequency, $\tilde{\rho}$, and the estimates obtained by the individual filter pairs, i.e.

$$\tilde{\sigma}^2 = \left[\sum_{i=1}^{N-1} q_i^2 \right]^{-1} \sum_{i=1}^{N-1} q_i^2 \left[2^{i+0.5} \frac{q_{i+1}}{q_i} - \tilde{\rho} \right]^2 \tag{8.29}$$

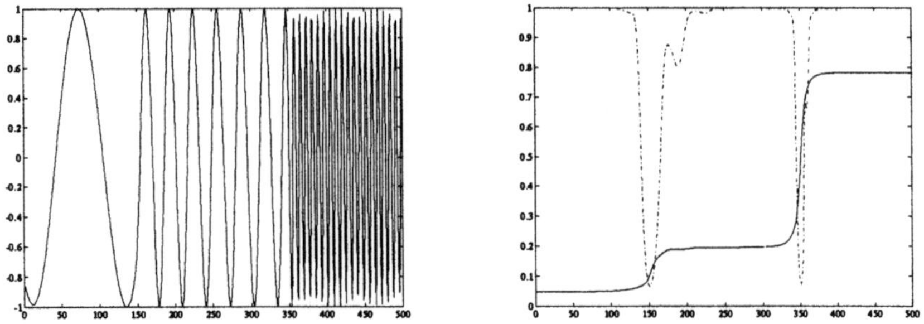

Figure 8.3 Left: A test signal constructed by sinusoids of a stepwise increasing frequency ($\pi/64$, $\pi/16$, $\pi/4$). Right: Estimated local frequency and certainty (dashed).

A certainty measure, c, based on the local spectrum variance can be defined

$$c = \frac{1}{1 + \tilde{\sigma}^2} \tag{8.30}$$

This is the certainty measure used in the examples below.

8.3 EXPERIMENTAL RESULTS

The first example contains a non-stationary signal, i.e. a signal with space-variant spectrum. It is constructed from three sinusoids having frequencies $\pi/64$, $\pi/16$ and $\pi/4$, see Figure 8.3. Note that the certainties of the estimates are high in the constant frequency regions and low in the regions where transitions between two frequencies occur.

The test signal in the second example is constructed as a sum of two components, one constant frequency component, $f = \pi/4$, plus a chirp signal which has linearly increasing frequency from $f = \pi/8$ to $f = 3\pi/8$. This means that the single frequency model is not valid. In this case the algorithm estimates the mean of the two frequency components. Due to interference between the frequencies, the mean frequency is difficult to measure in some spatial positions. In these singular points the certainty of the estimate is close to zero, see Figure 8.4 (left). A smoother estimate can be obtained using normalized averaging, (see [75] and Chapter 9. With this technique an adaptive smoothing can be performed based on the accompanying certainty values. A new estimate of the certainty after this operation is also provided, see the dashed line.

The third example contains a test signal with one constant mean frequency with exponentially increasing bandwidth from $B = 0.25$ to $B = 4$ octaves, see Figure 8.4 (right). As the bandwidth gets broader, the notion of instantaneous frequency for the signal becomes weaker. This is why the estimates get more and more noisy towards the right hand side in the middle right figure, and why the certainty is decreasing in the bottom right figure. Also plotted in these figures are the frequency estimates after using normalized averaging.

Figure 8.6 contains a two-dimensional example. Top left: A test image that contains a circular region with constant frequency surrounded by a region with a frequency gradient. In all two-dimensional examples six quadrature filters were used. The outputs from the filters in this example are shown in Figure 8.4.

A second two-dimensional example is shown in Figure 8.7. This image has regions from two different test images that are overlapping in radial sections. One component is a narrow-band noise process and the other component has increasing frequency toward the the center. In the overlapping regions the algorithm estimates the mean frequency component. If the frequency components differ significantly, this is captured in the certainity estimate.

For yet another example of local frequency estimation, see Figure 13.5.

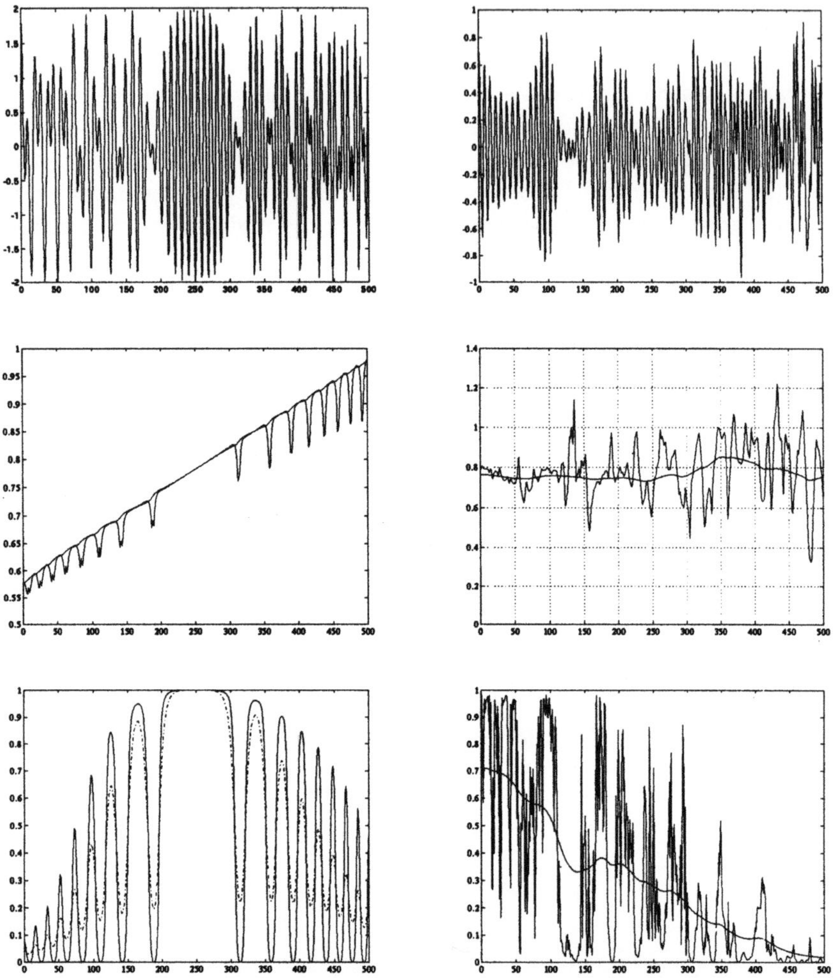

Figure 8.4 **Left** Top: A test signal containing one constant and one linearly increasing frequency component. Middle: Estimated local frequency before and after normalized averaging (the smoother one is after the averaging). Bottom: Estimated certainty and the the certainty after the normalized averaging step (dashed). **Right** Top: A test signal containing constant mean frequency but with exponentially increasing bandwidth. Middle: Estimated local frequency before and after normalized convolution (the smoother one is after the averaging). Bottom: Estimated certainty and the certainty after the normalized averaging step (the smoother one).

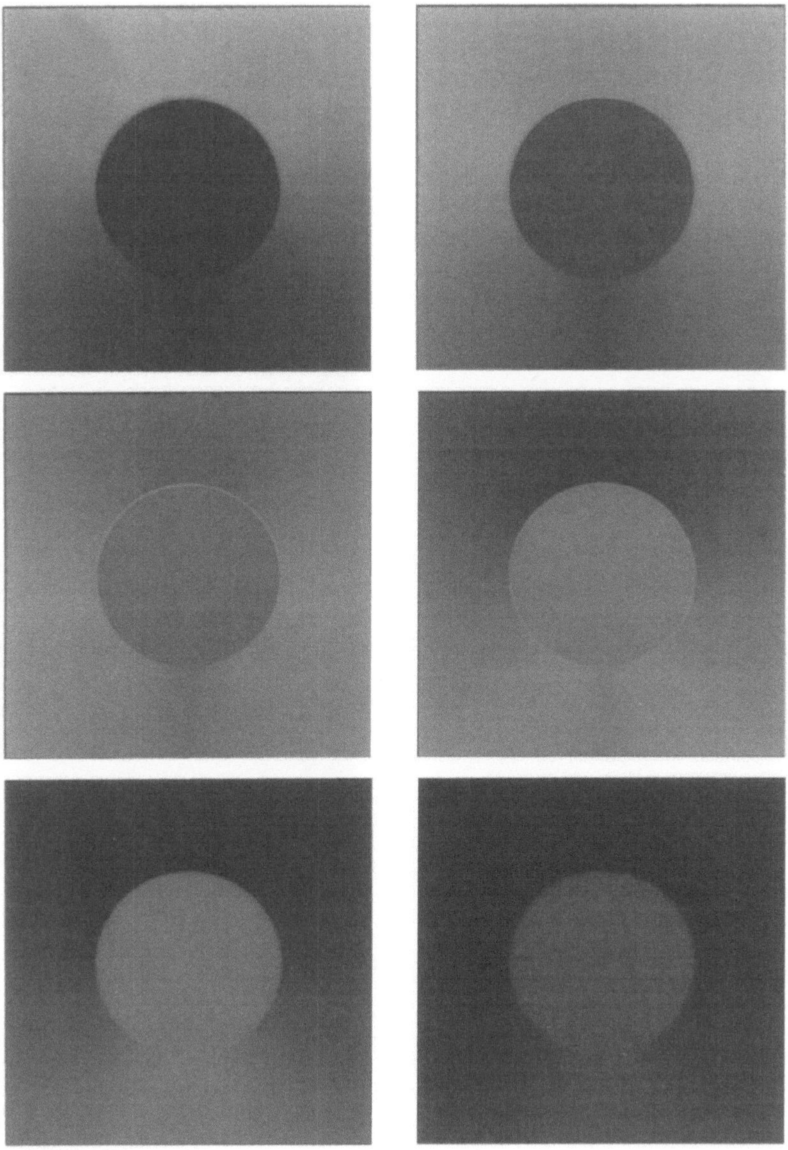

Figure 8.5 Filter magnitude outputs from the six quadrature filters used. The center frequencies of the filters are π, $\pi/\sqrt{2}$, $\pi/2$, $\pi/2\sqrt{2}$, $\pi/4$, $\pi/4\sqrt{2}$. All filters have a relative bandwidth of $B = 2\sqrt{2}$.

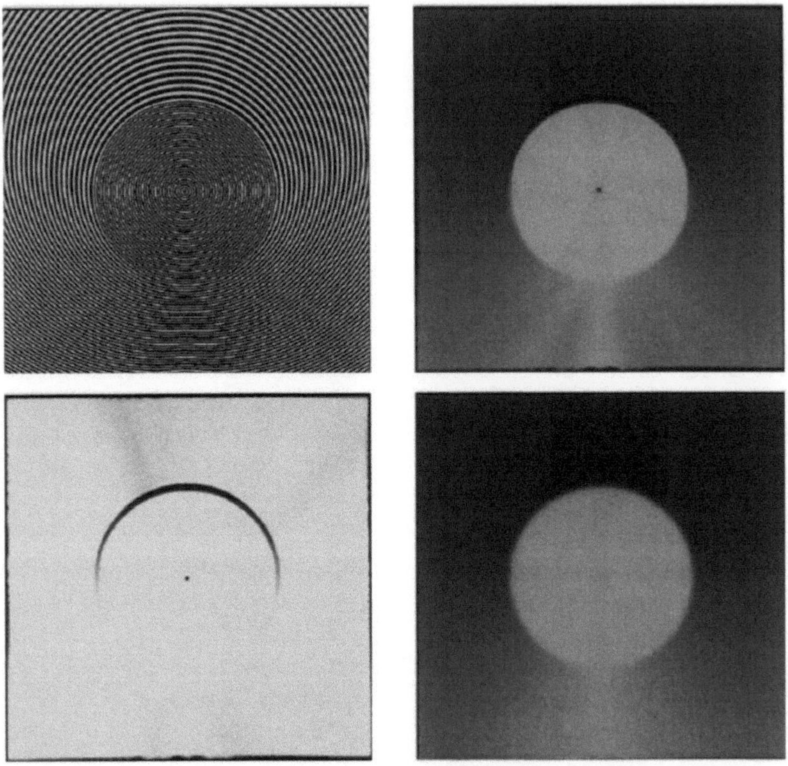

Figure 8.6 Top left: Test image. **Top right** Estimated local frequency. High intensity corresponds to high frequency. **Bottom left** Certainty measure of the estimated frequency. **Bottom right** Updated frequency estimate using normalized convolution.

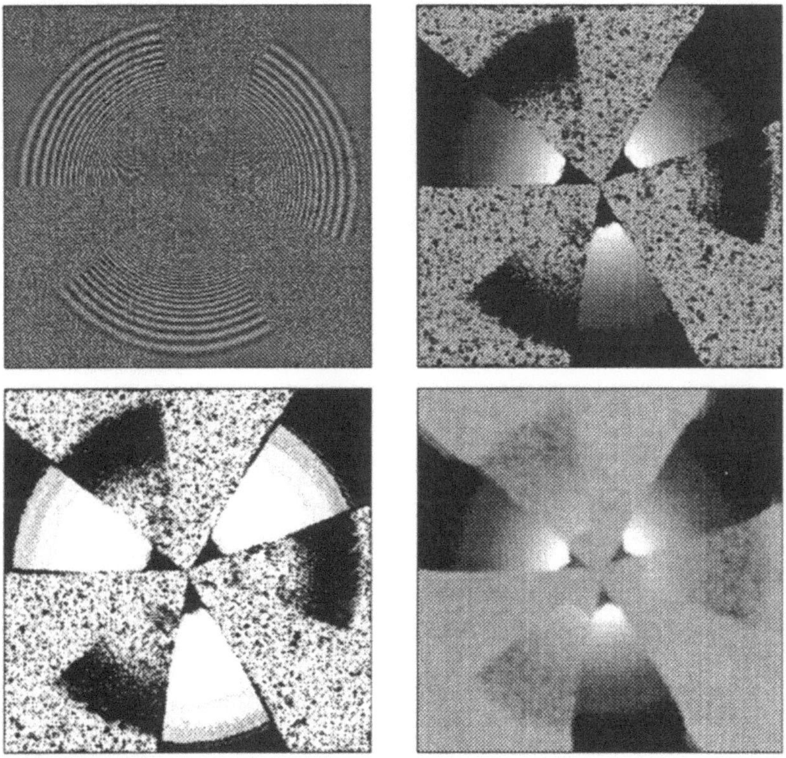

Figure 8.7 Top left: Test image. **Top right** Estimated local frequency. High intensity corresponds to high frequency. **Bottom left** Certainty measure of the estimated frequency. **Bottom right** Updated frequency estimate using normalized averaging.

9

REPRESENTATION AND AVERAGING

Information representation is an important and complex issue in all multi-level signal processing systems [42]. In this chapter we discuss specific properties that are claimed to be of major importance for representation of information in visual processing systems.

The last part of this chapter discusses averaging of information. Averaging is used to obtain, from a set of descriptors, a statement which can be assigned a higher degree of certainty than the individual statements. This has implications on the way information should be represented.

9.1 BACKGROUND

In general, the input information to a computer vision system consists of Cartesian sampled images arrays. This representation is convenient for efficient storage of information in regular data arrays such as in a computer. The gray-level, or color, information of the individual sample points or picture elements (pixels) are easily retrieved by simple pointer indexing.

A central debate in the field of computer vision concerns what type of information should be extracted in order to be able to get a compact and useful description of visual information. Exactly what is meant by a useful description will of course be formulated in different ways for different systems. Clearly, the standard description of an image as an array of pixels is redundant since a skilled artist can capture the essence of a complex image with only a few strokes of a pen on a paper.

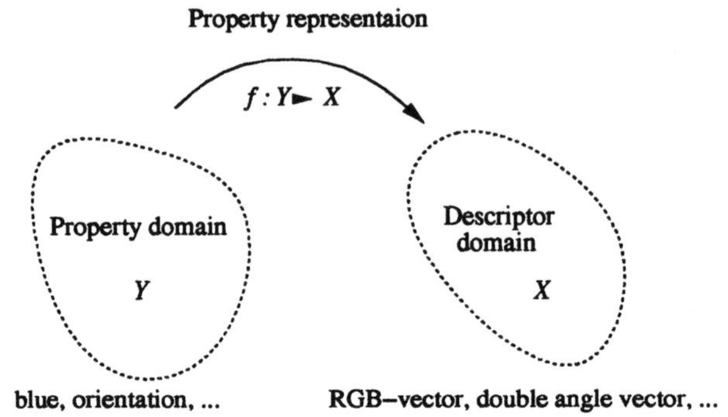

Figure 9.1 The property domain contains the basic properties that are to be described, e.g. color, size, or orientation. A specific property representation maps an basic concept onto a descriptor in the description domain. The descriptors are mathematical objects, e.g. scalars and vectors, upon which operations such as summation can be performed. The set of properties is denoted Y and the set of descriptors is denoted X.

9.2 INVARIANCE AND EQUIVARIANCE

The issues of invariance and equivariance were touched upon in Chapter 7, in a discussion of the properties of local phase, and in Chapter 6, in the context of designing a suitable representation of local orientation. This section gives a strict definition of these concepts.

Let Y be a set of properties and let f be a representation for these properties, i.e. $f : Y \to X$, where X is the set of descriptors. A property may, in general, change in any possible way, but if the variation can be described as a transformation $\mathbf{A} : Y \to Y$, the representation can be described in terms of how $\mathbf{A}$ is reflected in a transformation of the descriptors. The representation f is called *equivariant* with respect to a transformation $\mathbf{A}$ if there is a transformation $\mathbf{A}'$, not equal to the identity transformation, such that

$$f(\mathbf{A}y) = \mathbf{A}'f(y), \qquad y \in Y. \tag{9.1}$$

Given a representation f, the set of all transformations to which f is equivariant to is called the equivariance class of f. This implies that changes of a property y caused by any transformation in that class are reflected in variations of the descriptor $f(y)$. Furthermore, f is called *invariant* with respect to a transformation $\mathbf{B}$ if

$$f(\mathbf{B}y) = f(y), \qquad y \in Y. \tag{9.2}$$

Given a representation f, the set of all transformations $\mathbf{B}$ that f is invariant with respect to is called the invariance class of f. Hence, changes of a property, y, caused by a transformation in this class will not change the descriptor $f(y)$.

The concepts of equivariance and invariance in the context of computer vision were first described in [141] and are further developed in [146].

9.3 SIGNALS AND CERTAINTIES

We stress that an important common attribute of a good representation is that it should keep statement and certainty of statement separate. Consequently, a descriptor should contain two parts: a statement and a measure of confidence of that statement. From a philosophical point of view there should be no argument that "knowing" and "not knowing" are different situations regardless of what is to be known. Such thoughts are by no means new [95, 42, 45, 143, 144, 67] and can, depending on point of view, be said to be related to probability theory, fuzzy set theory, quantum mechanics and evidence calculus. However, we feel that the vision community would benefit from an increased awareness of the importance of these ideas.

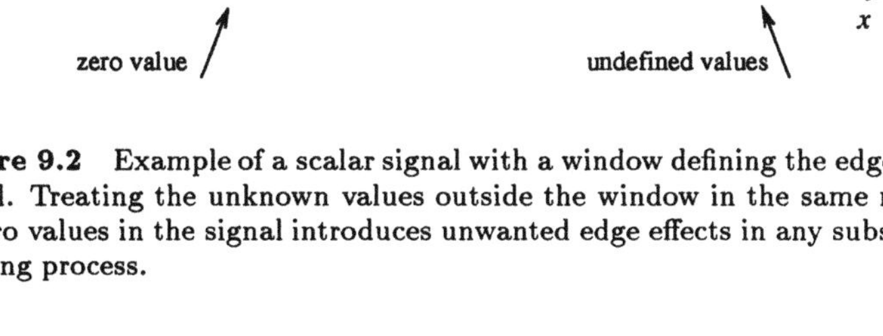

Figure 9.2 Example of a scalar signal with a window defining the edge of the signal. Treating the unknown values outside the window in the same manner as zero values in the signal introduces unwanted edge effects in any subsequent filtering process.

Consider the following simple example. Velocity and its direction are commonly represented by vectors. Only in regions of the image containing structure is it possible to measure the local image velocity. In a region having constant gray-level value, it is impossible to find motion and thus impossible to measure it. If the velocity vectors are set to zero in these regions, this will cause artifacts such as discontinuities in the motion field. Thus, if velocity is estimated from an image sequence, it cannot be fully described by vectors as a sole description.

A representation having a similar ambiguity is the standard scalar representation of gray-level images. The value zero is commonly used to represent both 'black' and 'outside the image'. This causes most operators to produce undesired edge effects. A data representation which has certainty and statement separated allows for setting the *certainty* to zero and avoids the need to decide the gray-level value outside the image.

In practice, estimation or generation of additional certainty information is not unusual. As an example, range data normally consists of two parts; a scalar value defining the distance and an energy measure used to identify points, so called drop-outs, where the range camera has failed to estimate the distance [87].

We argued for the importance of distinguishing between zero value of a descriptor and not knowing the value. In the subsequent discussion, descriptors have an additional scalar certainty value denoted by c. In general, operations act on the *product* of the descriptor and its certainty. Figure 9.3 presents what we call the *scalar representation.* Figure 9.4 illustrates another descriptor, the *argument representation.* For example, depth or intensity can be represented using the scalar descriptor, and orientation represented using the argument descriptor. For spatial frequency both types of descriptors may be of interest.

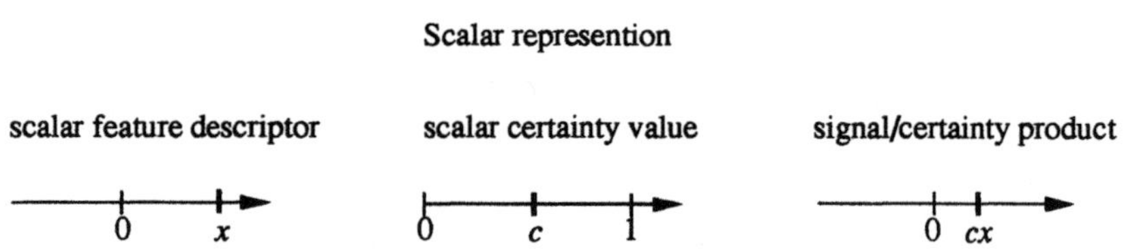

Figure 9.3 Example of a scalar descriptor with an additional statement of certainty. This representation may be used for representing gray-level information in an image. In regions of the image where data is missing, the certainty is set to zero

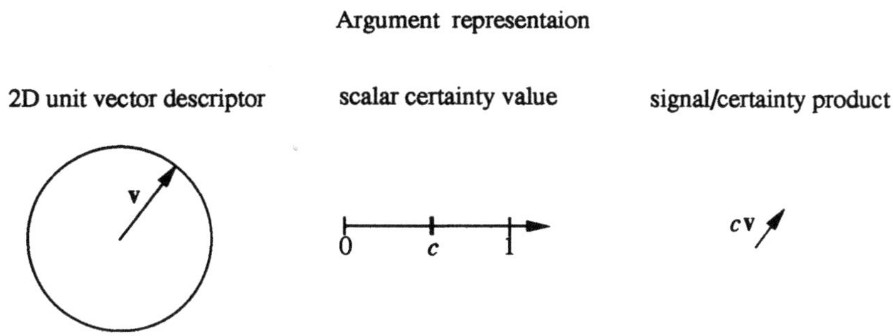

Figure 9.4 A 2D vector descriptor with corresponding certainty value. The property is given by the argument of the vector.

9.4 VECTOR AND TENSOR REPRESENTATION OF LOCAL ORIENTATION

Chapter 6 discussed two ways of obtaining a continuous representation for local orientation in two dimensions: a double angle *vector representation*, first described by Granlund using Gabor functions [42] and later an algorithm using quadrature filters by Knutsson [65], and a *tensor representation* introduced by Knutsson [67]. In Figure 9.5 both the vector representation and the tensor representation are shown. In this section it is shown that they are intimately related.

9.4.1 Geometric interpretation

A geometric interpretation of the relationship between the eigenvectors of a 2D tensor and a double angle description can be made. Suppose that we have a tensor:

$$\mathbf{T} = \lambda_1 \mathbf{e}_1 \mathbf{e}_1^T + \lambda_2 \mathbf{e}_2 \mathbf{e}_2^T \tag{9.4}$$

where λ_n, $\mathbf{e}_n$ denote the eigenvalues and eigenvectors of $\mathbf{T}$ respectively. When the tensor is of rank 1, only one of the eigenvalues is greater than 0,

$$\mathbf{T} = \lambda_1 \mathbf{e}_1 \mathbf{e}_1^T \quad \leftrightarrow \quad \lambda_1 \begin{pmatrix} e_{1x}^2 & e_{1x}e_{1y} \\ e_{1x}e_{1y} & e_{1y}^2 \end{pmatrix}, \tag{9.4}$$

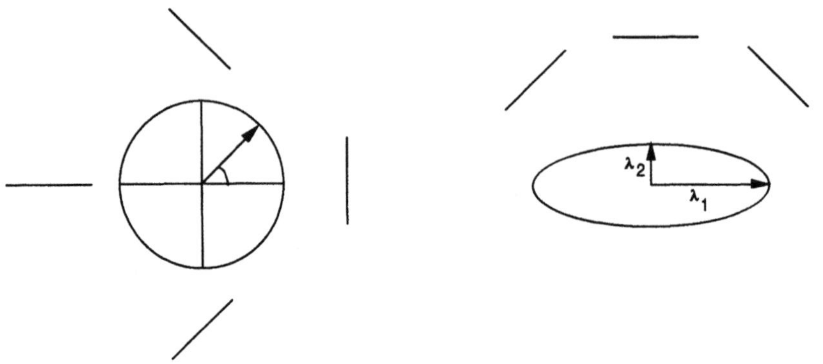

Figure 9.5 **Left**: A double angle vector representation of local orientation. **Right**: A tensor representation of local orientation. The largest eigenvector of the tensor indicates the local orientation (in single angle), and the relation between the eigenvalues expresses the certainty of the statement. The larger λ_1 is compared to λ_2 the higher the certainty ($\lambda_1 \geq \lambda_2$).

In this special case, proving the validity of the double angle method for extracting the direction of the eigenvectors is straightforward since

$$\begin{aligned} 2\varphi_1 &= 2\arg\,(e_{1x}, e_{1y}) && (9.6)\\ &= \arg\,(e_{1x}^2 - e_{1y}^2, 2e_{1x}e_{1y}) && (9.7)\\ &= \arg\,(\cos(\varphi_1)^2 - \sin(\varphi_1)^2, 2\cos(\varphi_1)\sin(\varphi_1)) && (9.8)\\ &= \arg\,(\cos(2\varphi_1), \sin(2\varphi_1)). && (9.9) \end{aligned}$$

This relationship applies separately for both the tensor $\lambda_1 \mathbf{e}_1 \mathbf{e}_1^T$ and the tensor $\lambda_2 \mathbf{e}_2 \mathbf{e}_2^T$ in Equation (9.3). However, the reason that this simple formula holds in general, i.e. even if the tensor has more than one eigenvalue greater than zero, is that the two eigenvectors are actually anti-parallel in the 2φ-domain. Figure 9.6 shows this relationship. The left side of this figure illustrates the direction of the two eigenvectors of a tensor; on the right the corresponding double angle representations of these vectors show that they are actually anti-parallel. This means that the resultant vector in the 2φ-domain has the same direction for all λ_2. Only its length, $(\lambda_1 - \lambda_2)$, varies. The figure also illustrates that the method collapses when $\lambda_1 = \lambda_2$. However, in that case the tensor is completely round, which means that *all* vectors are eigenvectors.

We have shown that, in two dimensions, the double angle vector descriptor and the tensor descriptor give the same result. The only difference is that the tensor descriptor has an additional statement of the local energy, the norm of the tensor. Both methods provide certainty estimates. In the tensor case, the

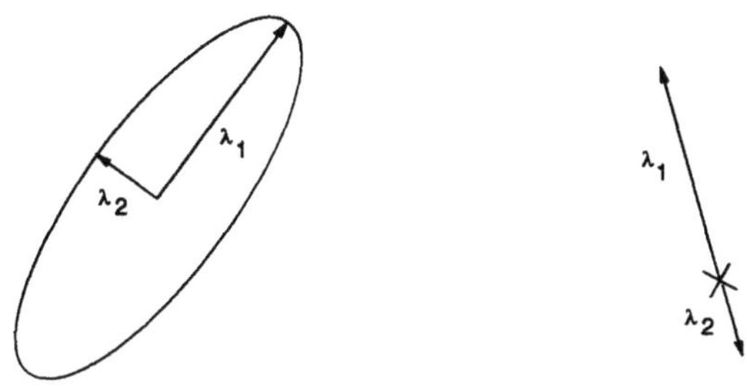

Figure 9.6 **Left:** Geometric representation of a tensor. **Right:** The two eigenvectors expressed in double angle. Note that they are anti-parallel.

relation between the eigenvalues gives an estimate of the certainty [14]. In the double angle methods these estimates are mixed together as the length of the produced vector. The similarities between the methods show, however, that the double angle concept is a fundamental form for description of directional data in 2D. In higher dimensions, however, the tensor formulation must be used.

9.5 AVERAGING

The purpose of averaging a descriptor field is to obtain new local statements which can be assigned a higher degree of certainty. Averaging of scalars, vectors and higher order tensors is simple. Since both averaging and change of representation base are linear operations, scalar averaging applied to the component fields is invariant to the chosen basis. Averaging of a signal by convolution with a lowpass filter function, a, can be written as

$$\mathbf{T}_a(\boldsymbol{\xi}) = \sum_{\mathbf{x}} a(\mathbf{x}) \cdot \mathbf{T}(\boldsymbol{\xi} - \mathbf{x}). \tag{9.9}$$

where $\mathbf{T}$ is the input signal in the form of a scalar, vector or a higher order tensor, $\mathbf{x}$ is the local filter coordinate and $\boldsymbol{\xi}$ is the global signal coordinate. As long as the basic operation is understood, explicitly indicating the dependence on the spatial coordinates $\boldsymbol{\xi}$ and $\mathbf{x}$ serves no purpose. Therefore the standard convolution notation, $*$, will be used,

$$\mathbf{T}_a = a * \mathbf{T} \tag{9.10}$$

Because of the linear properties of averaging mentioned above, each of the components in a vector or tensor field can be interpreted as a separate scalar field which is convolved with a scalar smoothing operator, e.g. a Gaussian filter. When averaging a vector field, vectors in a neighborhood pointing in much the same direction sum up to a large vector, while inconsistently pointing vectors sum up to a shorter vector (Figure 9.7). As previously mentioned, averaging is only a meaningful operation if the chosen representation is appropriate.

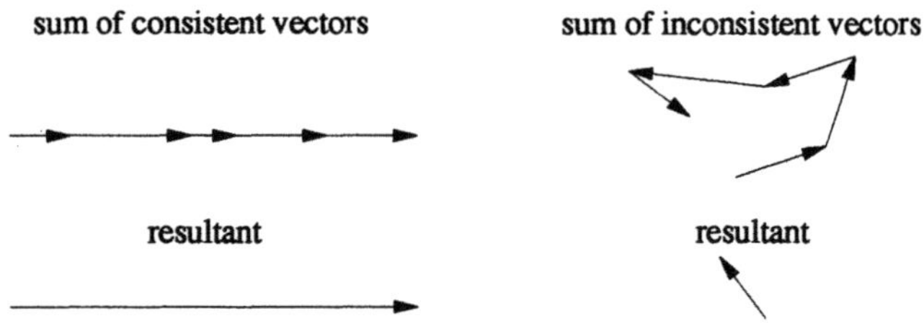

Figure 9.7 Sums of consistent and of inconsistent vectors.

Figure 9.8 shows a stylized example of averaging a tensor field. The tensors are all aligned in almost the same direction after the smoothing operation. The more inconsistent the orientation field is, the "thicker" the averaged tensors will be. Averaging a field with random orientations will produce a field with only almost "round" tensors where the radius is a measure of local energy (Figure 9.9). Another example of averaging can be found in Figure 9.10, in which a neighborhood with an energy gradient is averaged.

9.6 NORMALIZED AVERAGING

The length of the signal/certainty product vector illustrated in Figure 9.4 represents certainty. A common characteristic of a descriptor field is local stability, i.e. its rate of variation is lower than the rate of change in the described signal. As an example, we do not expect the local orientation to vary, from one pixel to another, from vertical to horisontal orientation. However, the estimation

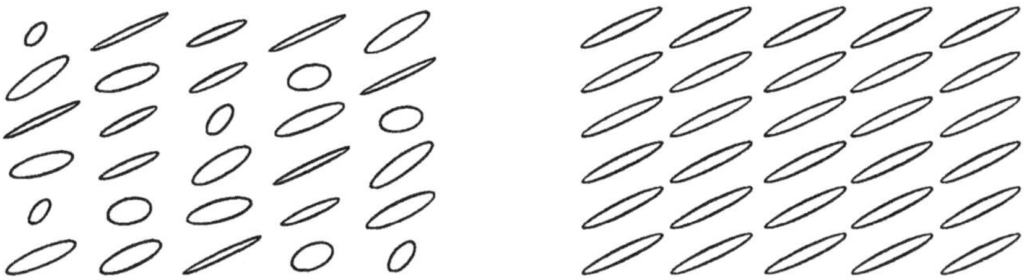

Figure 9.8 A stylized example illustrating what the averaging of a tensor field means. Left: input field. Right: output field.

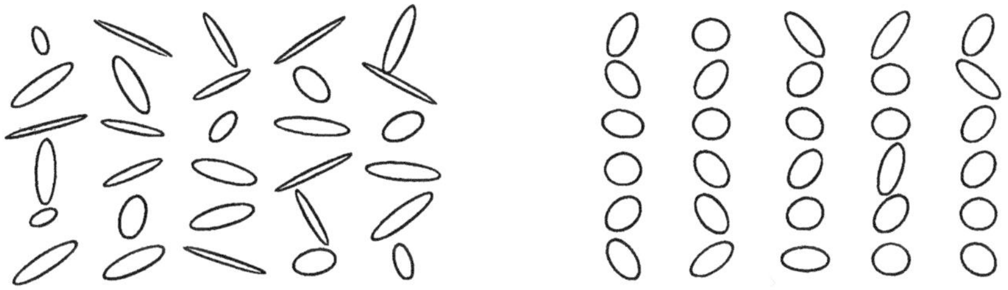

Figure 9.9 Averaging example where the field contains inconsistent data. The output tensors are almost "round" which indicates that the local structure is isotropic. Left: input field. Right: output field.

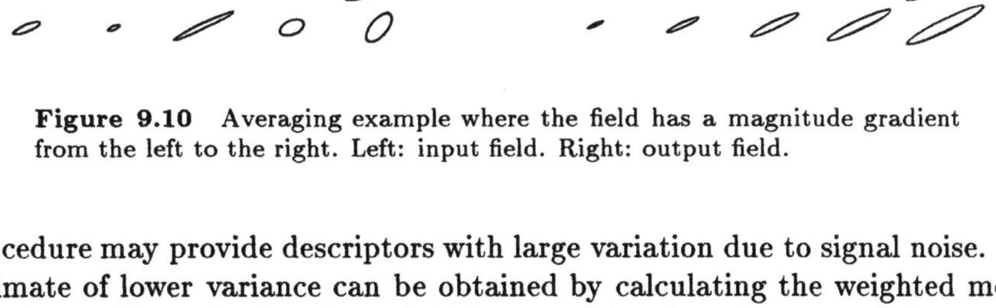

Figure 9.10 Averaging example where the field has a magnitude gradient from the left to the right. Left: input field. Right: output field.

procedure may provide descriptors with large variation due to signal noise. An estimate of lower variance can be obtained by calculating the weighted mean descriptor. Since the descriptor field is assumed to have low rate of change the averaging procedure will not significantly reduce the spatial resolution of the information.

The mean vector, $\mathbf{v}_m$, of a set of vectors, $\mathbf{v}_i = c_i\hat{\mathbf{v}}_i$, is

$$\mathbf{v}_m = \frac{\sum_i \mathbf{v}_i}{\sum_i \|\mathbf{v}_i\|} = \frac{\sum_i c_i\hat{\mathbf{v}}_i}{\sum_i c_i} \quad . \tag{9.11}$$

Since the norm of the mean vector, $\|\mathbf{v}_m\|$, varies between 0 and 1 we call this procedure *normalized averaging.* The maximum value, $\|\mathbf{v}_m\| = 1$, is obtained when the variance of the neighborhood is zero.

Performing normalized averaging using convolution is a special case of a filtering technique termed *normalized convolution*, which is a general filtering technique for filtering incomplete and uncertain data [75, 138]. In normalized convolution, data as well as operators are accompanied by a scalar component representing the appropriate weight and corresponding to the certainty assigned to the data or operator values. Data is represented by a signal part, in this section denoted $\mathbf{T}$, and a scalar certainty value, denoted c. Operators have the same structure: a basis function, $\mathbf{B}$, and a scalar function, denoted a. This scalar function is called an *applicability* function and is used for spatially localizing the basis function. In this special case of normalized convolution denoted *normalized averaging*, the basis function is reduced to a constant function equal to one.

Let normalized averaging be defined by:

$$\mathbf{U}_N = \frac{a * c\mathbf{T}}{a * c} \tag{9.12}$$

where $\mathbf{T}$ is a tensor describing the input data, c denotes the corresponding scalar certainty values, and a is the applicability function. The appropriate choice of this function depends, of course, on the application. The applicability functions used in our experiments belong to the following family of functions:

$$a = \begin{cases} r^{-\alpha} \cos^{\beta}(\frac{\pi}{2}\frac{r}{r_{max}}) & r < r_{max} \\ 0 & \text{otherwise} \end{cases} \tag{9.13}$$

where r denotes the distance from the neighborhood center and α and β are positive integers.

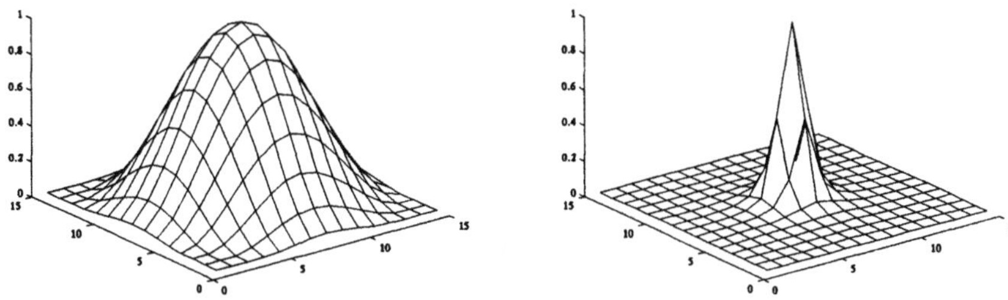

Figure 9.11 **Left**: Example of an applicability function with $\alpha = 0, \beta = 2, r_{max} = 8$ **Right**: An applicability function having $\alpha = 3, \beta = 0, r_{max} = 8$.

It is evident that all components of **T** will be subject to the same transformation and it is consequently sufficient to study the case where **T** is a scalar g (for gray scale). The expression for normalized averaging reduces to

$$u_N = \frac{a * cg}{a * c}. \tag{9.14}$$

where cg is the certainty value times a scalar descriptor, e.g. a gray-level value in an image.

Note that if all certainties are equal, Equation (9.14) is reduced to standard convolution and the standard representation of a gray-level image:

$$\frac{a * cg}{a * c} = \frac{ca * g}{ca * 1} = a' * g \tag{9.15}$$

where a' is the normalized applicability function. In other words, this is interpolation by convolution using a kernel maintaining the local DC-component, a technique widely used for resampling of images.

To illustrate the above method, consider an example using normalized averaging to obtain an efficient interpolation algorithm in a missing sample situation. Figure 9.12 (top left) shows a sparsely and *irregularly* sampled test-image, constructed by using gated white noise, with a threshold chosen so that only 10% of the data remains. Any attempt to reconstruct this image by simple smoothing will fail because of the variation of the sample density.

Figure 9.12 (top right) shows the result of smoothing done using filter of Figure 9.11 (left). Figure 9.12 (bottom left) shows the result using the same filter as applicability function in normalized averaging: the local normalization performed compensates effectively for the sample density variations. Although the result is satisfactory, an even better result can be obtained using a more localized applicability function. The parameter values for the applicability function used to produce the result shown in Figure 9.12 (bottom right) are $\alpha = 3$ and $\beta = 2$. Using such a filter results in an adaptive smoothing of the reconstructed image. The more samples missing, the greater the smoothing performed.

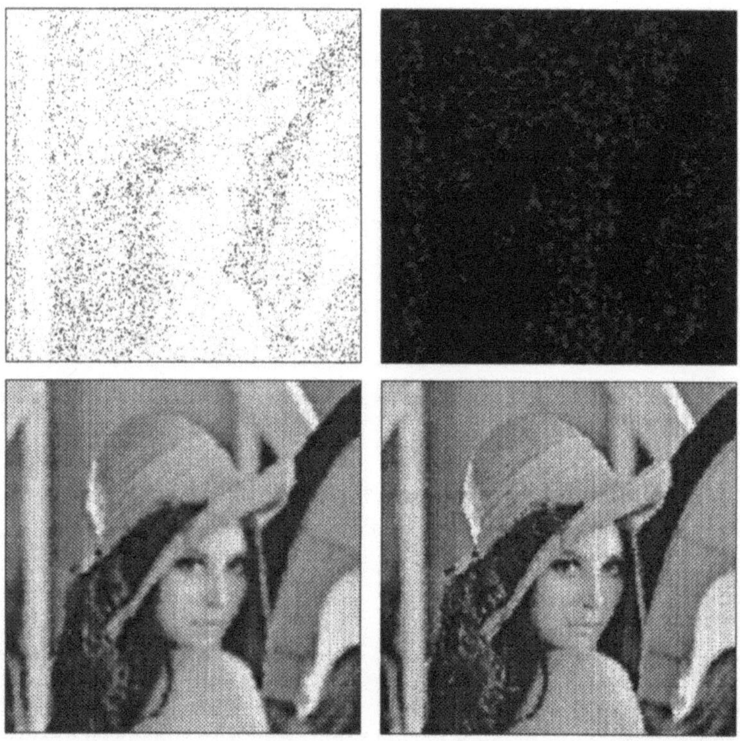

Figure 9.12 **Top left:** The famous Lena-image degraded to a gray-level test image containing only 10% of the original information. **Top right:** Interpolation using standard convolution with a normalized smoothing filter (see Figure 9.11, left). **Bottom left:** Interpolation using normalized averaging with the same filter used as applicability function. **Bottom right:** Normalized averaging using a more local applicability function (see Figure 9.11, right).

10

ADAPTIVE FILTERING

This chapter presents a computationally efficient technique for adaptive filtering of N-dimensional signals. The approach is based on the local signal description given by the orientation tensor discussed in chapter 6. The adaptive filter is synthesized as a tensor-controlled weighted summation of shift-invariant filters.

Background

Image filtering techniques for production of an, in some respect, improved image can be divided in two broad classes - *Restoration* and *Enhancement.* Techniques for image restoration are intended to minimize some generally objective error criterion. Techniques for image enhancement are intended to improve the visual quality of an image; error criteria for this are largely subjective and, as a consequence, no formal theory has emerged for producing "good quality" images.

The variety of existing image-degrading mechanisms has over the years resulted in a multitude of algorithms for image restoration and enhancement. Examples of some of the more commonly used methods are *Averaging, Inverse filtering, Wallis filtering* and *Median filtering.* In contrast to the two-dimensional situation relatively few algorithms for 3-dimensional signals have been put presented, [3, 39, 46, 71]. For a review of enhancement techniques see [29]. However, it is not the purpose of this chapter to discuss various existing algorithms but to present an adaptive filtering technique that has proven useful in a wide range of situations.

The present technique is related to a model for adaptive filtering presented by Abramatic and Silverman [1]. Basic principles for adaptive *anisotropic* filtering using local filter synthesis, presently commonly referred to as steerable or controllable filters [39, 111, 3], were first published for 2-dimensional signals by Knutsson, Wilson and Granlund in [76]. The technique is inspired by mechanisms displayed by biological vision systems and stresses the importance of locally one dimensional structures.

This technique was later extended and applied to 3-dimensional signals such as volumes or image sequences, [71, 47]. The material in this chapter contains original work extending this technique to signals of any dimension. The technique is applicable in image restoration as well as enhancement tasks.

10.0.1 Wiener filtering

In the case of stationary signals corrupted by additive stationary uncorrelated noise the Fourier transform of the optimal linear estimator is given by Equation (10.1).

$$W(\mathbf{u}) \;=\; \frac{H^2(\mathbf{u})}{H^2(\mathbf{u}) \;+\; \eta^2(\mathbf{u})} \tag{10.1}$$

where:

$H^2(\mathbf{u})$ is the spectrum of the signal.

$\eta^2(\mathbf{u})$ is the noise spectrum.

This is the classical solution to the restoration problem and W($\mathbf{u}$) is termed the *Wiener filter.* This approach has been successfully used in many problems involving one-dimensional time signals. However, image or image sequence signals are typically highly *non*-stationary; as a consequence, the Wiener filtering approach has proven unsatisfactory in most image applications. Wiener filtering normally causes an unacceptable blurring of edges and fine details. It is clear that efficient restoration of "image type" signals requires a non-stationary model. In the following a local adaptive filtering scheme, suitable for restoration of non-stationary signals, is developed.

10.1 ADAPTIVE FILTERING

Adaptive filtering implies a shift-variant approach, i.e. the filter coefficients are dependent on the position of the the filter in space or time or both. An adaptive filter can symbolically be expressed as:

$$F = F(\mathbf{u}, \boldsymbol{\xi}) \tag{10.2}$$

It is important to note that the operation implied by Equation (10.2) cannot be expressed in terms of standard convolution. As a consequence a shift-variant filter cannot be given a strict frequency domain interpretation. This fact would seem to render common Fourier-based filter design techniques inapplicable. However, for a certain class of space-variant filters it is possible to employ Fourier based techniques and maintain a concept of frequency. This class of filters can, in broad terms, be defined by the following two requirements:

1. The filter must be local.
2. The filter must vary slowly with position.

In the following it is assumed that the space-variant filters involved in the analysis belong to this class.

10.1.1 Local Wiener filtering

The power of a local approach is nicely demonstrated by the following simple example. Let the spectrum in Equation (10.1) be local instead of global. The result could be interpreted as an adaptive filter, a local Wiener filter, $W_l(\mathbf{u}, \boldsymbol{\xi})$, that is "optimal" for each neighborhood.

$$W_l(\mathbf{u}, \boldsymbol{\xi}) = \frac{H^2(\mathbf{u}, \boldsymbol{\xi})}{H^2(\mathbf{u}, \boldsymbol{\xi}) + \eta^2(\mathbf{u}, \boldsymbol{\xi})} \tag{10.3}$$

Before discussing the problems with this approach a few observations are appropriate.

In broad terms, the local Wiener filter will tend to have a value close to unity where the local signal spectrum is above the level of the noise while suppressing the rest. In this way the filter will adapt to the signal, e.g. if the signal is anisotropic the filter will tend to be anisotropic in a similar way.

The main problem with the local Wiener filter approach is that the filter would need to be computed individually for each position, ξ. The magnitude of the required computational effort renders the approach unsuitable for implementation.

10.1.2 Adaptive filter synthesis

The remainder of this chapter presents a computationally efficient technique for obtaining an adaptive filter displaying many of the desirable features of the local Wiener filter . In this technique the adaptive filter output is synthesized as a weighted summation of shift-invariant filter outputs using adaptive weighting coefficients. Figure 10.1 shows a schematic diagram of the adaptive filter synthesis technique including the production of the orientation tensor.

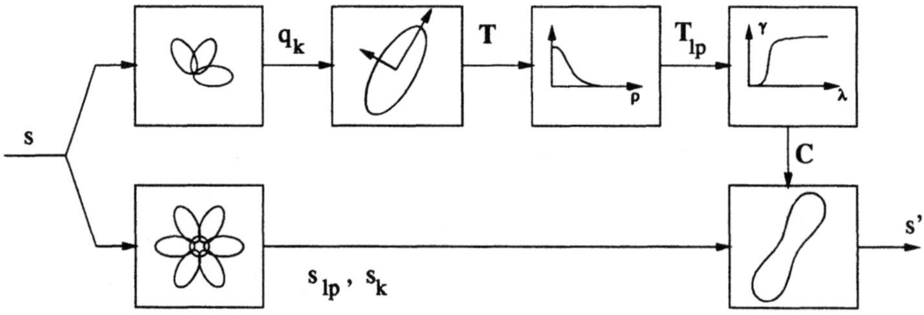

Figure 10.1 Schematic diagram of the adaptive filter synthesis technique including the production of the orientation tensor.

10.2 TENSOR CONTROLLED FILTER

In the following the adaptivity of the filter is mediated through a local control tensor $\mathbf{C}$, i.e.

$$F = F(\mathbf{u}, \mathbf{C}) \tag{10.4}$$

The adaptive filter is a sum of a position invariant low-pass filter and a spherically separable adaptive filter. For the present purpose it is sufficient that the angular part of the adaptive filter be controlled by $\mathbf{C}$, i.e.

$$F(\mathbf{u}, \mathbf{C}) = F_{lp}(\rho) + R(\rho)\, D(\hat{\mathbf{u}}, \mathbf{C}) \tag{10.5}$$

The basis for the adaptive filtering strategy is the simple relation between the control tensor $\mathbf{C}$ and the angular part of the adaptive filter $D(\hat{\mathbf{u}}, \mathbf{C})$ given by:

$$D(\hat{\mathbf{u}}, \mathbf{C}) = \sum_{n=1}^{N} \gamma_n\, (\hat{\mathbf{e}}_n \cdot \hat{\mathbf{u}})^2 \tag{10.6}$$

where γ_n and $\hat{\mathbf{e}}_n$ are the eigenvalues and eigenvectors of $\mathbf{C}$.

10.3 THE CONTROL TENSOR

The information contained in the orientation tensor $\mathbf{T}$ provides a good basis for control of an adaptive filter. The orientation tensor is processed in two steps to produce a suitable control tensor. 1. The tensor is filtered using a low-pass kernel. 2. The tensor eigenvalues are remapped.

10.3.1 Tensor low-pass filtering

The first step in producing the control tensor, **C**, is to low-pass filter the orientation tensor, **T**. The low-pass filtered tensor, $\mathbf{T}_{lp}$, is obtained by convolving the orientation tensor field with a smooth weighted-average kernel, h_{lp}.

$$\mathbf{T}_{lp} = h_{lp} * \mathbf{T} \tag{10.7}$$

The low-pass operation ensures that the requirement of a slowly varying adaptive filter is met. This type of local averaging in the representation space originates from [65].

The general effects of tensor averaging are discussed in Chapter 9; a few comments are, however, appropriate. Because of the properties of the tensor representation the low-pass tensor, $\mathbf{T}_{lp}$, should be thought of as representing the range of events within the neighborhood rather than the "mean" event. To clarify, consider the following two simple examples:

- If the neighborhood is the sum of a simple signal and isotropic noise the effects of the noise on the orientation estimate will be reduced by the low-pass filter. The presence of the noise will be manifested in that an isotropic term is added to the low-pass tensor.
- If the neighborhood consists of several perpendicular orientations the low-pass orientation tensor will represent all orientations in proportion to magnitude and relative occurrence. If all orientations are equal in magnitude and occurrence all eigenvalues of the low-pass tensor will be equal, i.e. the low-pass tensor will be isotropic.

10.3.2 Tensor mapping

The next step toward producing a suitable control tensor is a point-by-point mapping of the low-passed orientation tensor $\mathbf{T}_{lp}$. It should be noted that the purpose of the mapping functions is to provide the degrees of freedom that has been found useful in practise and that the exact mathematical formulation of the functions is of secondary importance.

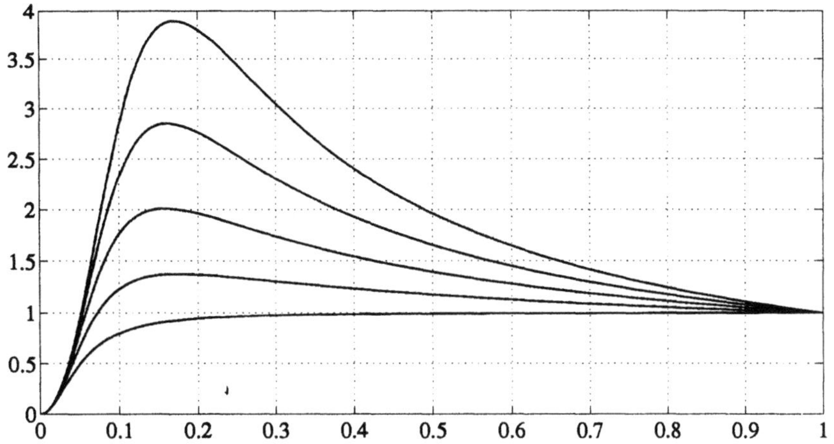

Figure 10.2 Plot of $m(x, \sigma\ ;\ \alpha, \beta, j)$ as a function of x in the range $0 < x < 1$, $\beta = 2.0$, $j = 1$ and $\sigma = 0.05$. The plot shows the function for five different values of α : 0 (bottom curve), 0.25, 0.5, 0.75 and 1.0 .

Two mapping functions

The two functions $m(x, \sigma\ ;\ \alpha, \beta, j)$ and $\mu(x\ ;\ \alpha, \beta, j)$ that will be used for mapping the orientation tensor are presented below. In both cases x is the position-dependent variable, while α, β and n are constant parameters. The use of the parameters α and β will be explained in the proper context. σ is intended to be an estimate of the noise and could be constant or vary with position. The estimation of σ, in different situations, is discussed below. The parameter j is convenient when iterative procedures are employed and normally equals the number of iterations.

$$m(x, \sigma\ ;\ \alpha, \beta, j) = \left[\frac{x^{\beta}}{x^{\beta+\alpha} + \sigma^{\beta}} \right]^{1/j} \tag{10.8}$$

The m-function is intended to control the overall high-pass content of the adaptive filter. Plots of this function for a few different parameter settings are shown in Figure 10.2.

$$\mu(x\ ;\ \alpha, \beta, j) = \left[\frac{(\, x\ (1-\alpha)\,)^{\beta}}{(\, x\ (1-\alpha)\,)^{\beta} + (\, \alpha\ (1-x)\,)^{\beta}} \right]^{1/j} \tag{10.9}$$

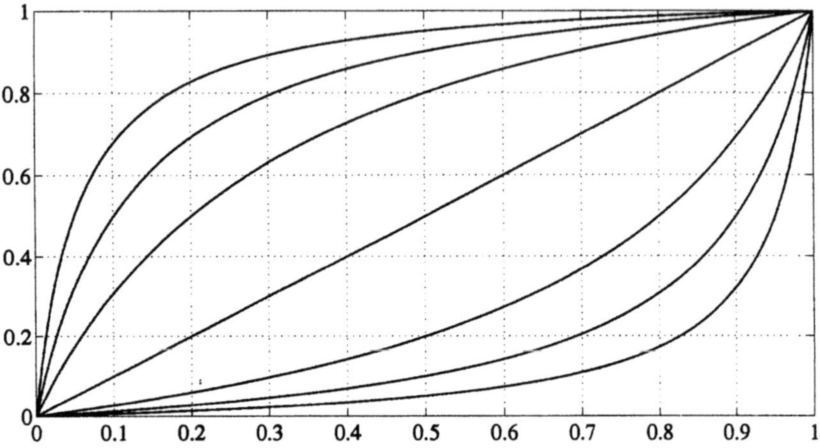

Figure 10.3 Plot of $\mu(x\ ;\ \alpha,\beta,j)$ as a function of x in the range $0 < x < 1$, $\beta = 1.0$, $j = 1.0$. The plot shows the function for seven different values of α : 0.05 (top curve), 0.1, 0.2, 0.5, 0.8, 0.9 and 0.95 .

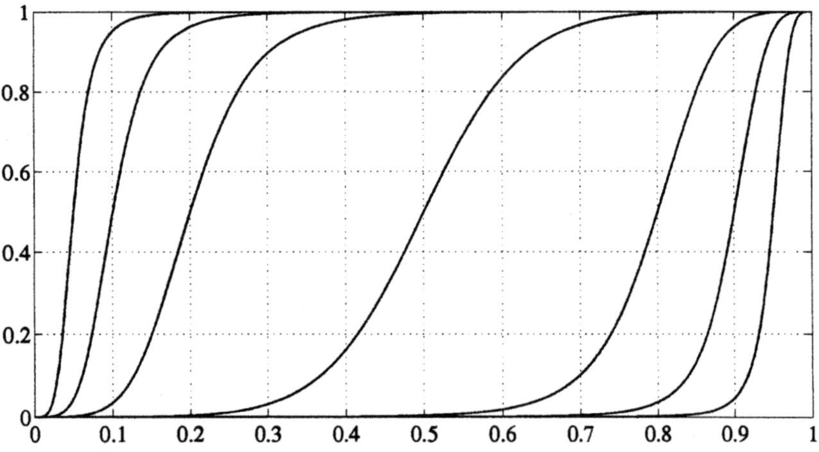

Figure 10.4 plot of $\mu(x\ ;\ \alpha,\beta,j)$ as a function of x in the range $0 < x < 1$, $\beta = 4.0$, $j = 1.0$. The plot shows the function for seven different values of α : 0.05 (top curve), 0.1, 0.2, 0.5, 0.8, 0.9 and 0.95 .

The μ-function is intended to control the orientational variation in the high-pass region of the adaptive filter. Plots of this function for different parameter settings are shown in Figures 10.3, 10.4 and 10.5.

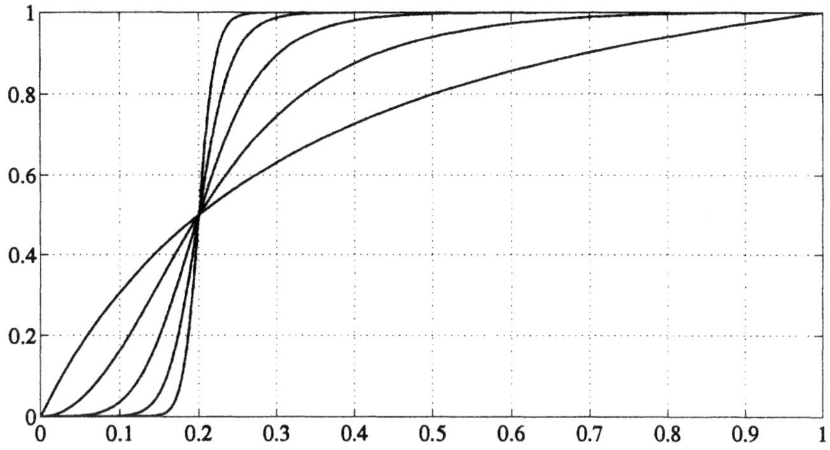

Figure 10.5 Plot of $\mu(x\ ;\ \alpha, \beta, j)$ as a function of x in the range $0 < x < 1$, $\alpha = 0.2$, $j = 1.0$. The plot shows the function for five different values of β : 1, 2, 4, 8 and 16.

10.3.3 Control of high-pass content

The largest eigenvalue, γ_1, of the control tensor will determine the overall high-pass content of the adaptive filter. More precisely, γ_1 will determine the local scaling of the entire high-pass part of the adaptive filter. γ_1 is obtained using the m-function.

$$\gamma_1 = m(\ \|\mathbf{T}_{lp}\|\,, \sigma\ ;\ \alpha_1, \beta_1, j) \tag{10.10}$$

The parameter α can be used for dynamic range compression. For example, if α is set to unity the amplification of the signal will be inversely proportional to the estimated local signal amplitude, $\|\mathbf{T}\|$. This will tend to equalize the local signal amplitude in all regions where the signal is above the noise. In general the dynamic range of the result signal will tend to be $D^{1-\alpha}$ where D is the dynamic range of the input signal.

The local estimated noise level, σ will function as a "soft threshold" and, in broad terms, define if the neighborhood should be considered to be noise or signal. The best adaptive filtering results are obtained when σ is close to the local noise level. The parameter β serves to set the slope of the m-function in the noise-signal transition area.

Local noise estimation

It is not possible to provide a general procedure for obtaining local noise estimates; the following should merely be taken as "hints" relevant to a few specific cases. Three types of noise commonly found in image data are **Additive noise, Poisson noise** and **Multiplicative noise.** It is also not uncommon for a signal to be corrupted by more than one noise source. For these classes of noise a reasonable estimate of the local noise level can be obtained from:

$$\sigma^2 = \sigma_a^2 + \sigma_p^2 s_{lp} + \sigma_m^2 s_{lp}^2 \tag{10.11}$$

where

s_{lp} is the output from the low-pass filter $F_{lp}(\rho)$

σ_a, σ_p and σ_m are constants corresponding to the level of additive, Poisson and multiplicative noise respectively.

Using Equation (10.11) as part of the noise estimation procedure is consistent with the requirement of a slowly varying adaptive filter. If the noise level has a known dependence on the signal level, s, estimation of local noise can be done in a similar manner. The essential problem of estimating the constants in Equation (10.11) still remains but this is a matter that has to be done using whatever knowledge is avaliabe in each case.

10.3.4 Control of filter shape

The ratios between the eigenvalues of a tensor can be thought of as the shape of the tensor. It is convenient to let the shape of the control tensor $\mathbf{C}$ depend only on the shape of the orientation tensor $\mathbf{T}_{lp}$.

This can be done by generating shape coefficients, μ_n, which are functions of ratios of consecutive eigenvalues of $\mathbf{T}_{lp}$.

$$\mu_n \;=\; \mu(\frac{\lambda_n}{\lambda_{n-1}}\;;\;\alpha_n,\beta_n,j) \qquad\qquad n = 2\ldots N \tag{10.12}$$

where:

λ_n are the eigenvalues of $\mathbf{T}_{lp}$.

The parameter α determines the value of $\frac{\lambda_n}{\lambda_{n-1}}$ for which the μ-function is $\frac{1}{2}$.

The parameter β is used to set the slope in the transition area.

10.3.5 The control tensor

$$\mathbf{C} \;=\; \gamma_1\;[\hat{\mathbf{e}}_1\hat{\mathbf{e}}_1^T \;+\; \mu_2\;[\hat{\mathbf{e}}_2\hat{\mathbf{e}}_2^T \;+\ldots+\; \mu_N\;[\hat{\mathbf{e}}_N\hat{\mathbf{e}}_N^T]\ldots]\;] \tag{10.13}$$

where $\hat{\mathbf{e}}_n$ are the eigenvectors of $\mathbf{T}_{lp}$.

Equation (10.13) can be rewritten as a weighted sum of outer products of the eigenvectors of the low-pass filtered orientation tensor.

$$\mathbf{C} \;=\; \sum_{n=1}^{N} \gamma_n\;\hat{\mathbf{e}}_n\hat{\mathbf{e}}_n^T \tag{10.14}$$

where

$$\gamma_n \;=\; \gamma_1\;\prod_{j=2}^{n}\mu_n \tag{10.15}$$

It is evident, from Equation (10.14), that the control tensor $\mathbf{C}$ has the same eigensystem as $\mathbf{T}_{lp}$ and that γ_n are the eigenvalues of $\mathbf{C}$. Note that the ordering of the eigenvectors is also preserved, i.e. the n:th eigenvector of $\mathbf{C}$ is identical to the n:th eigenvector of $\mathbf{T}_{lp}$. This is due to the fact that $\mu_n < 1$ if $\frac{\lambda_n}{\lambda_{n-1}} < 1$ and that γ_n is obtained as $\gamma_{n-1}\,\mu_n$.

Tensor scaling case

The coefficients, γ_n ; $n > 1$, are related to the "case coefficients" P_n in Equation (6.88). For the particularly simple case when $\alpha_2 ... \alpha_N = 0.5$ and $\beta_2 ... \beta_N = 1$ in Equation (10.9) all μ-functions become identity mappings and

$$\mu_n = \frac{\lambda_n}{\lambda_{n-1}} \qquad n = 2 \ldots N \tag{10.16}$$

which gives the following relation between the filter synthesis coefficients γ and the case coefficients P_j.

$$\gamma_n = \frac{\lambda_n}{\lambda_1} = \sum_{j=n}^{N} P_j \qquad n = 2 \ldots N \tag{10.17}$$

For this case the controlling tensor is simply an adaptively scaled version of the orientation tensor and is given by

$$\mathbf{C} = \frac{\gamma_1}{\lambda_1} \mathbf{T}_{lp} \tag{10.18}$$

10.4 ADAPTIVE FILTER SYNTHESIS

In this section it is shown how the adaptive filter can be synthesized as a weighted sum of fixed filters. More precisely the adaptive filter will be synthesized as:

$$\begin{cases} F(\mathbf{u}, \mathbf{C}) = F_{lp}(\rho) + a_{hp} \, F_{hp}(\mathbf{u}, \mathbf{C}) \\ F_{hp}(\mathbf{u}, \mathbf{C}) = \sum_k c_k \, F_k(\mathbf{u}) \end{cases} \tag{10.19}$$

Since the process is linear there is no need to actually produce the synthesized filter – the output from the synthesized filter, s', can be obtained by simply weighting the outputs from the fixed filters.

$$s' = s_{lp} + a_{hp} \sum_k c_k s_k \tag{10.20}$$

where

s_{lp} is the output from the low-pass filter F_{lp}.

s_k is the output from high-pass filter F_k.

The low-pass filter $F_{lp}(\rho)$

The low-pass filter should provide adequate filtering in regions where the local signal energy (or more precisely γ_1) is close to zero. The filter function $F_{lp}(\rho)$ can be chosen according to the specific needs in each particular case. The low-pass filter is normally designed to preserve the local mean of the signal, i.e. $F_{lp}(0) = 1.0$; if a zero mean result is desirable the low-pass filter can be omitted.

The high-pass amplification factor: a_{hp}

In the case of signal reconstruction the high-pass amplification factor, a_{hp}, is normally set to unity. In image enhancement cases it may, however, be appropriate to use significantly higher values.

The high-pass filter set: $F_k(\mathbf{u})$

The fixed high-pass filter set, $F_k(\mathbf{u})$, consists of K spherically separable filters. The required number of filters is dependent on the dimensionality of the signal.

$$F_k(\mathbf{u}) = R(\rho)\, D_k(\hat{\mathbf{u}}) \tag{10.21}$$

The radially dependent filter function, $R(\rho)$, is obtained by first choosing an "all-pass" filter function, $F_{ap}(\rho)$, that is appropriate for the regions where the local signal-to-noise ratio is very high, i.e $\gamma_1 \simeq 1$. $R(\rho)$ is then given by subtracting the low-pass filter function from $F_{ap}(\rho)$.

$$R(\rho) = F_{ap}(\rho) - F_{lp}(\rho) \tag{10.22}$$

The only restriction when choosing this function is that $F_{ap}(0) = F_{lp}(0)$. This ensures that $R(0) = 0$ which is required for the resulting spherically separable filters, $F_k(\mathbf{u})$, to be realizable; $F_k(\mathbf{u})$ will otherwise be discontinuous at $\mathbf{u} = 0$.

The set of orientationally-dependent filter functions, $D(\mathbf{u})$ is given by:

$$D_k(\hat{\mathbf{u}}) = (\hat{\mathbf{n}}_k \cdot \hat{\mathbf{u}})^2 \tag{10.23}$$

The filter directing vectors, $\hat{\mathbf{n}}_k$, and the number of filters are, in two, three, and four dimensions, identical to what was presented in the orientation estimation chapter, Equations (6.36), (6.37) and (6.38).

This set of filters spans the same function space as the union of the *spherical harmonic functions* of order 0 and order 2. This implies that the adaptive filter could be equally well synthesized using spherical harmonic functions or indeed any set of functions spanning the same space. The present choice, however, simplifies the calculations.

The signal-dependent weighting coefficients: c_k

It will be shown that the signal-dependent weighting coefficients, c_k, can be obtained by simply calculating the scalar product between the controlling tensor $\mathbf{C}$ and the filter associated tensors $\mathbf{M}_k$, i.e.

$$c_k = \mathbf{C} \bullet \mathbf{M}_k \tag{10.24}$$

The filter-associated dual tensors $\mathbf{M}_k$ are the same as in the orientation estimation case and are given by

$$\mathbf{M}_k = \alpha \, \hat{\mathbf{n}}_k \hat{\mathbf{n}}_k^T - \beta \mathbf{I} \tag{10.25}$$

where

α is $\frac{4}{3}$ in two, $\frac{5}{4}$ in three and 1 in four dimensions.

β is $\frac{1}{3}$ in two, $\frac{1}{4}$ in three and $\frac{1}{6}$ in four dimensions.

10.4.1 The tensor controlled filter

Combining Equations (10.19), (10.21) and (10.23) the adaptive high-pass filter can be expressed

$$F_{hp}(\mathbf{u}, \mathbf{C}) = R(\rho) \sum_k \mathbf{C} \bullet \mathbf{M}_k \, (\hat{\mathbf{n}}_k \cdot \hat{\mathbf{u}})^2 \tag{10.26}$$

The square of the scalar product can be rewritten as the scalar product of two outer product tensors, i.e

$$F_{hp}(\mathbf{u},\,\mathbf{C}) \;=\; R(\rho) \sum_{k} \mathbf{C} \bullet \mathbf{M}_k \; (\mathbf{N}_k \bullet \mathbf{U}) \tag{10.27}$$

where $\mathbf{N}_k = \hat{\mathbf{n}}_k \hat{\mathbf{n}}_k^T$ and $\mathbf{U} = \hat{\mathbf{u}}\hat{\mathbf{u}}^T$.

Since $\mathbf{C}$ does not depend on k, Equation (10.27) can be rewritten as

$$F_{hp}(\mathbf{u},\,\mathbf{C}) \;=\; R(\rho)\, \mathbf{C} \bullet \sum_{k} \mathbf{M}_k \; (\mathbf{N}_k \bullet \mathbf{U}) \tag{10.28}$$

By design, $\mathbf{M}_k$ are the dual elements of $\mathbf{N}_k$, implying that the sum in Equation (10.28) equals $\mathbf{U}$, see Equation (6.41). The expression thus simplifies to

$$F_{hp}(\mathbf{u},\,\mathbf{C}) \;=\; R(\rho)\, \mathbf{U} \bullet \mathbf{C} \tag{10.29}$$

Decomposing $\mathbf{C}$ and expressing $\mathbf{U}$ as an outer product yields

$$F_{hp}(\mathbf{u},\,\mathbf{C}) \;=\; R(\rho)\, \hat{\mathbf{u}}\hat{\mathbf{u}}^T \bullet \sum_{n=1}^{N} \gamma_n \hat{\mathbf{e}}_n \hat{\mathbf{e}}_n^T \tag{10.30}$$

Moving the outer product $\hat{\mathbf{u}}\hat{\mathbf{u}}^T$ inside the sum and expressing the tensor scalar products as squared vector scalar products yields the final result.

$$F_{hp}(\mathbf{u},\,\mathbf{C}) \;=\; R(\rho) \sum_{n=1}^{N} \gamma_n (\hat{\mathbf{e}}_n \cdot \hat{\mathbf{u}})^2 \tag{10.31}$$

Equation (10.31) shows that the control tensor and the adaptive filter are closely linked. The number of degrees of freedom is the same and the orientation and shape of the control tensor are directly inherited by the adaptive filter. The filter and the control tensor can in fact be thought of as completely isomorphic.

10.4.2 Range of adaptive filter shape

The shape of the adaptive filter is rotation-invariant and depends only on the eigenvalues of the control tensor. The range of different possible shapes of the adaptive filter does, of course, depend on the signal space dimensionality. The possible extremes in filter shape for two, three and four dimensions are listed below. The "names" given to the different cases refer to the the shape of the autocorrelation function of signals that have maximum gain. The exact filter shape is in each case given by Equation (10.31) and the list is only intended to serve as an intuitive guide. The interpretation of the filter shape when all dimensions are spatial is given first. The corresponding interpretation if one dimension represents time is also given.

Two-dimensional shapes:

1. Low-Pass $\gamma_1 = 0 \quad \gamma_2 = 0$
2. Oriented - Moving point $\gamma_1 = 1 \quad \gamma_2 = 0$
3. Isotropic - Velocity all-pass $\gamma_1 = 1 \quad \gamma_2 = 1$

Three-dimensional shapes:

1. Low-Pass $\gamma_1 = 0 \quad \gamma_2 = 0 \quad \gamma_3 = 0$
2. Plane - Moving line $\gamma_1 = 1 \quad \gamma_2 = 0 \quad \gamma_3 = 0$
3. Line - Moving point $\gamma_1 = 1 \quad \gamma_2 = 1 \quad \gamma_3 = 0$
4. Isotropic - Velocity all-pass $\gamma_1 = 1 \quad \gamma_2 = 1 \quad \gamma_3 = 1$

Four-dimensional shapes:

1. Low-Pass $\gamma_1 = 0 \quad \gamma_2 = 0 \quad \gamma_3 = 0 \quad \gamma_4 = 0$
2. 3D Hyperplane - Moving plane $\gamma_1 = 1 \quad \gamma_2 = 0 \quad \gamma_3 = 0 \quad \gamma_4 = 0$
3. 2D Hyperplane - Moving line $\gamma_1 = 1 \quad \gamma_2 = 1 \quad \gamma_3 = 0 \quad \gamma_4 = 0$
4. Line - Moving point $\gamma_1 = 1 \quad \gamma_2 = 1 \quad \gamma_3 = 1 \quad \gamma_4 = 0$
5. Isotropic - Velocity all-pass $\gamma_1 = 1 \quad \gamma_2 = 1 \quad \gamma_3 = 1 \quad \gamma_4 = 1$

10.4.3 Filter examples

The class of low-pass filter functions, $F_{lp}(\rho)$, and high-pass functions, $F_{hp}(\rho)$, used in the examples presented below are given by Equations (10.32) and (10.33). Figure 10.6 indicates the radial behavior of the adaptive filter for this choice of $F_{lp}(\rho)$ and $F_{hp}(\rho)$. One reason for this particular choice of $F_{hp}(\rho)$ is that in high signal-to-noise regions it is often desirable to maintain the relative frequency content. A second reason for this choice is discussed in Section 10.4.4.

$$F_{lp}(\rho) = \begin{cases} \cos^2(\frac{\pi\rho}{2\rho_{lp}}) & 0 \leq \rho < \rho_{lp} \\ 0 & \rho_{lp} \leq \rho \end{cases} \tag{10.32}$$

$$F_{hp}(\rho) = \begin{cases} 1 & 0 \leq \rho < \rho_{hp} \\ \cos^2(\frac{\pi(\rho-\rho_{hp})}{2\rho_{lp}}) & \rho_{hp} \leq \rho < \rho_{hp} + \rho_{lp} \\ 0 & \rho_{hp} + \rho_{lp} \leq \rho \end{cases} \tag{10.33}$$

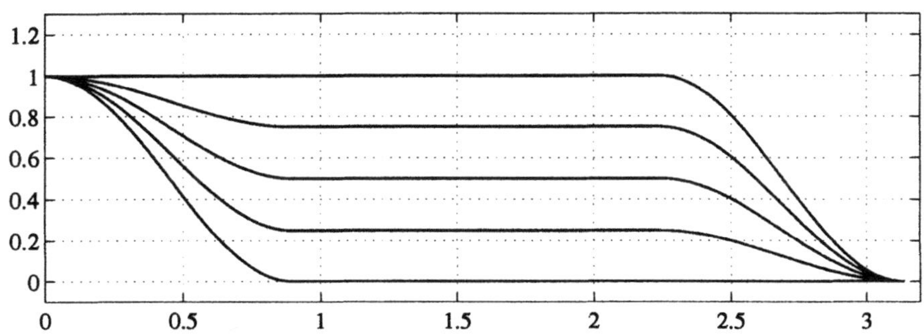

Figure 10.6 Plot showing the range of possible behavior of the adaptive filter as a function of ρ, $\rho_{lp} = 0.9$, $\rho_{hp} = 2.24$. For $\mathbf{u} = \rho\hat{\mathbf{e}}_n$ the level of the "plateau" is given by γ_n. The plot shows the radial behavior for five different "plateau" levels: 0, 0.25, 0.5, 0.75 and 1.0.

An understanding of the general behavior of the adaptive filter can be gained by studying Figures 10.6 and 10.7 together. If $\rho_{lp} < \rho < \rho_{hp}$ the adaptive filter will be constant on rays through the origin and the variation on a circle in any plane through the origin will be sinusoidal.

Figure 10.8 illustrates the degrees of freedom of a two-dimensional adaptive filter. In three dimensions it is not possible to fully represent the adaptive

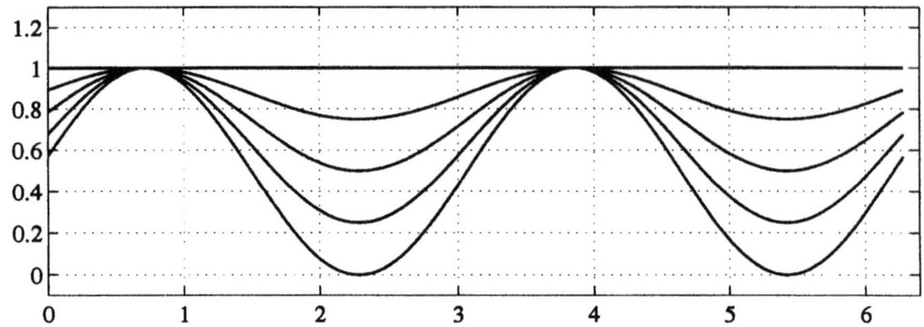

Figure 10.7 Plot indicating the range of angular variation of the high pass part of the adaptive filter. The ratio between the minimum and maximum levels is given by γ_{n+1}/γ_n for a circle in the plane spanned by $\hat{\mathbf{e}}_n$ and $\hat{\mathbf{e}}_{n+1}$ The plot shows the angular variation for five different min/max ratios: 0, 0.25, 0.5, 0.75 and 1.0.

filter function with an image. However, the isosurface plots in Figure 10.9 are intended to facilitate understanding of the three-dimensional adaptive filter range.

10.4.4 Iterating the adaptive filter

It is sometimes advantageous to pass the signal through the adaptive filter more than once. The effects of iterating the adaptive filter can be studied in Figures 10.10, 10.11 and 10.12.

Assuming that the control tensor variations are low-pass, the resulting adaptive filter, $F_J(\mathbf{u}, \mathbf{C})$, can be approximated by simply raising the single-iteration filter function to the number of iterations, J.

$$F_J(\mathbf{u}, \mathbf{C}) = \left[F_{lp} + a_{hp} R(\rho) \sum_{n=1}^{N} \gamma_n (\hat{\mathbf{e}}_n \cdot \hat{\mathbf{u}})^2 \right]^J \tag{10.34}$$

A few special cases deserve particular attention. For simplicity a_{hp} is assumed to be unity.

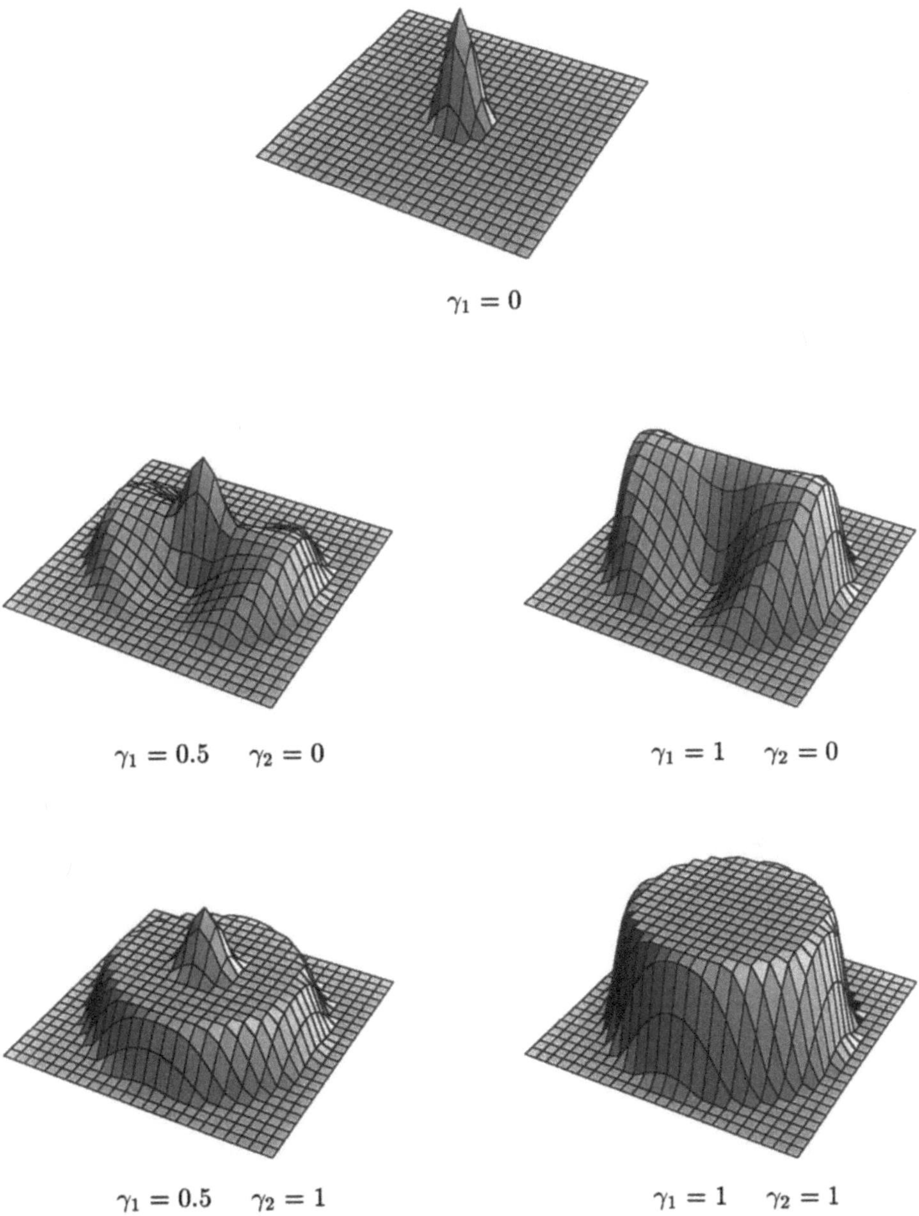

Figure 10.8 Plots of the 2D adaptive filter in the Fourier domain for different values of γ_1 and γ_2.

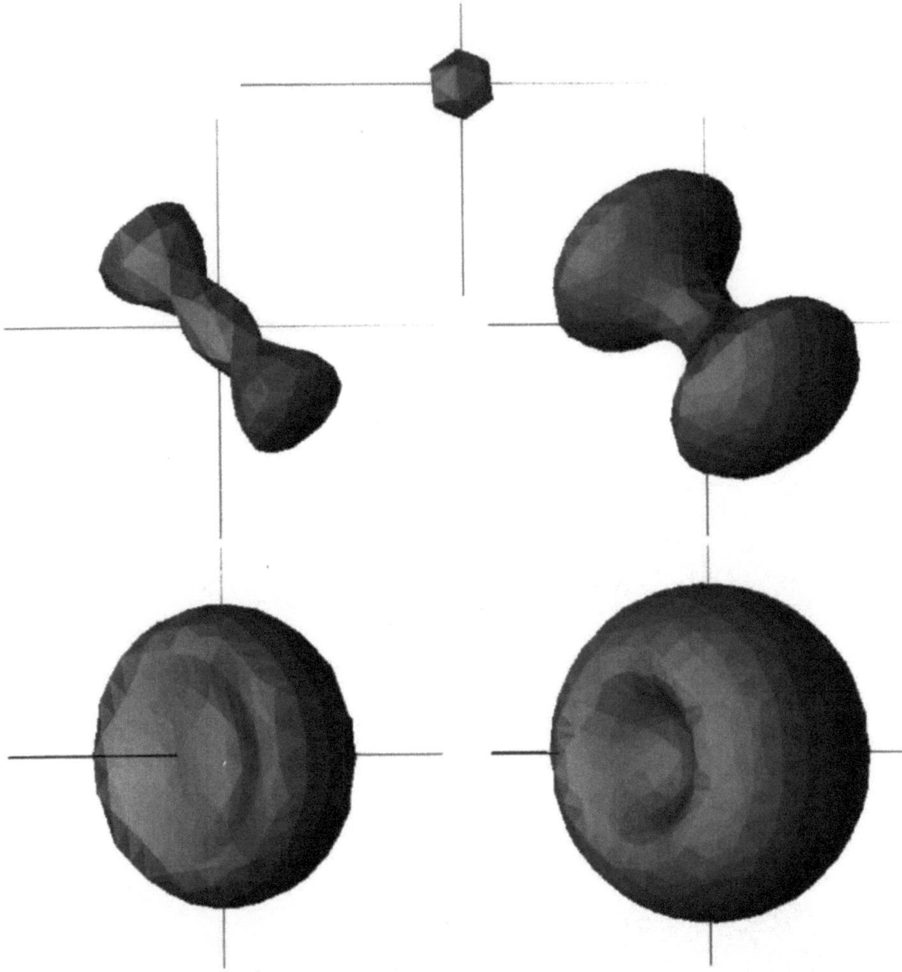

Top:	$\gamma_1 = 0$	$\gamma_2 = 0$	$\gamma_3 = 0$	Low-Pass
Middle left:	$\gamma_1 = 0.5$	$\gamma_2 = 0$	$\gamma_3 = 0$	"Plane"
Middle right:	$\gamma_1 = 1$	$\gamma_2 = 0$	$\gamma_3 = 0$	"Plane"
Bottom left:	$\gamma_1 = 1/2\sqrt{2}$	$\gamma_2 = 1/2\sqrt{2}$	$\gamma_3 = 0$	"Line"
Bottom right:	$\gamma_1 = 1/\sqrt{2}$	$\gamma_2 = 1/\sqrt{2}$	$\gamma_3 = 0$	"Line"

Figure 10.9 Isosurface plots of the 3D adaptive filter in the Fourier domain illustrating the filter shapes for different relations of the eigenvalues of the control tensor **C**. For a better understanding of the filter functions compare this figure to Figure 10.8.

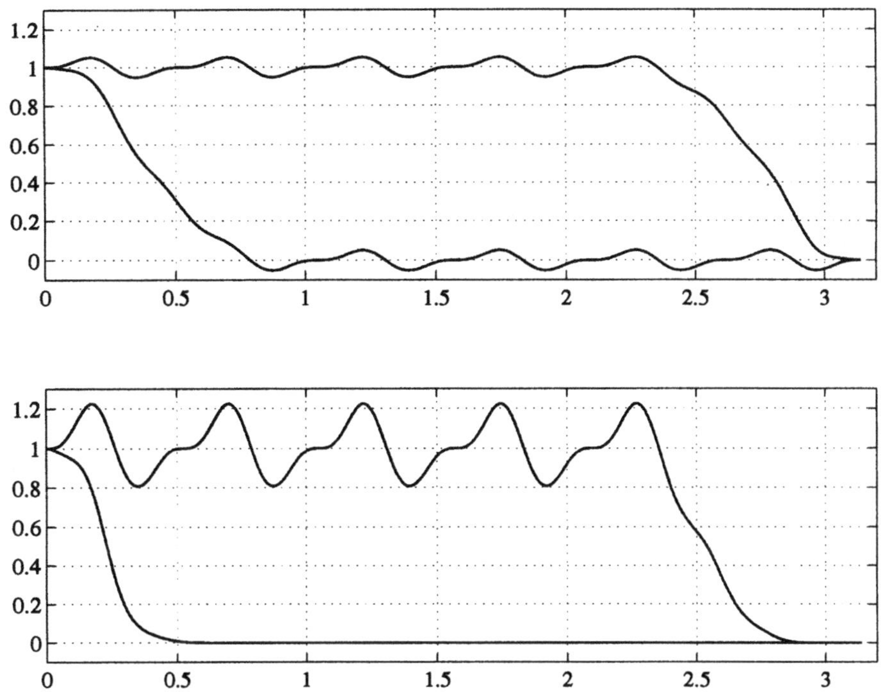

Figure 10.10 Plot showing the effect of iteration on filter ripple and filter shape for low-pass and broadband type filters. **Top:** One iteration. **Bottom:** Four iterations.

Iteration of a low-pass filter:

If $\gamma_1 \simeq 0$, implying a very low local signal-to-noise ratio, the adaptive filter will be low-pass.

$$F_{Jlp}(\mathbf{u}, \mathbf{C}) = F_{lp}^J \qquad \gamma_1 = 0 \tag{10.35}$$

Equation (10.35) implies that a more effective low-pass filtering effect can be obtained by iterating the adaptive filter. The effect will be an improved noise suppression in areas where $\gamma_1 \simeq 0$.

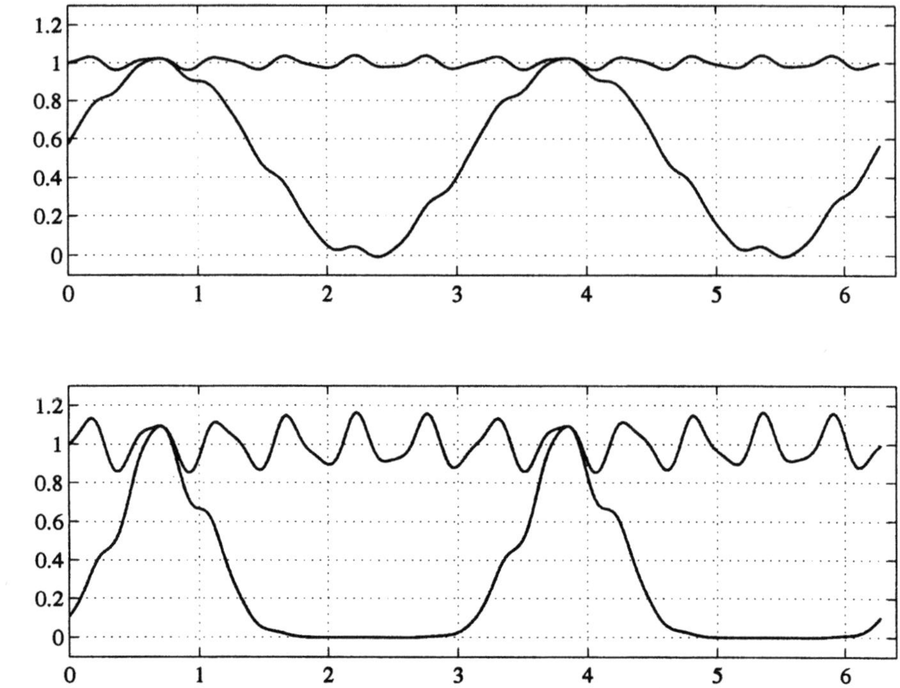

Figure 10.11 Plot showing the effect of iteration on filter ripple and filter shape for oriented and isotropic filters. **Top:** One iteration. **Bottom** Four iterations.

Iteration of a distinct filter:

A distinct filter is defined by: $\gamma_1, \ldots \gamma_m = 1$ and $\gamma_n = 0$ for $n > m$. For signals in the pass band, $\rho_{lp} < \rho < \rho_{hp}$, the iterative distinct filter, $F_{Jd}(\mathbf{u}, \mathbf{C})$, can be written:

$$F_{Jd}(\mathbf{u}, \mathbf{C}) = \left[\sum_{n=1}^{m} (\hat{\mathbf{e}}_n \cdot \hat{\mathbf{u}})^2 \right]^J \tag{10.36}$$

As in the low-pass case a "filter sharpening" effect is obtained. By iterating a distinct filter a more angular selective filter can be obtained. The gain of the simple filter will, as with the one-pass filter, be unity for signals in the subspace spanned by $\hat{\mathbf{e}}_1 \ldots \hat{\mathbf{e}}_m$ but will decay faster as the angle between the subspace

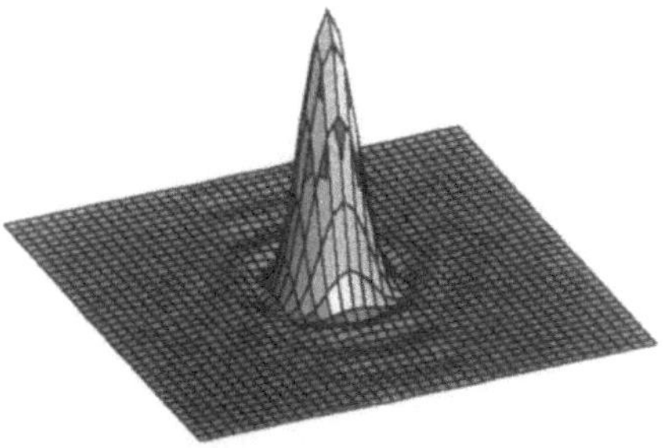

Low-Pass filter, 1 iteration

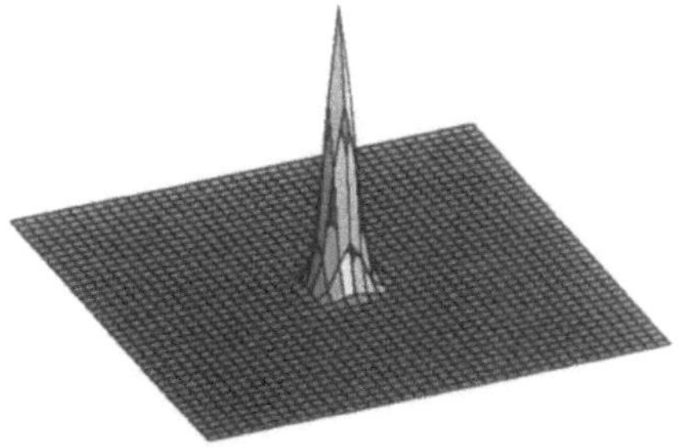

Low-Pass filter, 4 iterations

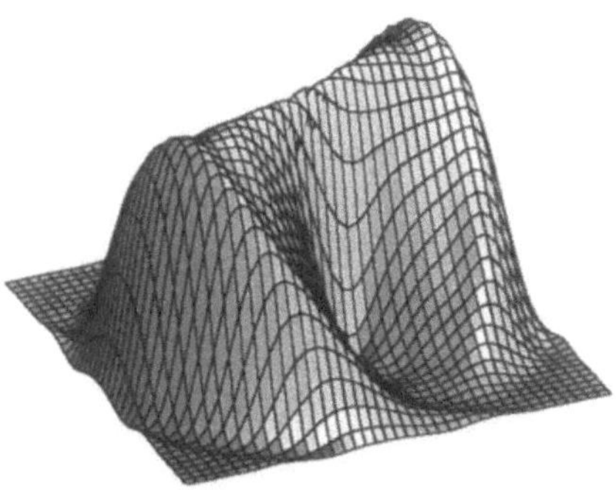

"Line" filter, 1 iteration

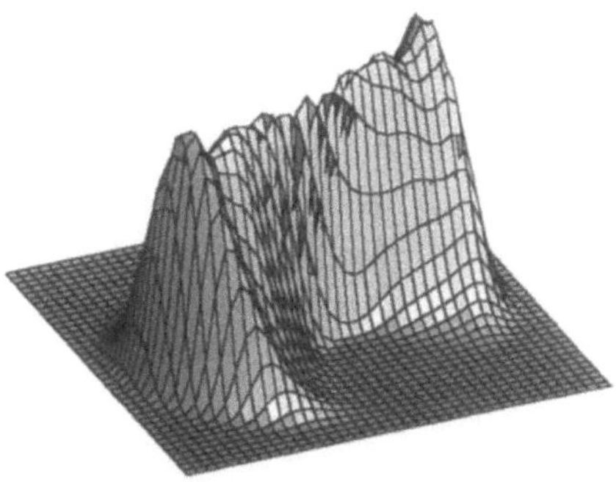

"Line" filter, 4 iterations

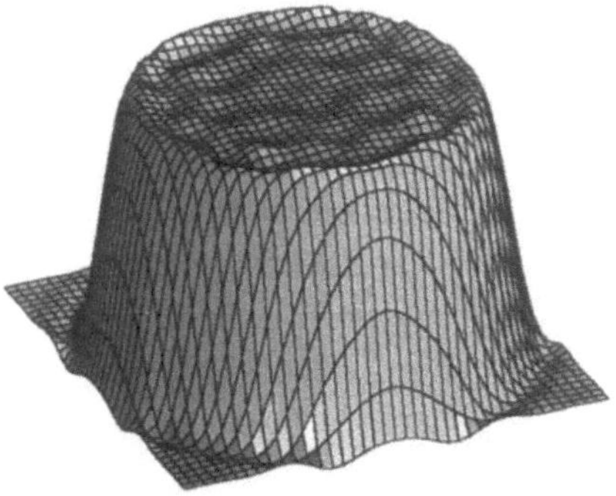

Isotropic filter, 1 iteration

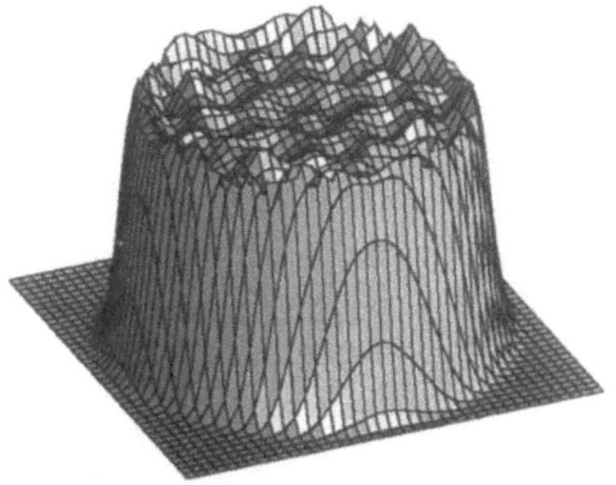

Isotropic filter, 4 iterations

Figure 10.12 **Left:**Plots of Fourier transforms of the 15 by 15 coefficient filter used in the 2D examples. **Right:**Plots of the corresponding Fourier transforms after four iterations.

and the signal increases. If $m = N$ then $\hat{\mathbf{e}}_1 \ldots \hat{\mathbf{e}}_m$ spans the entire signal space and the adaptive filter gain in the pass band will be unity in all orientations.

Iteration of a general filter:

For signals in the pass band, $\rho_{lp} < \rho < \rho_{hp}$, the general iterative filter can be written:

$$F_{Jpb}(\mathbf{u}, \mathbf{C}) = \gamma_n^J \qquad \mathbf{u} = \rho\,\hat{\mathbf{e}}_n, \quad \rho_{lp} < \rho < \rho_{hp} \tag{10.37}$$

Using Equations (10.8), (10.9) and (10.15) to express γ_n yields

$$\gamma_n^J = m(x, \sigma\ ;\ \alpha_1, \beta_1, j)^J \prod_2^N \mu(x\ ;\ \alpha_n, \beta_n, j)^J \tag{10.38}$$

The reason for introducing the parameter j in the mapping functions m and μ is now apparent. If j is set to the number of iterations, J, Equation (10.38) reduces to

$$\gamma_n^J = m(x, \sigma\ ;\ \alpha_1, \beta_1, 1) \prod_2^N \mu(x\ ;\ \alpha_n, \beta_n, 1) \tag{10.39}$$

This implies that pre-setting j to the number of iterations J will keep the gain constant in the orientations given by $\hat{\mathbf{e}}_n$ for signals in the pass band, i.e.

$$F_{Jpb}(\mathbf{u}, \mathbf{C}) = F(\mathbf{u}, \mathbf{C}) \qquad \mathbf{u} = \rho\,\hat{\mathbf{e}}_n, \quad \rho_{lp} < \rho < \rho_{hp} \tag{10.40}$$

This relation is helpful for obtaining the desired result when iterative procedures are employed. It should be noted that Equation (10.40) is true only for filter sets where $R(\rho) = 1$ in the pass band; this is a second reason for the choice of $F_{hp}(\rho)$, Equation (10.33).

Effects on filter ripple

The effects on filter ripple have to be considered when an iterative filtering scheme is used. For example, assume that the ideal gain of the adaptive filter at a given frequency $\mathbf{u}$ is g. If the implemented filter deviates from the ideal with an amount Δ, the deviation after J iterations will be

$$\Delta_J = (g+\Delta)^J - g^J \tag{10.41}$$

Two cases are of particular interest. If $g = 1$ and $\Delta \ll 1$, the deviation will increase with the number of iterations.

$$\Delta_J \simeq J\Delta \tag{10.42}$$

If $g \simeq 0$ and $\Delta \ll 1$, the deviation will be reduced with the number of iterations.

$$\Delta_J \simeq \Delta^J \tag{10.43}$$

Both of these effects can be studied in Figures 10.10, 10.11 and 10.12. The six instances of a two-dimensional adaptive filter shown in Figure 10.12 were synthesized using four (one low-pass and three high-pass) 15×15 coefficient filters. The same four filters were used in the two-dimensional adaptive filtering examples presented in Section 10.5.

10.5 EXAMPLES OF FILTERING

Orientation estimation filters

The quadrature filters used for producing the orientation tensor were optimized using a lognormal reference function. The parameters for the two-dimensional filters were: $\rho_i = \frac{\pi}{2\sqrt{2}}$, $B = 4$ and $A = 1$. The kernel size used was 15×15 coefficients. The parameters for the three-dimensional filters were: $\rho_i = \frac{\pi}{2\sqrt{2}}$, $B = 2$ and $A = 1$. The kernel size used was $9 \times 9 \times 9$ coefficients. The number of coefficient were in both cases chosen so that the optimized filters had a reasonable fit to the reference function.

Tensor smoothing filter

The filter used for tensor low-pass filtering in the two-dimensional as well as in the three-dimensional examples were truncated isotropic Gaussian filters.

The spatial standard deviations used in the two-dimensional examples were 4.0 (inter pixel distances) for the texture generation, 2.5 for the face image and 1.5 for the boat image. The filters were implemented using 15 × 15, 11 × 11 and 9 × 9 coefficients respectively. The spatial standard deviation used in the three-dimensional examples was 1.3 (inter voxel distances). The filter was implemented using 7 × 7 × 7 coefficients.

Filters for adaptive filtering

The filter sets used for adaptive filtering were designed according to Equations (10.32) and (10.33). The parameters for the two-dimensional filters were: $\rho_{lp} = 0.9$ and $\rho_{hp} = 2.24$. The kernel size used was 15 × 15 coefficients. The parameters for the three-dimensional filters were: $\rho_{lp} = 1.0$ and $\rho_{hp} = 2.14$. The kernel size used was 9 × 9 × 9 coefficients. The number of coefficient were in both cases chosen so that the optimized filters had a good fit to the reference function.

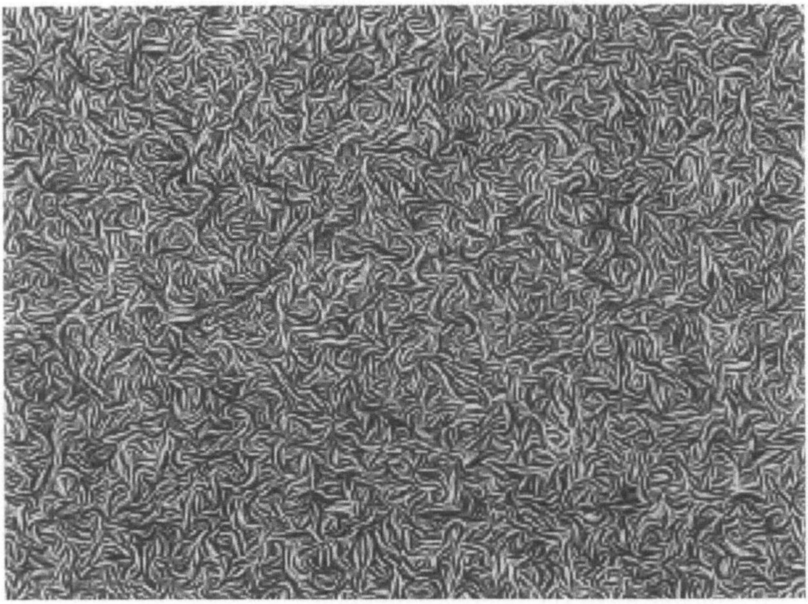

Figure 10.13 Example of texture generation.

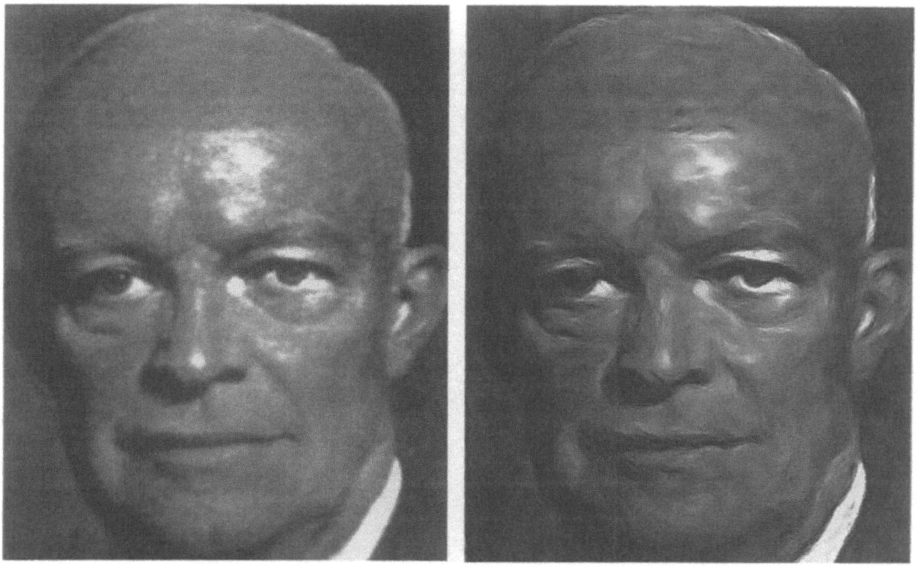

Figure 10.14 **Left:** D. D. Eisenhower. **Right:** Image obtained after adding noise and iterating the adaptive filtering procedure four times. See text for details.

10.5.1 Examples of two-dimensional filtering

Four two-dimensional examples intended to demonstrate a few possible modes of operation of the adaptive filtering technique are given below.

Texture generation

Figure 10.13 shows a texture field generated by iterated adaptive filtering. The control tensor was obtained by estimating local orientation in a white noise image and smoothing the result using a relatively large low-pass filter. The parameter σ in the m-function was set to zero implying maximally anisotropic filtering everywhere. The white noise image was then used as input to the adaptive filter and iterated 8 times.

Illustration of tensor mediated information

Figure 10.14 is intended to facilitate understanding of the type of information that is mediated by the orientation tensor. The control information was ob-

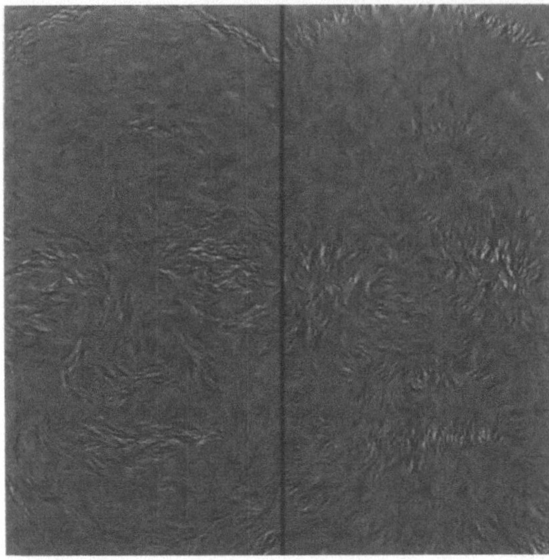

Figure 10.15 **Left:** The filtered noise component of the result in Figure 10.14. **Right:** The filtered noise with orientation information altered by 90 deg.

tained from the left image. The right image was obtained by iterating the adaptive filter four times using a 0 dB SNR image, consisting of the left image plus white noise, as input (from [76]). The left half of Figure 10.15 shows the filtered noise component of the result in Figure 10.14. The right half shows the filtered noise when orientation information was deliberately altered by 90 deg (from [76]).

Enhancement examples

Two examples of results from using the adaptive filtering technique for enhancement are given. Figure 10.16 shows the original used in the first example and Figure 10.17 shows the result. The main purpose of this example is to display the achievable level of noise reduction.

The second example, found in the color plate section, Color image 2, shows the effect of dynamic compression. The input image is a picture of a landscape where visibility of the distant hills is fading. Using dynamic compression enables the weak signals to be significantly amplified without saturating the strong signals from the nearby objects. The dynamic compression factor, α in the m-function, was in this example set to 0.7 reducing the dynamic

Figure 10.16 Original noisy image.

Figure 10.17 Result after adaptive filtering.

range to about 1/3 in logarithmic terms. The parameter σ was set to 0.01 of the maximum signal level. The remaining parameter settings are given by: $\beta_1 = 2$, $\alpha_2 = 0.7$, $\beta_2 = 2$ and $j = J = 1$.

10.5.2 Examples of three-dimensional filtering

Three examples of adaptive filtering of three-dimensional data is given below. The first two are computer generated test volumes, the third example is an ultrasonic sequence of a beating heart (from [47]).

The first test volume consists of a spherically symmetrical signal having a sinusoidal variation in the radial direction. The frequency of the sinusoid is high in the center of the volume and decreases toward the periphery. In this way the volume contains all possible three-dimensional plane orientations for a wide frequency range. Sixteen slices of the original volume is shown in Figure 10.18. The test volume was corrupted with white additive Gaussian noise to a signal-to-noise ratio of 0 dB, see Figure 10.20. The corresponding slices from the enhanced volume are shown in Figure 10.21. To illustrate the noise performance of the algorithm, the signal-to-noise ratios before and after the restoration are shown in Table 10.1 (from [47]). When regarded as an image

Input SNR	Output SNR
∞ dB	21.0 dB
20 dB	21.0 dB
10 dB	18.5 dB
0 dB	9.8 dB

Table 10.1 Table of SNR ratios before and after restoration of the volume in Figure 10.18

sequence the second test pattern consists of 10 points moving toward the center of the image. The velocity of the points varies from 0.24 to 9.4 pixels/frame. The spatiotemporal volume has been corrupted with white Gaussian noise to a SNR of -4 dB. Sixteen slices from the original sequence and sixteen slices from the filtered sequence are shown in Figure 10.19 and Figure 10.23.

Adaptive filtering of an ultrasonic sequence

The third example is an ultrasonic sequence of a beating heart. Figure 10.24 contains four frames from the original sequence. The low-signal-to-noise ratio of the original sequence is inherent to the imaging method. Adaptive filtering significantly reduces the noise, as is seen in Figure 10.25.

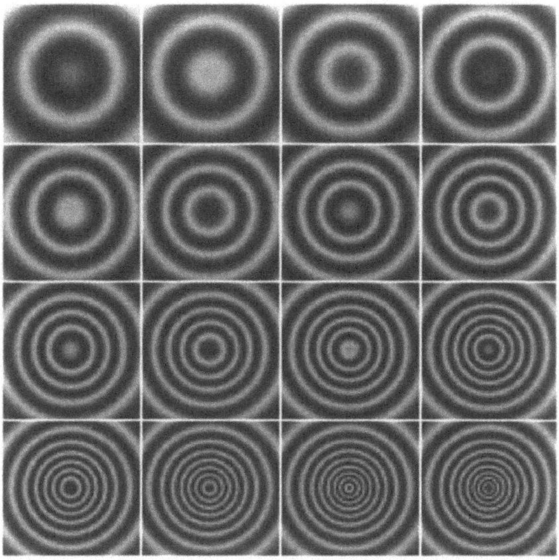

Figure 10.18 Sixteen slices from the spherically symmetric test volume.

Figure 10.19 Sixteen slices from the moving points test sequence.

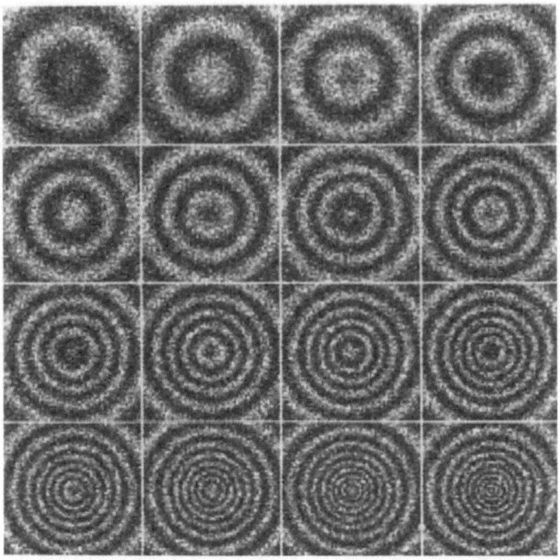

Figure 10.20 Sixteen slices from the test pattern corrupted with white additive Gaussian noise, SNR= 0dB.

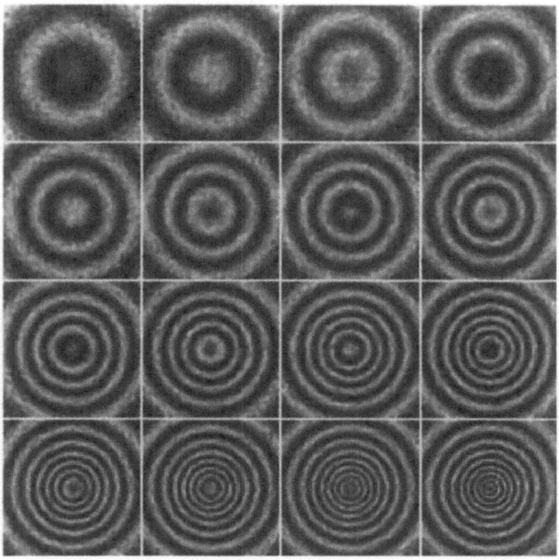

Figure 10.21 Result from three-dimensional adaptive filtering of the test pattern in Figure 10.20.

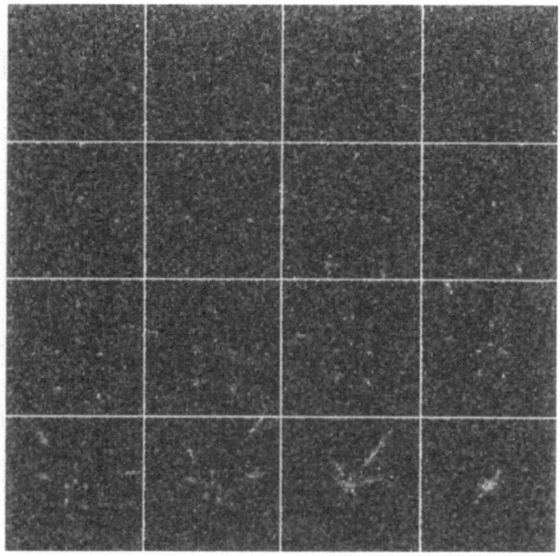

Figure 10.22 Sixteen slices from the test pattern corrupted with white additive Gaussian noise, SNR= -4dB.

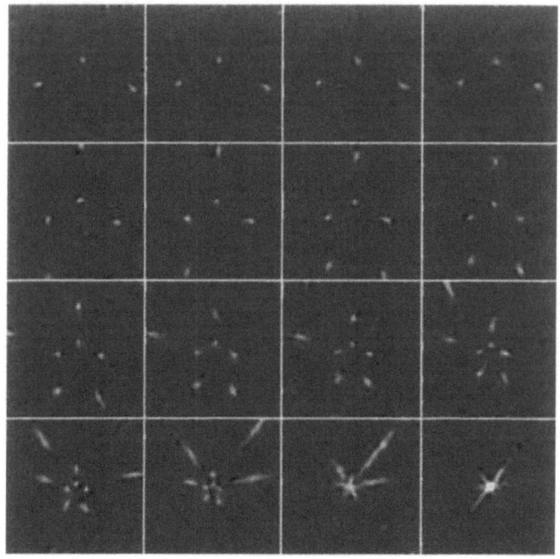

Figure 10.23 Result from three-dimensional adaptive filtering of the test pattern in Figure 10.22.

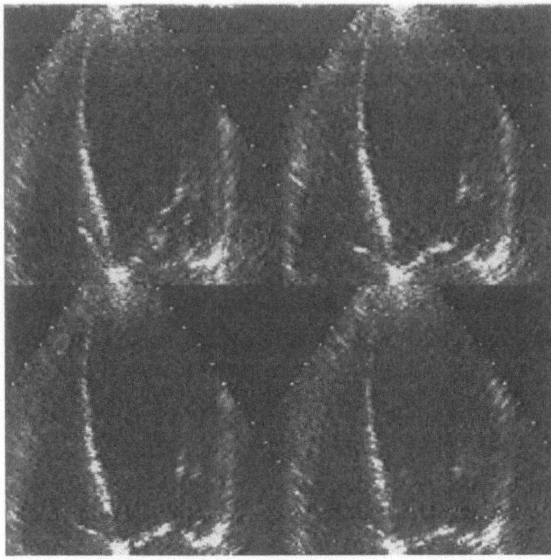

Figure 10.24 Four original frames from an ultrasonic image sequence of a beating heart.[1]

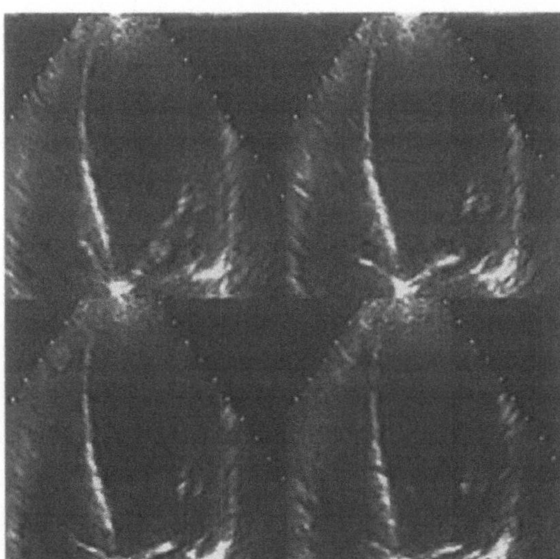

Figure 10.25 Result of three-dimensional adaptive filtering of the ultrasonic sequence in Figure 10.24.

[1] Provided by Tomas Gustavsson, Chalmers Institute of Technology, Gothenburg, Sweden.

11

VECTOR AND TENSOR FIELD FILTERING

11.1 INTRODUCTION

In the preceding chapters, we have been concerned with operations acting on multi-dimensional scalar images. Recall, for example, the orientation algorithm estimating the local orientation in Chapter 6 and the algorithm estimating local frequency in Chapter 8. In this chapter, we describe some algorithms acting on vectors and higher order tensors.

It is shown that the orientation algorithm presented in Chapter 6 can be extended to work on higher order fields. This can be useful, for example, for detection of borders between homogeneous vector or tensor fields. An application of this is described in Chapter 13, where texture borders are defined as sudden changes in a feature vector descriptor field. Examples of operators acting on vector data performing circular symmetry estimation and curvature estimation will also be discussed.

11.2 ORIENTATION IN VECTOR AND TENSOR FIELDS

Lines and edges are naturally defined in gray-level images. However, these concepts do not extend directly to higher order fields since variation of the local data may have more than one direction of change. For example, the gradient of a vector field is not a new vector, but a tensor of order two. In two dimensions, this tensor describes the direction of the maximum change and

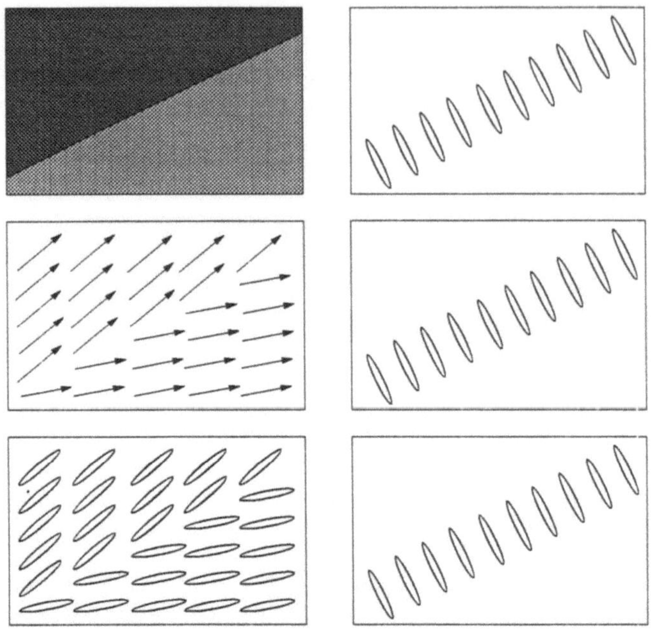

Figure 11.1 **Left**: Three *simple functions* showing edges in scalar, vector, and tensor fields. **Right**: The corresponding orientation tensor defining the location and orientation of the borders.

the (orthogonal) direction of minimum change. The corresponding eigenvalues describe the rate of change in each direction.

Although there are some difficulties in defining edges and lines in vector and higher order tensor fields, it can be done. If a neighborhood can be modeled as a *simple function*, the direction of change is uniquely defined even in higher order tensor fields. This function was defined for complex functions in Chapter 4. A higher order simple function is similarly defined by

$$\mathbf{S}(\boldsymbol{\xi}) = \mathbf{G}(\boldsymbol{\xi}^T \hat{\mathbf{e}}) \tag{11.1}$$

where $\boldsymbol{\xi}$ is the spatial signal coordinate and $\hat{\mathbf{e}}$ is the direction of change in the signal. $\mathbf{S}$ is a simple function if there exists a function $\mathbf{G}$ so that

$$\begin{aligned} \mathbf{S} : \quad & \mathbb{R}^n \rightarrow \text{tensors} \\ \mathbf{G} : \quad & \mathbb{R} \rightarrow \text{tensors} \end{aligned}$$

where n is the spatial dimension of the signal. This definition ensures that the function values (scalars, vectors or tensors) are changing only when moving in the direction of $\hat{\mathbf{e}}$. The values are constant on all hyperplanes that are normal to this vector.

Three examples of simple functions and their stylized orientation estimates are found in Figure 11.1.

11.2.1 Estimation of local orientation in higher order fields

If the input signal $\mathbf{S}$ is simple, all the scalar component fields also are simple and defined by the same direction vector, $\hat{\mathbf{e}}$, as $\mathbf{S}$,

$$\mathbf{S}(\boldsymbol{\xi}) = \mathbf{G}(\boldsymbol{\xi}^T\hat{\mathbf{e}}) \quad \rightarrow \quad S_{mn\ldots o}(\boldsymbol{\xi}) = G_{mn\ldots o}(\boldsymbol{\xi}^T\hat{\mathbf{e}}) \tag{11.2}$$

where the right expression is a coordinate description of the tensors. Dividing the signal into scalar components makes it possible to apply well-known Fourier techniques to the signals. For each of the scalar components fields, $S_{mn\ldots o}$, we know that the Fourier transform is concentrated on a line in the Fourier domain. The direction of this line can be estimated using the orientation algorithm for scalar fields presented in Chapter 6,

$$\mathbf{T} = \sum_k q_k \mathbf{M}_k \tag{11.3}$$

where q_k are the filter response from quadrature filter k

Each scalar component field is either a constant function or a simple function. These simple functions are constant on hyperplanes that are normal to the same vector as the one defining the direction of the higher order simple function. Thus, estimating the orientation in each component field gives

$$a_{mn\ldots o}\hat{\mathbf{T}} = \sum_k q_{kmn\ldots o}\mathbf{M}_k \tag{11.4}$$

where $q_{kmn\ldots o}$ are the filter magnitude responses from quadrature filter k when acting on the scalar component fields $S_{mn\ldots o}$ and where $a_{mn\ldots o}$ are scalars ≥ 0.

A linear summation of the filter output magnitudes gives

$$\sum_{mn\ldots o} a_{mn\ldots o}\hat{\mathbf{T}} = b\hat{\mathbf{T}} \tag{11.5}$$

where b is a scalar value ≥ 0. Although this equation gives a correct estimate of local orientation, the strength may be affected by a rotation of the input signal. Applying an orthonormal operator $\mathbf{A}$ to the input field $\mathbf{G}$ should not affect the strength, b, of the estimate. This requirement is met by the following equation

$$b\hat{\mathbf{T}} = \sum_k \Big(\sum_{mn\ldots o} q^2_{kmn\ldots o} \Big)^{\frac{1}{2}} \mathbf{M}_k \tag{11.6}$$

A vector field example

In this section we apply the algorithm presented in the previous chapter on a vector test image containing a circular vector edge (Figure 11.2). This figure also shows two component fields. The image size is 128 × 128 pixels.

In the vector case, $\mathbf{S}$ has only one index, S_l. Equation (11.6) then reduces to

$$\mathbf{T} = \sum_{k=1}^{4} \Big(\sum_{l=1}^{2} q^2_{kl} \Big)^{\frac{1}{2}} \mathbf{M}_k \tag{11.7}$$

where q_{kl} denotes the filter magnitude response of filter k obtained from filtering on vector component S_l of the input data.

Figure 11.3 shows the result of applying the orientation algorithm to the respective scalar component fields. To the right in the figure the norm of the component orientation tensors are shown, and to the left the fields of eigenvectors multiplied by the corresponding largest eigenvalue in double angle are shown. The quadrature filters used in this example have the size 15 × 15. Figure 11.4 shows the combined estimate, $\mathbf{T}$.

Another example using orientation estimation in vector fields is shown in Figure 13.5 where this technique is used for texture segmentation.

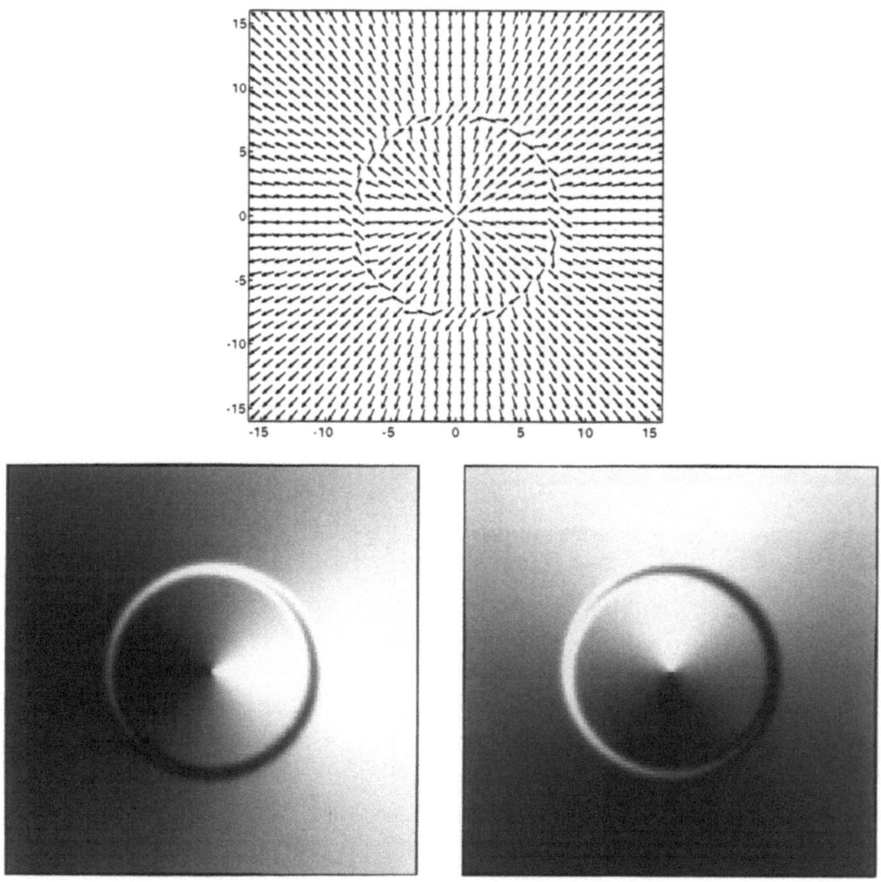

S_1 S_2

Figure 11.2 **Top**: A vector test pattern containing a circular vector edge. **Left**:Horizontal vector component of the test image. **Right**: Vertical vector component of the test image. Dark means negative values, gray means zero and bright means positive values.

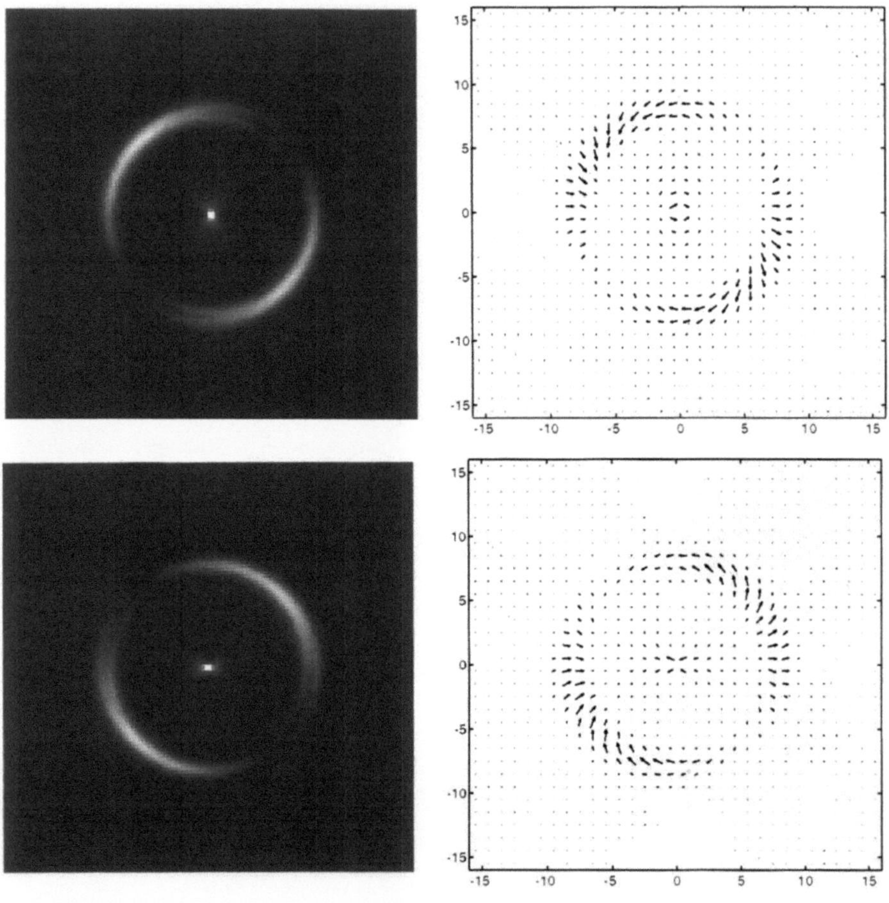

Figure 11.3 Orientation estimation on the individual components.

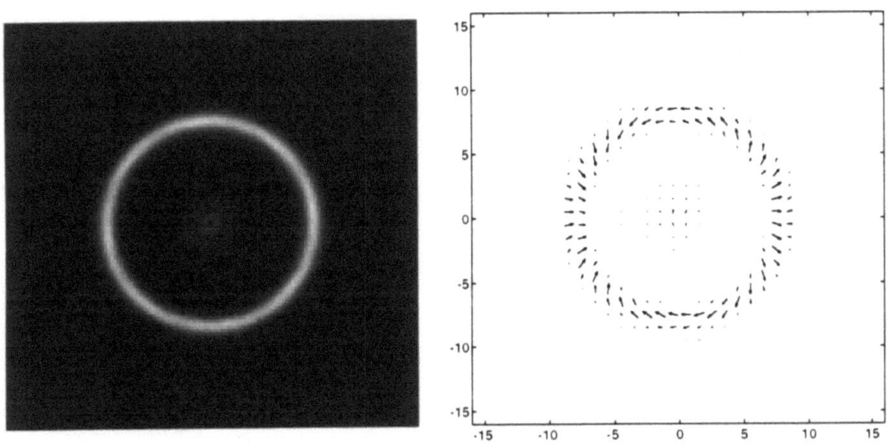

Figure 11.4 **Left**: Norm of the estimated orientation tensor field. **Right**: Eigenvector times the corresponding largest eigenvalue in double angle.

11.3 ROTATIONAL SYMMETRIES

The vectors described in this and in the preceding sections are two-dimensional, i.e. they can be described by pairs of scalar numbers. In two dimensions it is often convenient to work with an analogous description using complex numbers.

$$\mathbf{v} = v_1\mathbf{x}_1 + v_2\mathbf{x}_2 \longleftrightarrow v_1 + iv_2 \tag{11.8}$$

where $\mathbf{x}_1$ and $\mathbf{x}_2$ are basis vectors. The representation of vectors as complex numbers is used in this chapter.

The class of different symmetries is large. Here we shall limit ourselves to the study of polar separable functions, whose angular variation is defined by circular harmonic functions.

$$e^{ik\varphi} \tag{11.9}$$

The angular modulation speed, k, determines the family of symmetries to which the filter is sensitive. Two important ones are 1st order symmetries ($k = 1$) also termed *parabolic* symmetries, and 2nd order symmetries ($k = 2$) also termed *circular* symmetries. The names of the symmetries refer to the gray-level patterns the filters match to in terms of the double angle orientation representation used, although they are applied on vector data produced by the two-dimensional orientation algorithm described in Chapter 6.

11.3.1 2nd order rotational symmetries

The simplest rotationally symmetry objects are circles with centers at the origin. Figure 11.5 (left) shows such a gray-level neighborhood. The corresponding orientation representation is described in terms discussed earlier:

$$\mathbf{z}_1 = ce^{i2\varphi} \tag{11.10}$$

where c is a constant magnitude. It has been stated that maximally different or *maximally incompatible* patterns should be maximally different in the representation space [42]. Now, the notion of compatibility of patterns is highly subjective, but in certain simple cases there are "natural" ways of defining pairs of maximally different patterns. Conceptually, the most incompatible pattern

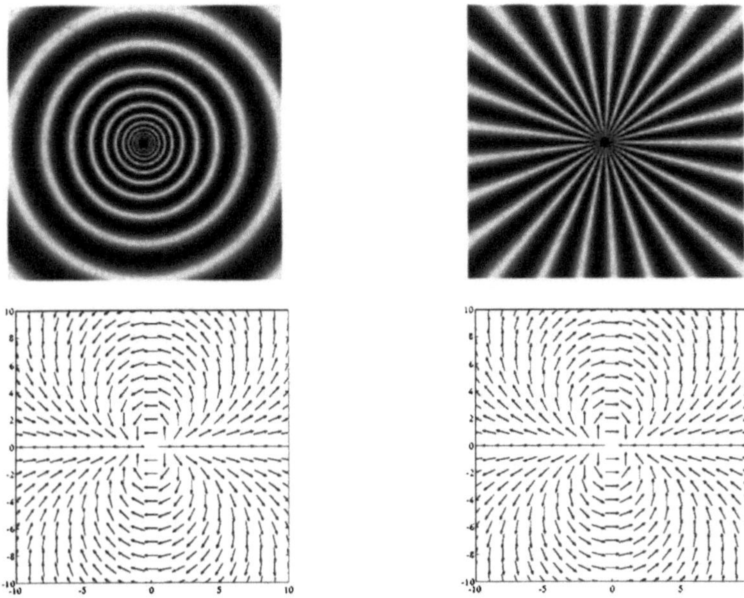

$\mathbf{z}_1 \;=\; ce^{i2\varphi}$ $\qquad$ $\mathbf{z}_3 \;=\; ce^{i2(\varphi+\frac{\pi}{2})}$

Figure 11.5 Two circular symmetries and the corresponding local orientation representation. Note that the right vector field is -1 times the left one, i.e. every vector are differs maximally between the two fields.

to a circle is a star, since the star structure is orthogonal to the circular structure (Figure 11.5, right). The vectorial orientation representation of the star pattern is thus generated by adding $\frac{\pi}{2}$ to φ (Figure 11.5, right) ,

$$\mathbf{z}_3 \;=\; ce^{i2(\varphi+\frac{\pi}{2})} \tag{11.11}$$

Note that this gives vector representations that are maximally different for the two patterns. The vectors are in every point opposite to each other.

A filter function matched to the signal $\mathbf{z}_1$ would be a filter having the same angular variation as $\mathbf{z}_1$, with some suitable lowpass magnitude function limiting the spatial extent of the filter,

$$\mathbf{b}_2 = a(r)e^{i2\varphi} \tag{11.12}$$

where the subscript indicates the order of the symmetry. Filtering a signal $\mathbf{z} = z(r)e^{i\varphi_z}$ with this filter function can be expressed in terms of the scalar product between the signal and the filter,

$$\langle \mathbf{z}, \mathbf{b}_2 \rangle = \int \mathbf{z}(\mathbf{x})\mathbf{b}_2^*(\mathbf{x})d\mathbf{x} \tag{11.13}$$

where $*$ denotes complex conjugate. The convention used is that the scalar product is linear in the first argument and anti-linear in the second. Assuming some suitable discretization of the pattern and the filter this scalar product may be expressed by a product sum,

$$\langle \mathbf{z}, \mathbf{b}_2 \rangle = \sum_k \mathbf{z}_k \mathbf{b}_2^* = \sum_k c_k a_k e^{i((\varphi_z)_k - (\varphi_b)_k)} \tag{11.14}$$

where φ_b denotes the phase of the filter $\mathbf{b}_2$. Note that the complex conjugate of the filter is used in the product sum. If the signal has the same angular variation speed as the filter function but differs by a phase shift of α,

$$\varphi_z = 2\varphi + \alpha \tag{11.16}$$

$$\varphi_b = 2\varphi \tag{11.17}$$

then the filter output from Equation (11.14) gives

$$\sum c_k a_k e^{i(2\varphi_k + \alpha - 2\varphi_k)} = e^{i\alpha} \sum c_k a_k = s e^{i\alpha} \tag{11.17}$$

This shows that the filter, $\mathbf{b}_2$, proposed provides an unambiguous discrimination between the whole class of different second order symmetries.

$$\mathbf{z} = c e^{i(2\varphi+\alpha)}, \qquad \alpha \in [0, 2\pi[\tag{11.18}$$

The argument of the complex filter output describes the type of pattern. Figure 11.6 shows gray-level images corresponding to second order symmetry patterns described by $\alpha = \frac{\pi}{4}$ and $\alpha = \frac{3\pi}{4}$,

$$\mathbf{z}_2 = c e^{i2(\varphi + \frac{\pi}{4})} \quad \text{and} \quad \mathbf{z}_4 = c e^{i2(\varphi + \frac{3\pi}{4})} \tag{11.19}$$

We can see that the descriptor allows us to discriminate between various classes of patterns. In addition to circular and radial patterns, we can discriminate between left-handedness and right-handedness. The pattern may in any given case be any combination of the these four, excepting a combination of two opposites that would cancel each other, such as a left-handed and right-handed pattern of the same magnitude. The resulting vector representation describing circular symmetries is shown in Figure 11.7.

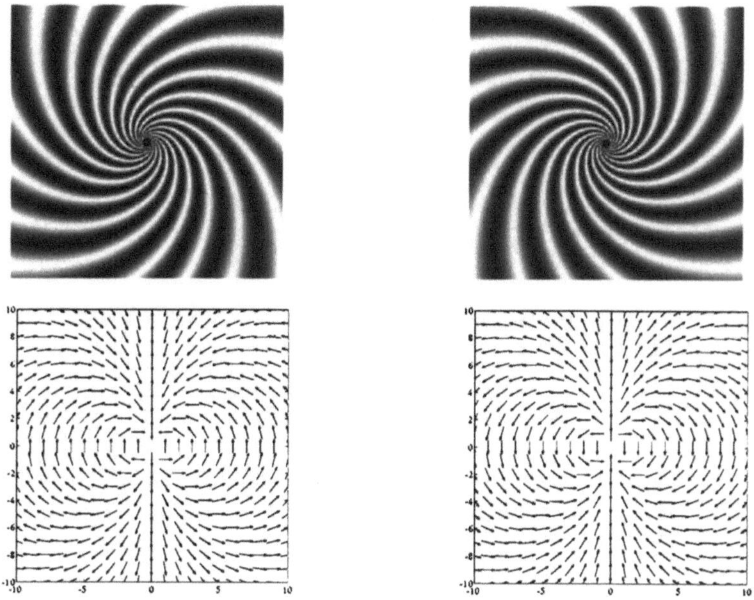

$$\mathbf{z}_2 = ce^{i2(\varphi+\frac{\pi}{4})} \qquad\qquad \mathbf{z}_4 = ce^{i2(\varphi+\frac{3\pi}{4})}$$

Figure 11.6 Two circular symmetries and their corresponding local orientation representation. Note that the right vector field is -1 times the left one, i.e. every vector differs maximally between the two fields.

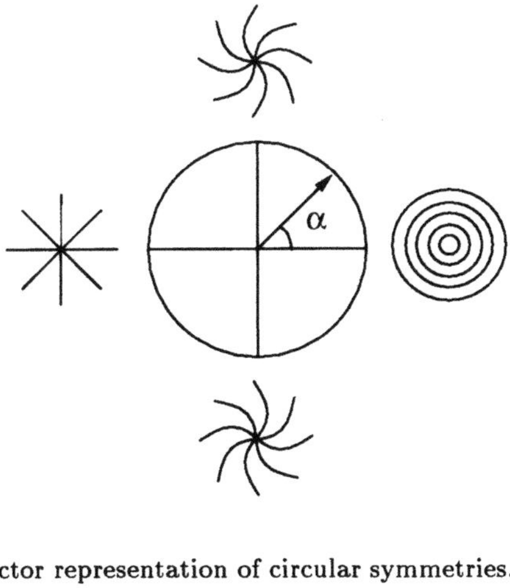

Figure 11.7 Vector representation of circular symmetries.

11.3.2 1st order symmetries

For 1st order symmetries or parabolic symmetries we will simply assume that we have patterns that, in the vectorial representation given earlier, can be written as

$$\mathbf{z} = ce^{i\varphi+\alpha} \tag{11.20}$$

Gray-level patterns producing this type of local orientation description are found in Figure 11.8 with $\alpha = \pi$ (left) and $\alpha = \frac{3\pi}{2}$.

What distinguishes Equation (11.20) from the corresponding equation in the preceding section, Equation (11.10), is the fact that the vector rotates with a rate of φ, rather than 2φ. As in the preceding case, the detection of these and intermediary classes of patterns can be achieved through filtering with a complex filter that matches the signal in Equation (11.20),

$$\mathbf{b}_1 = a(r)e^{i\varphi}, \tag{11.21}$$

where the subscript indicates the order of the symmetry. As in the previous section, the filtering is expressed as a scalar product between the signal, $\mathbf{z} = z(r)e^{i\varphi_z}$, and the filter. In the discrete case we get,

$$\langle \mathbf{z}, \mathbf{b}_1 \rangle = \sum_k \mathbf{z}_k \mathbf{b}_1^* = \sum_k c_k a_k e^{i((\varphi_z)_k - (\varphi_b)_k)} \tag{11.22}$$

where φ_b denotes the phase of the filter $\mathbf{b}_1$. If the signal has the same angular variation speed as the filter function but differs by a phase shift of α,

$$\varphi_z = \varphi + \alpha \tag{11.24}$$
$$\varphi_b = \varphi \tag{11.25}$$

then the filter output from Equation (11.22) gives

$$\sum c_k a_k e^{i(\varphi_k + \alpha - \varphi_k)} = e^{i\alpha} \sum c_k a_k = se^{i\alpha} \tag{11.25}$$

This shows that the filter, $\mathbf{b}_1$, proposed provides an unambiguous discrimination between the whole class of different first order symmetries.

$$\mathbf{z} = ce^{i(\varphi+\alpha)}, \qquad \alpha \in [0, 2\pi[\tag{11.26}$$

The operation presented in this section describes curvature, more specifically convexity and concavity, and the direction towards the center of concavity or outward from the center of convexity. The vector representation used to describe parabolic symmetries is illustrated in Figure 11.9.

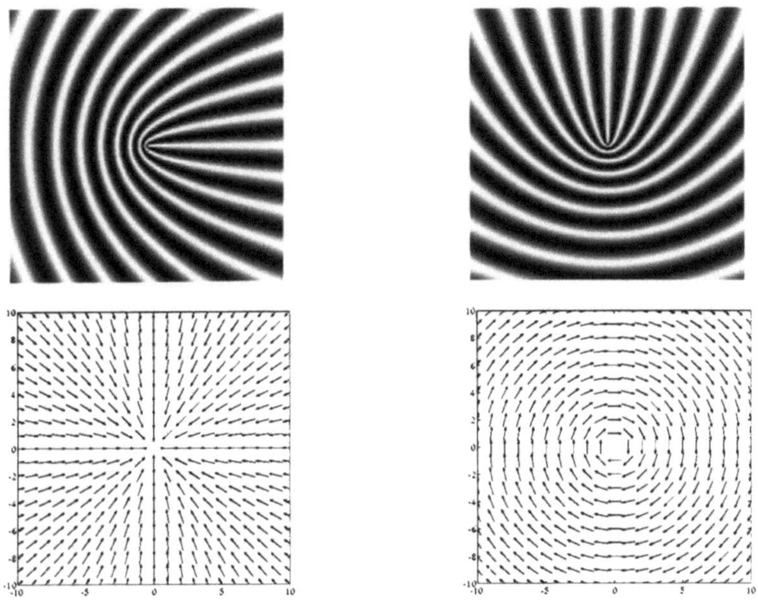

$$\mathbf{z}_3 = ce^{i\varphi+\pi} \qquad \mathbf{z}_2 = ce^{i\varphi+\frac{3\pi}{4}}$$

Figure 11.8 Two different parabolic symmetries and the corresponding local orientation representations.

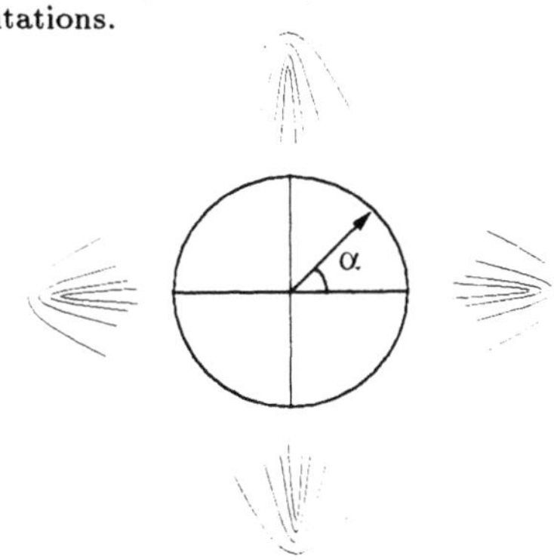

Figure 11.9 Vector representation of parabolic symmetries.

11.3.3 Symmetry filters

As mentioned above, the filter needed for estimation of a symmetry is a filter that is the complex conjugate of the pattern. Figure 11.10 plots complex conjugate of the parabolic and rotational filters.

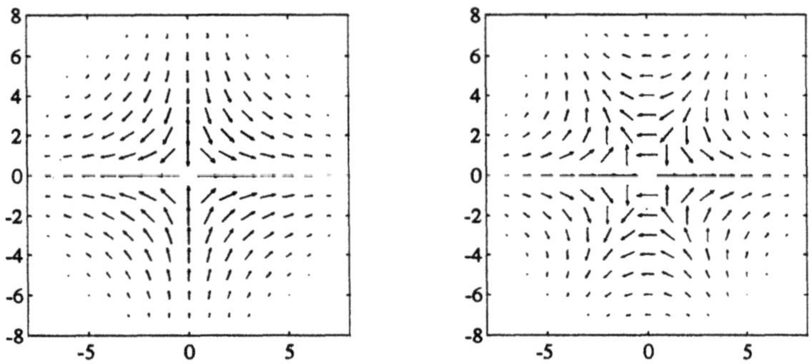

Figure 11.10 Filters for detecting symmetries. Left: Parabolic symmetries, $\mathbf{b}_1^* = a(r)e^{-i\varphi}$. Right: Circular symmetries, $\mathbf{b}_2^* = a(r)e^{-i2\varphi}$.

11.3.4 Increasing the model selectivity

The issue addressed in this section is the possibility to increase operator feature selectivity. The standard windowing approach for convolution operators necessarily makes such operators sensitive to a broad spectrum of signals, often much broader than that for which the operator was originally intended.

Consider for example the following example based on Figure 11.11. First, note that the star pattern and the line are non-orthogonal in the signal space, i.e. the line is a part of the star pattern. It follows that, in general, any *linear* operator designed to detect the line will have an expected response greater than zero for the star pattern and vice versa. In many situations a more specific operator would be preferable, e.g. when a line is not considered to be evidence of a star pattern.

The symmetry operators described in the previous section are sensitive to *changes* in vector fields. In the discussion above, however, only the change

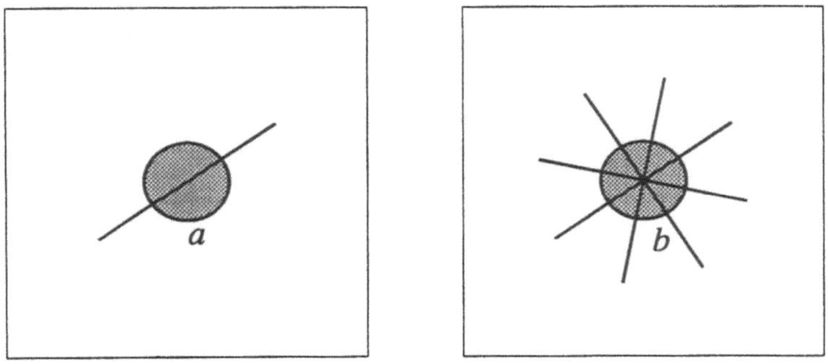

Figure 11.11 The left region, a, can be seen as a part of the more complex right region, b. The b-region can be seen as a linear combination of different oriented a-regions. This observation leads to the conclusion that, in general, any ***linear*** operation designed to detect patters like a will give some response on patterns like b and vice versa!

due to rotation is of interest. The magnitude of the vectors in the input field only reflects the *certainty* of the local orientation estimated. If this variation is not compensated for when estimating local symmetries, the operators respond to many structures they were not designed for. A straightforward normalization of the input orientation field,

$$\hat{\mathbf{z}} = \frac{\mathbf{z}}{\|\mathbf{z}\|} = \frac{ce^{i\varphi}}{c} \tag{11.27}$$

gives a field that is undefined in positions where the certainty, c, is zero. Inserting the value zero, or any other value in order to make the signal well-defined, introduces artificial step edges. Experimental results have shown that operators in general give strong erroneous responses to these new structures, unfortunately often much stronger than the responses to the structures of interest.

A solution to this problem is described in a patent from 1986 [73]. The patent describes a method termed a "consistency" operation. The name refers to that the operator only should respond to signals that are consistent with the signal model. The method is based on a combination of a set of convolutions. The following four filter results are needed:

$$\mathbf{s}_1 = \sum_k \mathbf{b}_k^* \mathbf{z}_k \tag{11.29}$$

$$\mathbf{s}_2 = \sum_k \mathbf{b}_k^* \|\mathbf{z}_k\| \quad (11.30)$$

$$\mathbf{s}_3 = \sum_k \|\mathbf{b}_k\| \mathbf{z}_k \quad (11.31)$$

$$\mathbf{s}_4 = \sum_k \|\mathbf{b}_k\| \|\mathbf{z}_k\| \quad (11.32)$$

where $\| \cdot \|$ denotes the magnitude of the filter or the data. The first term, $\mathbf{s}_1$, corresponds to standard convolution between the filter and the data. The forth term, $\mathbf{s}_4$, can be regarded as the local signal energy under the filter. Regarding the second and third term, the interpretation is somewhat harder. The filter is weighted locally with corresponding data producing a weighted average operator, where the weights are given by the data giving a "data dependent mean operator". For the third term it is vice versa. The mean data is calculated using the operator certainty as weights producing "operator dependent mean data".

The four filter results are combined into

$$\mathbf{s} = \frac{\mathbf{s}_4 \mathbf{s}_1 - \mathbf{s}_2 \mathbf{s}_3}{\mathbf{s}_4^{\gamma}} \quad (11.32)$$

where γ is a constant controlling the model selectivity. This value is normally set to one, $\gamma = 1$. The nominator in Equation (11.32) can consequently be interpreted as a standard convolution weighted with the local "energy" minus the "mean" operator acting on the "mean" data. The denominator is an energy normalization controlling the model versus energy dependence of the algorithm. With $\gamma = 1$, the output magnitude varies linearly with the magnitude of the input signal. Decreasing the value increases the selectivity.

Figure 11.12 (top left) shows a gray-level test image with size 128×128 and the corresponding estimates of local orientation using quadrature filters with size 15×15. The result of applying the filter corresponding to parabolic symmetries (Figure 11.10, left) is shown in Figure 11.13. The size of the parabolic filter is 15×15. Note that the filter is not only sensitive to the corners of the boxes where the orientation vector field is rotating but also to the edges due to the variation in the magnitude of the vectors. The other images in this figure shows that this can effectively be compensated for by applying the consistency scheme described above.

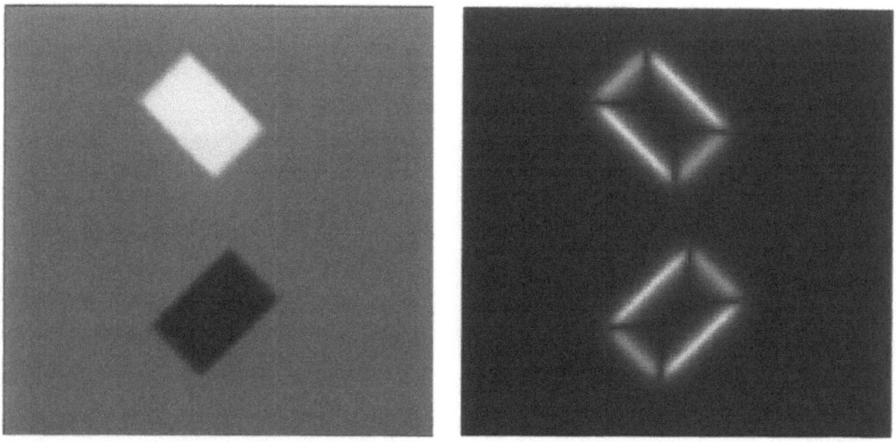

Figure 11.12 A gray-level test image containing two boxes is shown to the left. To the right, the local orientation is estimated. The brightness corresponds to the norm of the eigenvectors corresponding to the largest eigenvalue.

A more advanced method for filtering data with varying local certainty is *normalized convolution* [75, 138]. The above consistency operation is a special case of this method. Another special case of normalized convolution termed normalized averaging were touched upon in Chapter 9. Normalized convolution is a general method for filtering missing or uncertain data that can also handle the case of irregularly sampled signals. The method is based on notions from tensor theory and calculates implicitly local dual bases to the filters used taking the local certainty field into account. The theory needed for describing normalized convolution properly goes beyond the scope of this book. A detailed description of the theory can be found in [138].

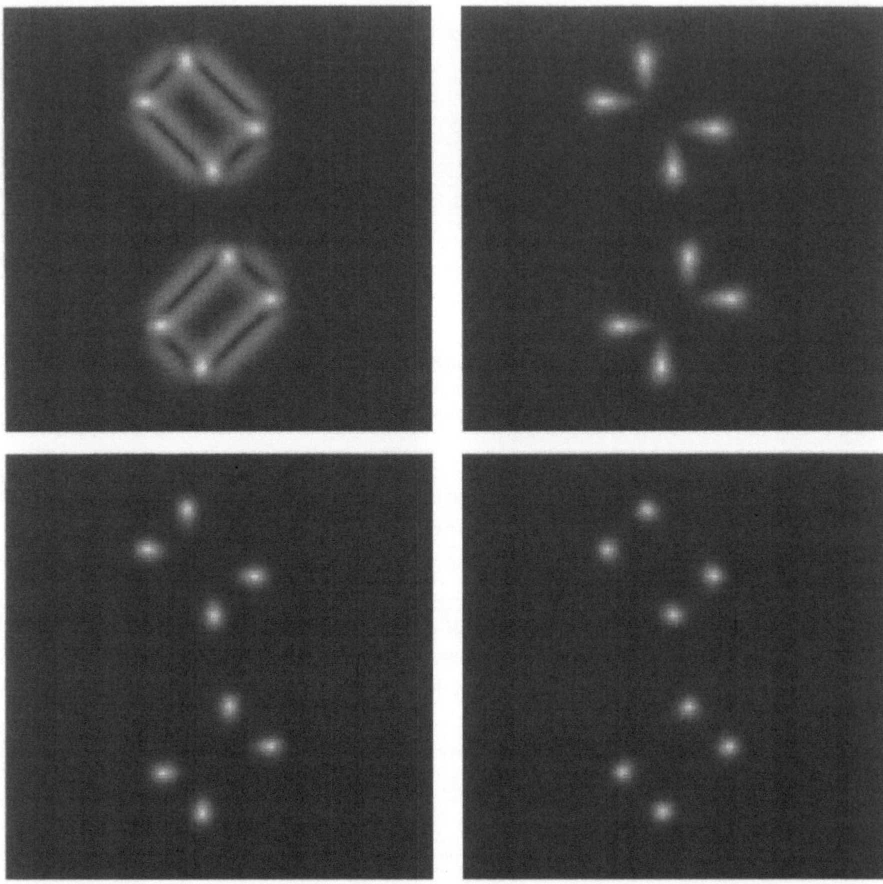

Figure 11.13 **Top**: Result after filtering the orientation vector field above with a parabolic symmetry filter (left). Note that the filters not only respond to the vector patterns they were designed for, but to many other structures as well. Increasing the model selectivity using a consistency operation is shown to the right where the selectivity parameter is set to one, $\gamma = 1$. **Bottom**: Same as the top right figure but with $\gamma = 0$ (left) and $\gamma = -2$ (right).

11.4 CURVATURE ESTIMATION

It is widely accepted that curvature is a relevant property of images and is thus useful for computer vision tasks. Moreover, the human visual system offers additional motivation for adopting curvature as an image feature since there are many indications of curvature detection being a mechanism in human perception. A well-known example of this was the demonstration by Attneave [4] that the approximation of the silhouette of a cat by straight lines ending in curvature maxima resulted in a figure that was still recognizable as a cat. Consequently curvature estimation forms an important tool in computer vision. Its use extends from guiding the selection of points in, for example, the polygonal approximation of curves [109], to matching features in the estimation of the three-dimensional orientation of objects from two-dimensional images [36]. The following presentation is based on original material from Bårman [8].

Curvature is a relatively simple concept as illustrated in Figure 11.14. Stated simply, curvature describes the local change in orientation of a curve and is commonly defined in terms of the tangent of the curve, i.e.

$$\kappa = \left|\frac{d\hat{\mathbf{t}}}{ds}\right| = \frac{\Delta\theta}{\Delta s} \tag{11.33}$$

where $\hat{\mathbf{t}}$ is a unit tangent vector, θ is the tangent angle of the vector and s denotes the arc length. This can, in terms of an image $f(\xi_1, \xi_2)$, be written as

$$\kappa = \frac{\left|(\frac{\partial f}{\partial \xi_1})^2 \frac{\partial^2 f}{\partial \xi_2^2} + (\frac{\partial f}{\partial \xi_2})^2 \frac{\partial^2 f}{\partial \xi_1^2} - 2\frac{\partial f}{\partial \xi_1}\frac{\partial f}{\partial \xi_2}\frac{\partial^2 f}{\partial \xi_1 \partial \xi_2}\right|}{((\frac{\partial f}{\partial \xi_1})^2 + (\frac{\partial f}{\partial \xi_2})^2)^{\frac{3}{2}}} \tag{11.34}$$

Here, the term "curvature" refers, for 2D images, to a feature of the curves in the image. If the image is interpreted as a surface where the pixel values represent "height values", curvature would then instead be a feature of that surface. The need for reliable estimation of curvature in computer vision has led to considerable theoretical and experimental investigations. Koenderink [79, 78] was one of the pioneers introducing the concepts of differential geometry into the field of computer vision.

Estimation schemes have traditionally been based on line and/or edge detection followed by thresholding before the actual curvature estimation takes place.

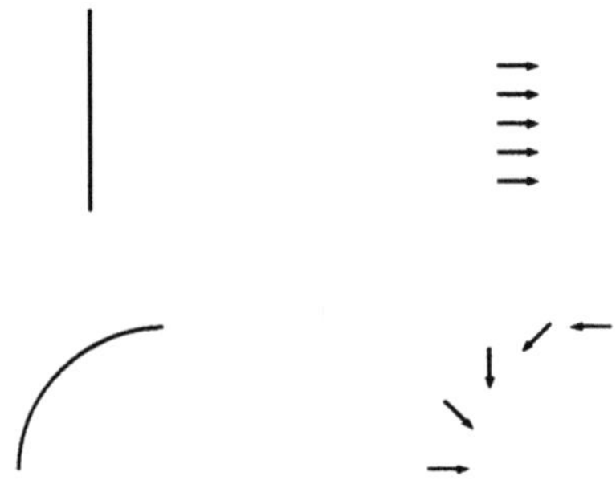

Figure 11.14 The orientation representation of a line (zero curvature) and an arc (curvature).

However, a significant drawback of these approaches concerns the use of thresholding, which tends to introduce some serious difficulties when considering real application problems:

- If noise is present, which is generally the case, it is hard or even impossible to select a proper threshold to discriminate the curve from the background.
- The conversion to a binary image introduces additional quantization noise since angular changes occur only in multiples of $\frac{\pi}{4}$.
- Small changes in the threshold level may cause dramatically different curvature estimates.

The curvature algorithm described here is in many respects different from other methods. The most significant difference is that the method gives a unified approach to the estimation and description of orientation and curvature. In addition, it avoids thresholding and premature decision making. Figure 11.15 shows the basic outline of the method. The input data to the curvature algorithm consists of an orientation description of the image, i.e. a complex-valued image in which each point represents an estimate of the dominant orientation in a local region about the point:

$$\mathbf{z} = f_1(\xi_1, \xi_2) = a + ib = re^{i2\varphi} \tag{11.35}$$

where the magnitude r is a certainty measure that the region has a dominant orientation and φ is the estimate of that orientation.

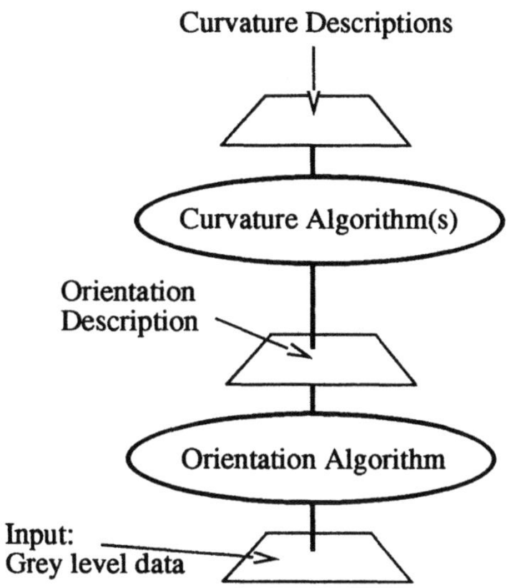

Figure 11.15 Estimation of curvature is an example of a second level operation.

It is not likely that a small neighborhood contains more than one type of curvature. This constraint makes it possible to express the angular function as

$$\varphi(\xi_1, \xi_2) = 2\pi(a\xi_1 + b\xi_2) \tag{11.36}$$

where $2\pi\sqrt{a^2 + b^2}$ is the magnitude of the curvature and $\arg(a + ib)$ is the direction of the tangent of the curve from which the curvature originates. Figure 11.16 gives an example of a neighborhood having the angular function given by Equation (11.36).

Instead of using the tangent angle of a curve as a definition of curvature, a more general definition of curvature in terms of the amount of directional change in a vector field will be used. This does not conflict with the more common tangent angle definition in Equation (11.33), since the orientation algorithm ensures that the vector field is produced by *one* curve. With the approximation of $\varphi(\xi_1, \xi_2)$ in Equation (11.36), curvature is reflected as a shift in the Fourier

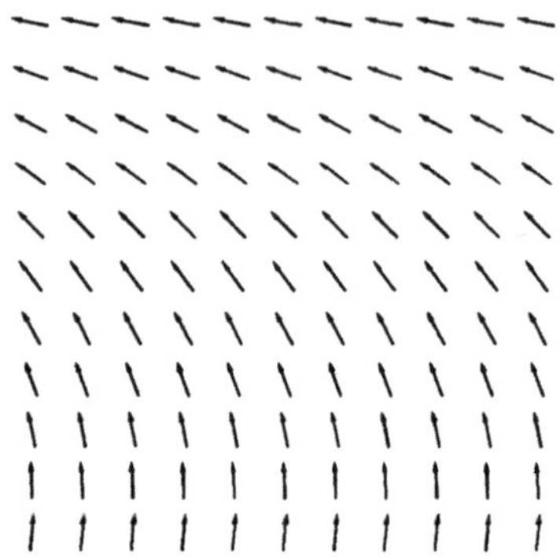

Figure 11.16 An example of a neighborhood consisting of a constant-valued magnitude and the angular function of Equation (11.36).

domain,

$$F_1(u_1, u_2) = \int_{-\infty}^{\infty}\int_{-\infty}^{\infty} r(\xi_1, \xi_2)e^{2\pi i(a\xi_1 + b\xi_2)}e^{-2\pi i(u_1\xi_1 + u_2\xi_2)}d\xi_1 d\xi_2 = $$
$$= R(u_1 - a, u_2 - b) \tag{11.38}$$

The correspondence between curvature estimation and shift estimation in local Fourier transforms of the orientation image is illustrated in Figure 11.17. The curvature direction can be calculated by estimation of the direction to the center of gravity of the amplitude spectrum $|F(u_1, u_2)|$ of the image.

In contrast to the orientation algorithm described in chapter 6, the curvature estimation requires filters spanning the whole frequency domain.

$$f_2(\xi_1, \xi_2) = \sum_{k=1}^{K} q_k(\xi_1, \xi_2)e^{i\varphi_k} \tag{11.38}$$

The double angle used in the orientation vector sum is in the context of curvature replaced by a 'single angle'. The resulting vector points in the direction of the shift, which is equivalent to the curvature direction (Figure 11.18).

A more detailed treatment of this curvature algorithm and experimental results on both synthetic and real world data can be found in [8]

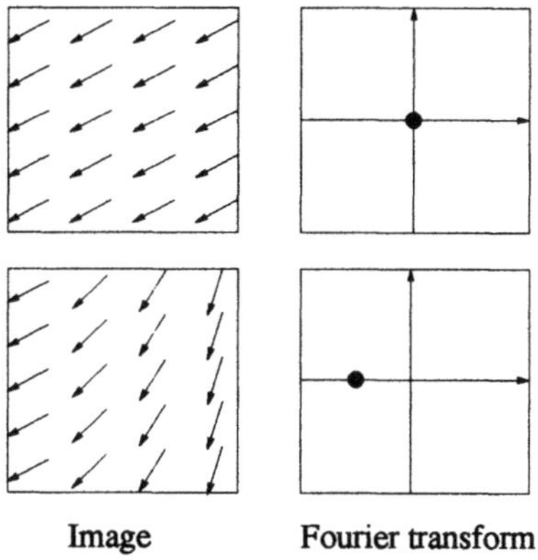

Figure 11.17 Left: Two fields of complex numbers. Right: Corresponding Fourier transforms.

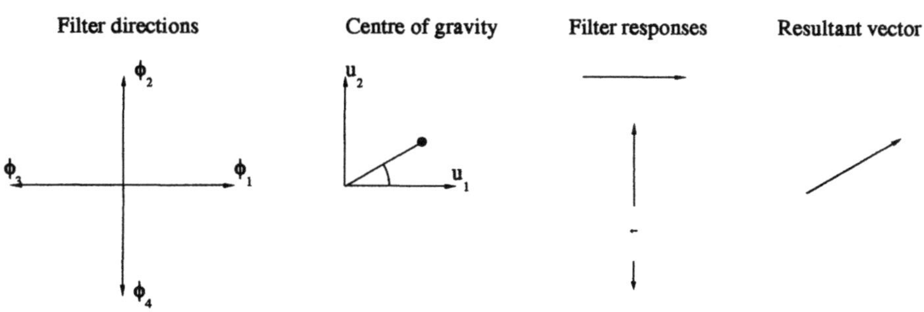

Figure 11.18 The curvature direction is estimated by summing the filters using a "single angle" convention in contrast to the "double angle" convention used, in Chapter 3, for local orientation estimation.

12

CLASSIFICATION AND RESPONSE GENERATION

The last stage of a computer vision system generally produces some action in the external world. It may be to control the motors of an industrial robot. It may be to point out defective units on the conveyor belt of a quality control system. Or it may be to point out suspected cancerous regions in an X-ray image. Sometimes the input to the system is used to improve a robot's knowledge and model of the world without any immediate external response being generated. Still, the conceptual purpose of a computer vision system is to produce responses, or at least to represent visual data in such a form that a correct decision is made.

Each of the tasks mentioned above is a problem that requires combining all the descriptive properties derived in the preceding chapters into an output control signal of suitable form. The general case requires using additional invariance mechanisms, taking into account geometry, object modeling, objects, and events in a 3-D or 4-D space, and other techniques. These mechanisms are beyond the scope of the present book. Although we cannot encompass all varieties of scenes, it is of interest to bring the task of computer vision to a logical completion in the combination of feature properties for response generation and classification.

In Chapter 1, referring to Figure 1.4, we saw a first discussion of the mechanisms related to identification of an object part having certain properties. We will continue this discussion along a fairly classical view of the problem, with a focus on segmentation of 2-D images. The purpose is to segment an image into regions that are in some sense homogeneous. This can be done either interactively by a user –*supervised* classification– or automatically –*unsupervised* classification.

12.1 INFORMATION NEEDED FOR CLASSIFICATION

Features

A classification must be performed with enough resolution to describe the image in sufficient detail. This often implies that it is performed pixelwise, i.e. for each pixel in the input image we determine a class membership. The corresponding pixel in a classification result image can be mapped to a color to indicate the class visually. In a satellite image this could mean that all water is given one color, e.g. blue, forest is colored green and arable land is colored yellow. As we have seen earlier for the pyramid feature representation, the information in a pixel at a high level in the pyramid may depend on pixels within a large region in the original image.

To perform the classification we need to determine a number of *features* at each position and resolution level in the input image. By features we mean values measured from the image such as local spatial frequency, local spatial orientation, intensity values in one or several spectral bands, shape measurements, as discussed in the preceding chapters.

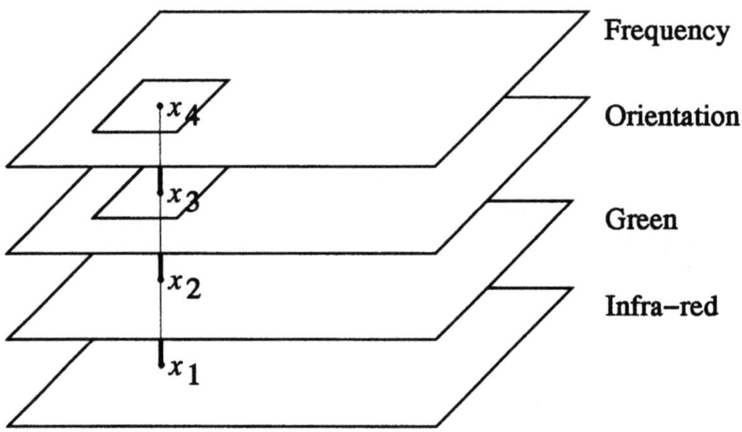

Figure 12.1 Feature images.

Example

Figure 12.1 shows an example situation. To distinguish between land and water in a satellite image, we might use the intensity in the infra-red (IR) band as a feature. Water emits less IR radiation than land and is thus much darker than land on an IR image. We can now use the following classification procedure: To each pixel with an intensity value less than some threshold T we assign the class `water` and assign the corresponding pixel in the output visualization image the color blue, and to each pixel having an intensity greater than T we assign the class `land` and give the corresponding pixel in the output visualization image the color green. Generally, this one feature is not enough to distinguish between all possible classes. We might want to subdivide the class `land` further into different types such as `forest` and `field`.

The feature space

To structure the ensuing discussion we define a few terms. The features can be said to span a *feature space* containing one dimension for each feature. Each pixel in the image corresponds to a point in this feature space. If the features are properly chosen, the pixels corresponding to homogeneous areas or objects form *clusters* in this space. From each point in the input image we extract several feature values such as intensity in the IR band, local spatial orientation, etc. These feature values become components in a *feature vector*. Figure 12.2 depicts a one-dimensional feature space with clusters of pixels from the two classes `land` and `water`. Note that the second axis, "number of pixels", is a measure of feature strength and *not* a dimension in the feature space.

Figure 12.3 illustrates a two-dimensional feature space with features *IR* and *GREEN* on the two axes. In this case the features are assumed to be *independent*, which makes it possible to represent them with a system having orthogonal axes.

In the classification phase, the class membership of a pixel is determined by determining its coordinates in the feature space, the feature vector $\mathbf{x} = (x_1, x_2, \ldots, x_n)^T$, and, by some rule, `class:=` $f(\mathbf{x})$, assign a class to this pixel.

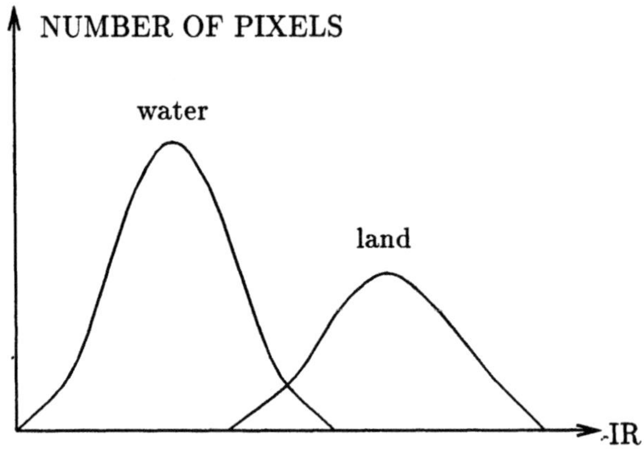

Figure 12.2 One-dimensional feature space.

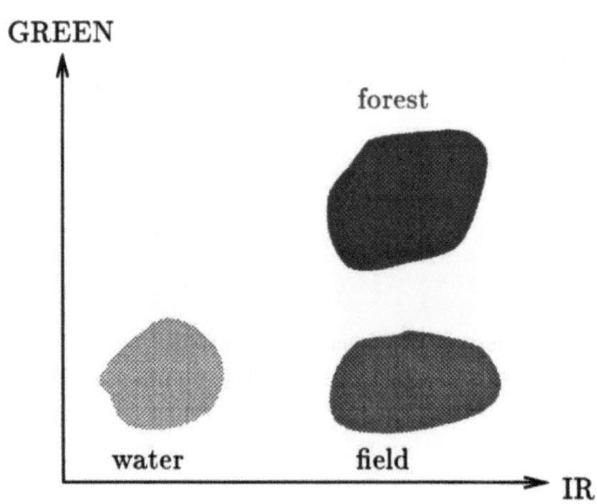

Figure 12.3 Two-dimensional feature space.

A simple classification rule

Once we have obtained a good set of features we reach the other crucial part of the classification procedure, namely, the rule by which points in the feature space are assigned to a class.

The general way to obtain such a rule is to subdivide the feature space into subspaces $\mathcal{S}_i$ with a set of boundaries mathematically defined as hyperplanes or other types of surfaces. The class membership can now be determined by deciding into which subspace the feature vector falls. There are several ways to perform this subdivision. How to do this in a general manner is the topic of most of the remainder of this chapter.

We look first at one of the simplest approaches, thresholding, which has already been mentioned. This is illustrated for the earlier example given in Figure 12.4. The generalization of thresholding is *box classification*, in which the feature space is subdivided into n-cubes. This approach extends the notion of orthogonality by assuming that features are independent.

The rule is now:

IF $l_{i1} < x_1 < u_{i1}$ AND $l_{i2} < x_2 < u_{i2}$ AND ... AND $l_{in} < x_n < u_{in}$
THEN this pixel belongs to class i

Values l_{ik} and u_{ik} refer to the lower and upper thresholds of feature k for class i. In this case, the subdivision of the feature space is done with planes determined by equations $x_i = const$ (see Figure 12.5).

12.2 LINEAR DISCRIMINANT FUNCTIONS

In general, features are not independent, which makes it necessary to divide the feature space into different regions using slanted hyperplanes or *discriminant functions*. In this section we consider discriminant functions which are either linear in the components of the feature vector **x** or linear in some fixed set of functions of **x**. Linear discriminant functions are attractive for several reasons.

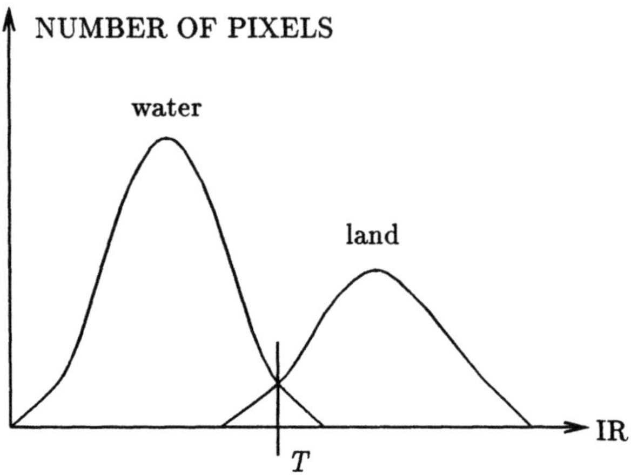

Figure 12.4 One-dimensional thresholding.

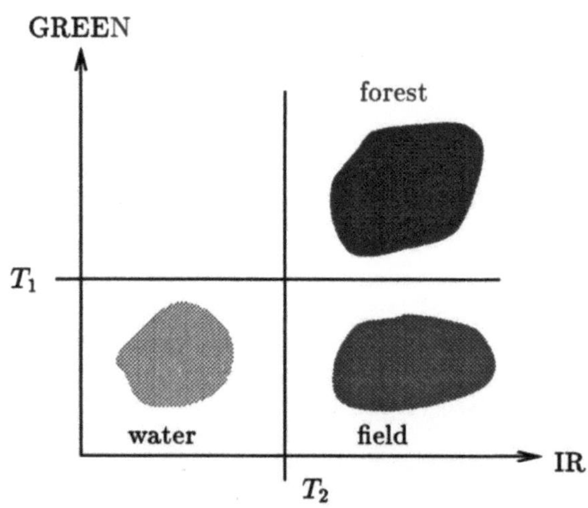

Figure 12.5 Box classification.

They allow a fairly compact treatment using matrix methods, and operations can be visualized geometrically. They are easy to compute as they can be implemented as convolutions or product sums, standard operations in today's procedures for signal processing and computer vision. They also allow us to give a unified treatment of all types of operations ranging from pre-processing to high level classification. We will follow the formalisms in Nilsson [107] and in Duda and Hart [37].

As an introductory example we consider a case with two feature variables, x_1 and x_2. One of the variables, viewed as a function of the other, forms the equation of a straight line

$$x_2 = w_1 x_1 + w_0. \tag{12.1}$$

For given values of w_1 and w_0, the equation describes pairs of values x_1 and x_2 which fit to a particular straight line. The same equation can be written

$$0 = -x_2 + w_1 x_1 + w_0. \tag{12.2}$$

We can now form a function which is the right-hand part of the preceding equation. We define

$$g(x_1, x_2) = -x_2 + w_1 x_1 + w_0. \tag{12.3}$$

This discriminant function has the interesting property that it is 0 for pairs of values x_1 and x_2, that are located on the straight line defined above. It also has the property that values of x_1 and x_2 located on one side of the line give a positive output from the function, while value pairs on the other side give a negative output. Figure 12.6 shows an example of this.

12.2.1 The two-category case for multi-dimensional features

A more common case bases the decision on a number of features. We assume that there are d feature components which form a d-dimensional feature vector $\mathbf{x}$. The discriminant function, which is a linear combination of the components of $\mathbf{x}$, can be written as

$$g(\mathbf{x}) = \mathbf{w}^T \mathbf{x} + w_0, \tag{12.4}$$

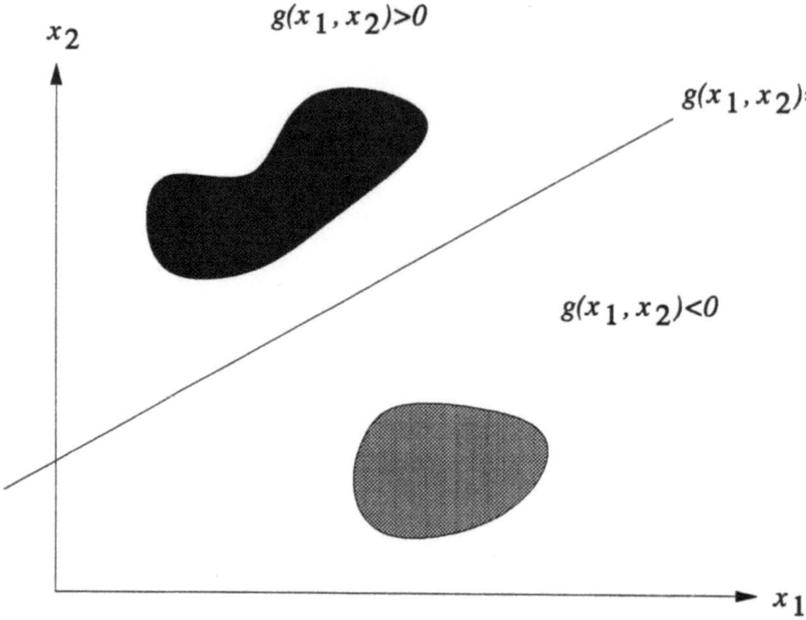

Figure 12.6 Example of a linear discriminant function in two dimensions.

where $\mathbf{w}$ is called the *weight vector* and w_0 the *threshold weight.* We will devise a decision rule for the two-category linear classifier as implied in the introductory discussion: Decide class ω_1 if $g(\mathbf{x}) > 0$ and class ω_2 if $g(\mathbf{x}) < 0$. We can see that the feature vector $\mathbf{x}$ is assigned to class ω_1 if the scalar product $\mathbf{w}^T\mathbf{x}$ exceeds the threshold $-w_0$.

The equation $g(\mathbf{x}) = 0$ defines the decision surface that separates points assigned to one class, ω_1, from points assigned to the other class, ω_2. As $g(\mathbf{x})$ is linear, this decision surface is a hyperplane. If $g(\mathbf{x}) = 0$, the feature vector $\mathbf{x}$ is right on the decision surface. If two vectors, $\mathbf{x}_1$ and $\mathbf{x}_2$, are both on the decision surface, then

$$\mathbf{w}^T\mathbf{x}_1 + w_0 = \mathbf{w}^T\mathbf{x}_2 + w_0, \tag{12.5}$$

which gives

$$\mathbf{w}^T(\mathbf{x}_1 - \mathbf{x}_2) = 0. \tag{12.6}$$

This implies that $\mathbf{w}$ is normal to any vector lying in the hyperplane, and we say that $\mathbf{w}$ is a normal vector of the hyperplane.

As we defined that $g(\mathbf{x}) > 0$ if $\mathbf{x}$ belongs to class ω_1, it follows that the normal vector $\mathbf{w}$ points into the decision region corresponding to ω_1. We sometimes say that an $\mathbf{x}$ corresponding to class ω_1 is on the positive side of the hyperplane.

The discriminant function $g(\mathbf{x})$ gives a measure of the distance between an arbitrary feature vector $\mathbf{x}$ and the hyperplane. One way to visualize this is to express $\mathbf{x}$ as

$$\mathbf{x} = \mathbf{x}_p + r\frac{\mathbf{w}}{||\mathbf{w}||}. \tag{12.7}$$

The point $\mathbf{x}_p$ is the projection of the point $\mathbf{x}$ onto the hyperplane and consequently lies in the hyperplane. The vector $r(\mathbf{w}/||\mathbf{w}||)$ is, as was shown earlier, normal to the hyperplane with r expressing the desired distance. Since $g(\mathbf{x}_p) = 0$, it follows that

$$\begin{aligned} g(\mathbf{x}) &= \mathbf{w}^T\mathbf{x} + w_0 = \mathbf{w}^T(\mathbf{x}_p + r\frac{\mathbf{w}}{||\mathbf{w}||}) + w_0 \\ &= g(\mathbf{x}_p) + r\frac{\mathbf{w}^T\mathbf{w}}{||\mathbf{w}||} = r\frac{\mathbf{w}^T\mathbf{w}}{||\mathbf{w}||} = r||\mathbf{w}||, \end{aligned} \tag{12.9}$$

which gives

$$r = \frac{g(\mathbf{x})}{||\mathbf{w}||}. \tag{12.9}$$

As one special case we find that the distance from the origin to the hyperplane is given by

$$r_0 = \frac{g(\mathbf{0})}{||\mathbf{w}||} = \frac{w_0}{||\mathbf{w}||}. \tag{12.10}$$

This indicates that if $w_0 > 0$, the origin is on the positive side of the hyperplane, and if $w_0 < 0$ it is on the negative side. If $w_0 = 0$, then $g(\mathbf{x})$ has the homogenous form $g(\mathbf{x}) = \mathbf{w}^T\mathbf{x}$, and the hyperplane passes through the origin. These relations are illustrated geometrically in Figure 12.7.

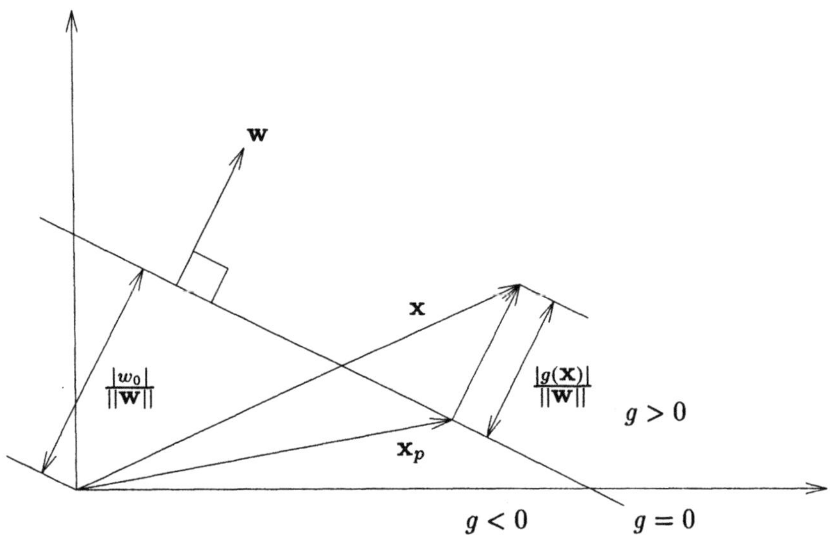

Figure 12.7 A linear decision boundary with different parameters indicated.

12.2.2 The multi-category case for multi-dimensional features

In general we want to separate events into more than two possible categories. There are several ways to devise such multi-category classifiers employing linear discriminant functions. One way is to reduce the problem to $c-1$ two-class problems, where a particular subclass is dealt with by a linear discriminant function that separates points assigned to class ω_i from those not assigned to ω_i. Another more costly approach is to use $c(c-1)/2$ linear discriminants, one for every pair of classes. Both of these approaches have some less attractive ambiguities, in that they may fail to classify points in certain regions.

We will choose the approach of defining c linear discriminant functions

$$g_i(\mathbf{x}) = \mathbf{w}_i^T\mathbf{x} + w_{i0}, \qquad i = 1, \ldots, c. \tag{12.11}$$

Every discriminant function has the same appearance as those discussed earlier, but the interpretation is slightly different. We will now assign a feature vector to ω_i if $g_i(\mathbf{x}) > g_j(\mathbf{x})$ for all $j \neq i$. In the case of ties, the classification is usually left undefined. This type of classifier is called a *linear machine*. A linear machine divides the feature space into c decision regions, where $g_i(\mathbf{x})$ is the largest discriminant function if $\mathbf{x}$ is in the region corresponding to class ω_i (Figure 12.8). If the regions corresponding to classes ω_i and ω_j are contiguous, the boundary between them is a portion of a new hyperplane defined by

$$g_i(\mathbf{x}) = g_j(\mathbf{x}) \tag{12.12}$$

or

$$g_{ij}(\mathbf{x}) = g_i(\mathbf{x}) - g_j(\mathbf{x}) = (\mathbf{w}_i - \mathbf{w}_j)^T\mathbf{x} + (w_{i0} - w_{j0}) = 0. \tag{12.13}$$

We notice that the linear discriminant function $g_{ij}(\mathbf{x})$, which is the difference between two linear discriminant functions of the new type, has the same character and interpretation as the linear discriminant functions discussed in the preceding section.

From Equation (12.13) it is apparent that $(\mathbf{w}_i - \mathbf{w}_j)$ is a normal vector to the hyperplane separating classes ω_i and ω_j. The distance between $\mathbf{x}$ and the separating hyperplane is in this case given by

$$r_{ij} = \frac{(g_i - g_j)}{||\mathbf{w}_i - \mathbf{w}_j||}. \tag{12.14}$$

It can be shown that the decision regions for a linear machine are convex. Every decision region must be singly connected, which makes the linear machine best suited for problems where the underlying probability densities are unimodal. This is often the case: classes can be divided into subclasses, which are themselves unimodal. As we proceed we will handle various phenomena in terms of linear discriminant functions; this provides a unifying way to handle most image operations.

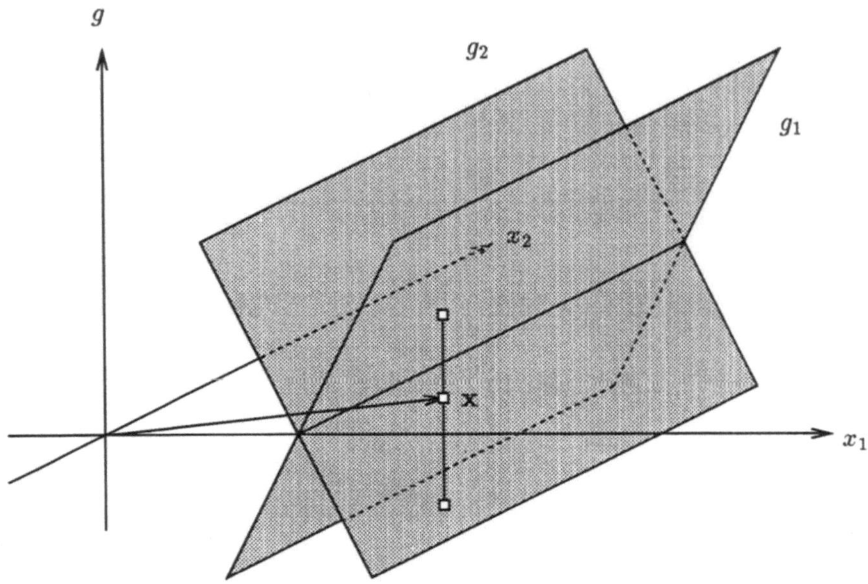

Figure 12.8 Two linear discriminant functions.

12.3 MINIMUM DISTANCE CLASSIFICATION

We will now approach linear discriminant functions from a slightly different direction, formulated as a minimum distance problem. Let us assume c different points $\mathbf{w}_1, \mathbf{w}_2, \ldots, \mathbf{w}_c$ in the feature space. We assume that these points are centers of different class clusters ω_i. The Euclidean distance between an arbitrary point $\mathbf{x}$ and $\mathbf{w}_i$ is given by

$$||\mathbf{x} - \mathbf{w}_i|| = \sqrt{(\mathbf{x} - \mathbf{w}_i)^T(\mathbf{x} - \mathbf{w}_i)}. \tag{12.15}$$

Associated with each point $\mathbf{w}_i$ is a class number i, $i = 1, \ldots, c$. A *minimum distance classifier* with respect to the points $\mathbf{w}_1, \mathbf{w}_2, \ldots, \mathbf{w}_c$ places each point $\mathbf{x}$ into class i_0 associated with the nearest point $\mathbf{w}_{i_0}$ of the points $\mathbf{w}_1, \mathbf{w}_2, \ldots, \mathbf{w}_c$.

For any $\mathbf{x}$, the quantities $||\mathbf{x} - \mathbf{w}_i||$, $i = 1, \ldots, c$, are calculated, and $\mathbf{x}$ is associated with the class corresponding to the smallest distance. We call the points $\mathbf{w}_1, \mathbf{w}_2, \ldots, \mathbf{w}_c$ *prototype points.*

We can simplify computations by comparing the squared distances $||\mathbf{x} - \mathbf{w}_i||^2$ for $i = 1, \ldots, c$. Squaring both sides of Equation (12.15) produces

$$\begin{aligned} ||\mathbf{x} - \mathbf{w}_i||^2 &= (\mathbf{x} - \mathbf{w}_i)^T(\mathbf{x} - \mathbf{w}_i) \\ &= \mathbf{x}^T\mathbf{x} - 2\mathbf{x}^T\mathbf{w}_i + \mathbf{w}_i^T\mathbf{w}_i. \end{aligned} \tag{12.17}$$

As $\mathbf{x}^T\mathbf{x}$ is constant for a given unknown feature vector $\mathbf{x}$, the minimum distance classification can be implemented by comparing the expressions

$$\mathbf{x}^T\mathbf{w}_i - \frac{1}{2}\mathbf{w}_i^T\mathbf{w}_i, \qquad i = 1, \ldots, c \tag{12.17}$$

and selecting the largest. Possible discriminant functions for this case are apparently

$$g_i(\mathbf{x}) = \mathbf{x}^T\mathbf{w}_i - \frac{1}{2}\mathbf{w}_i^T\mathbf{w}_i, \qquad i = 1, \ldots, c. \tag{12.18}$$

We obtain at the origin

$$g_i(\mathbf{0}) = -\frac{1}{2}\mathbf{w}_i^T\mathbf{w}_i = -\frac{1}{2}||\mathbf{w}_i||^2, \qquad i = 1, \ldots, c. \tag{12.19}$$

and

$$r_{i0} = \frac{g_i(\mathbf{0})}{||\mathbf{w}_i||} = -\frac{1}{2}||\mathbf{w}||. \tag{12.20}$$

is the distance from the origin to the hyperplane. On the hyperplane we have $g_i(\mathbf{x}) = 0$, which gives $\mathbf{x} = \frac{1}{2}\mathbf{w}_i$.

Hyperplanes for the minimum distance classifier are, consequently, located at a distance $\frac{1}{2}||\mathbf{w}_i||$ from the origin. It is apparent that a minimum distance classifier is in effect a linear machine or is implementable with linear discriminant functions. We have the discriminant function in the same form as discussed earlier if we select

$$w_{i0} = -\frac{1}{2}\mathbf{w}_i^T\mathbf{w}_i. \tag{12.21}$$

This is a somewhat forced selection, but it gives us a way of finding a direct correspondence with earlier versions. If we try to make an intuitive interpretation, it is apparent that the weight vectors $\mathbf{w}_i$ act as prototypes for different signal vector classes. When $\mathbf{w}_i$ and $\mathbf{x}$ are collinear, the scalar product between them will produce a large value. This will reduce the squared distance to a minimum. Note the difference between minimizing the distance $||\mathbf{x} - \mathbf{w}_i||$, which is obtained for $\mathbf{x} = \mathbf{w}_i$, and minimizing the discriminant function $g_i(\mathbf{x})$, which is obtained for $\mathbf{x} = \frac{1}{2}\mathbf{w}_i$.

12.4 CONVOLUTION OPERATIONS

Conventional convolution operations, in the form of product sums, can be viewed as linear discriminant functions. The ordinary form for a linear discriminant function

$$g(\mathbf{x}) = \mathbf{w}^T\mathbf{x} + w_0, \tag{12.22}$$

is equivalent to a convolution, where values in different points of the neighborhood are represented by the image density vector $\mathbf{x}$, which is multiplied by the weighting vector $\mathbf{w}$. In the case of convolution, the constant term w_0 is often zero, as we usually want a zero output from the convolution when all components in the image neighborhood vector are zero. As we have seen earlier, the values of the weighting function vector, $\mathbf{w}$, typically describe a function such as a weighted sine or a weighted cosine within the window under consideration.

We saw in Section 12.2 that we can view the output from the convolution, $g(\mathbf{x})$, as the non-normalized distance between the image vector point and the hyperplane corresponding to the convolution operator

$$g(\mathbf{x}) = r||\mathbf{w}||, \tag{12.23}$$

or

$$r = \frac{g(\mathbf{x})}{||\mathbf{w}||}. \tag{12.24}$$

We can see that a large output from the convolution implies a large distance between the image neighborhood vector point $\mathbf{x}$ and the hyperplane defined by the weighting vector $\mathbf{w}$. Zero output from the convolution implies, on the other hand, that the image neighborhood vector $\mathbf{x}$ is located on the hyperplane defined by $\mathbf{w}$.

12.4.1 Combination of convolution outputs

For a multi-dimensional signal vector, $\mathbf{x}$, a single convolution only establishes the distance between the vector and one single hyperplane. For a multi-dimensional signal vector a single convolution does not establish very much of its character. It is necessary to pinpoint its position and enclose it with a number of hyperplanes corresponding to different convolutions. The combination of hyperplanes is done in a nonlinear fashion; several varieties are used dependending on the desired effects.

Vector representation in terms of linear discriminant functions

One example of combination is the one used for orientation representation as described in Chapter 1 and Chapter 3. In this case the output from a discriminant function is a complex value $\mathbf{g}(\mathbf{x})$, since both the weighting coefficients and the input image data in each point are complex-valued. The real and imaginary components correspond to cosine and sine weighting functions respectively.

Four functions, $\mathbf{q}_1(\mathbf{x}), \ldots, \mathbf{q}_4(\mathbf{x})$, are typically computed, corresponding to four different orientations within the neighborhood as described earlier. These components are assembled into a description vector $\mathbf{z}(\mathbf{x})$, where

$$\mathbf{z}(\mathbf{x}) = \begin{pmatrix} z_1 \\ z_2 \end{pmatrix} = \begin{pmatrix} \|\mathbf{q}_1(\mathbf{x})\| - \|\mathbf{q}_3(\mathbf{x})\| \\ \|\mathbf{q}_2(\mathbf{x})\| - \|\mathbf{q}_4(\mathbf{x})\| \end{pmatrix}. \tag{12.25}$$

The implication of this operator in terms of linear discriminant functions is as follows. As the output from a particular convolution component is zero for image neighborhood vectors $\mathbf{x} = \mathbf{0}$, this implies that the hyperplanes corresponding to the convolutions go through the origin. We assume in the following discussion that the weight vectors $\mathbf{w}$ contain complex-valued components, whose real and imaginary parts correspond to cosine and sine convolution coefficients respectively. The essential non-linearity introduced is due to taking the magnitude of individual convolution components before the combination. Taking the magnitude results in the ambiguity that two points on opposite sides of the hyperplane, but at the same distance, yield the same result. An alternative non-linearity is to take the square of the value, which gives the same ambiguity but implies a different distance measure.

As the image neighborhood feature vector space has very high dimensionality of, say, 5×5 to 11×11, it is not possible to give any good geometric visualization of the situation. Figure 12.9 gives, however, an intuitive representation of some of the pertinent aspects.

The four weight vectors are all centered at the origin. As the convolution functions are pairwise orthogonal, the corresponding weight vectors are orthogonal in the hyperspace in question. This means that $\mathbf{w}_1$ is orthogonal to $\mathbf{w}_3$ and that vector $\mathbf{w}_2$ is orthogonal to $\mathbf{w}_4$. As adjacent vectors are not orthogonal but have some correlation, it follows that the angles between these are less than $90°$.

Figure 12.10 illustrates another property of the convolution weight vectors viewed as hyperplanes. If we consider two vectors that are orthogonal, we can consider the subspace that contains these two vectors. The output from a convolution indicates the distance from the hyperplane corresponding to the weight vector. When a signal vector reaches a hyperplane, the output from the scalar product is 0. This is because $\mathbf{w}^T\mathbf{x} = 0$, which implies that $\mathbf{x}$ and $\mathbf{w}$ are orthogonal. In fact, the weight vectors are located in the "complementary" hyperplane, as indicated in Figure 12.10. Consequently they are indeed orthogonal. A signal vector giving zero scalar product output is collinear with the complementary weight vector.

There is another vector of interest which is 0 as $\Sigma \mathbf{w}_i = 0$, and that is $\mathbf{x} = \mathbf{1}$. We can view this vector as being orthogonal to the other, i.e. normal to the paper.

We can see that it is possible to treat what we consider to be low-level operations with the same formalism as classification, which is often seen as a high level function. This allows us to handle descriptions in a consistent way independent of level.

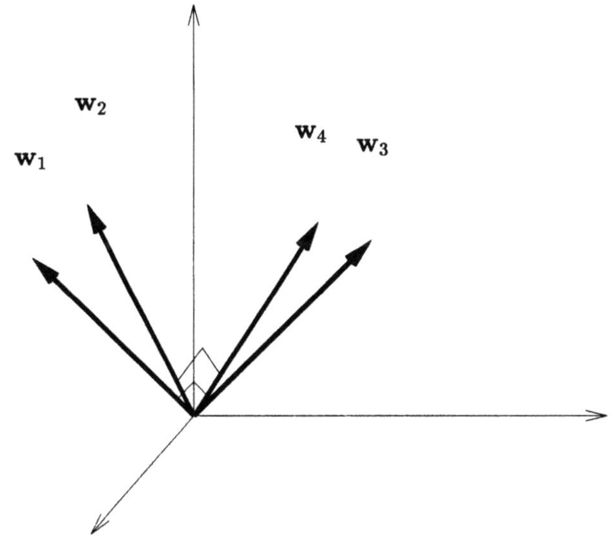

Figure 12.9 Intuitive representation of weight vectors corresponding to convolution for generation of the orientation vector estimate.

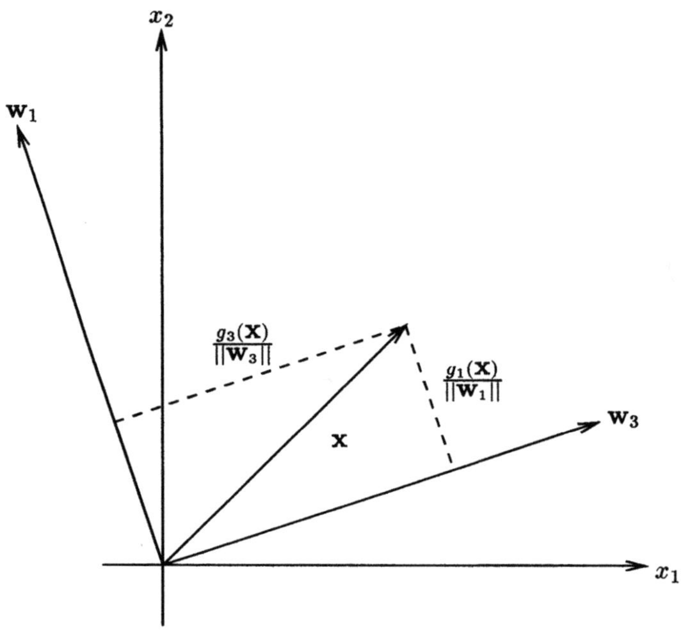

Figure 12.10 A two-dimensional subspace containing two weight vectors $\mathbf{w}_1$ and $\mathbf{w}_3$.

12.5 GENERALIZED DISCRIMINANT FUNCTIONS

We have seen earlier that the discriminant function $g(\mathbf{x})$ can be written as

$$g(\mathbf{x}) = w_0 + \Sigma_{i=1}^{d} w_i x_i,$$

with coefficients w_i which are components of the weight vector $\mathbf{w}$. It is possible to add additional terms containing products of pairs of components of $\mathbf{x}$. This gives us the quadratic discriminant function

$$g(\mathbf{x}) = w_0 + \Sigma_{i=1}^{d} w_i x_i + \Sigma_{i=1}^{d} \Sigma_{j=1}^{d} w_{ij} x_i x_j. \tag{12.26}$$

Since $x_i x_j = x_j x_i$ we need not distinguish between $w_{ij} = w_{ji}$. This quadratic discriminant function contains additional $d(d+1)/2$ coefficients; this allows us to produce more complex separating surfaces. The separating surface defined by $g(\mathbf{x}) = 0$ is now a second degree or *hyperquadric* surface.

Example 1. What is the decision surface of the discriminant function

$$g(\mathbf{x}) = -x_1^2 - x_2^2 + 9 \quad ?$$

The function $g(\mathbf{x}) = 0$ corresponds to the equation of a circle centered at the origin with radius 3. We can see that the function $g(\mathbf{x})$ is positive inside the circle and that it is negative outside. □

Example 2. What is the decision surface corresponding to the discriminant function

$$g(\mathbf{x}) = -x_1^2 - 4x_2^2 + 4x_1 + 32x_2 - 52 \quad ?$$

If we modify $g(\mathbf{x})$ to complete the square terms of each variable, we obtain

$$g(\mathbf{x}) = -(x_1 - 2)^2 - 4(x_2 - 4)^2 + 16.$$

We recognize this as an ellipse centered at $(2, 4)$, and with radii 4 and 2 respectively (Figure 12.11). □

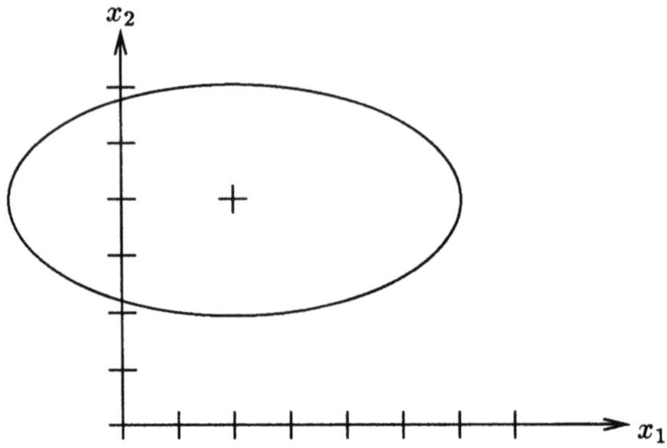

Figure 12.11 Illustration of decision function $g(\mathbf{x})$ discussed in Example 2.

The simple basic forms of discriminant functions often perform very well. Important special cases are linear functions of the variable $\mathbf{x}$ and simple quadratic functions of $\mathbf{x}$ corresponding to hyperspherical or hyperelliptical decision surfaces. These also correspond to two dominant classes of symmetry often found in data representing different phenomena: linear or one-dimensional symmetry and central or rotational symmetry. This relates to the descriptions in terms of such symmetries that was discussed in Chapter 3.

Example 3 Describe the decision regions that can be defined by decision functions

$$\begin{aligned}
g_1(\mathbf{x}) &= -x_1 + 4 \\
g_2(\mathbf{x}) &= x_2 - 4 \\
g_3(\mathbf{x}) &= x_2 - x_1 \\
g_4(\mathbf{x}) &= x_2 + x_1 - 8 \\
g_5(\mathbf{x}) &= -(x_1 - 4)^2 - (x_2 - 4)^2 + 1 \\
g_6(\mathbf{x}) &= -(x_1 - 4)^2 - (x_2 - 4)^2 + 4 \\
g_7(\mathbf{x}) &= -(x_1 - 4)^2 - (x_2 - 4)^2 + 9.
\end{aligned}$$

These seven discriminant functions produce the decision regions illustrated in Figure 12.12.

We see that it is possible to define the individual regions in the figure as functions of the signs of individual discriminant functions. In this way it is possible to adapt to data that exhibit a predominantly linear or angular distribution, or data that exhibit predominantly hyperspherical clusters. □

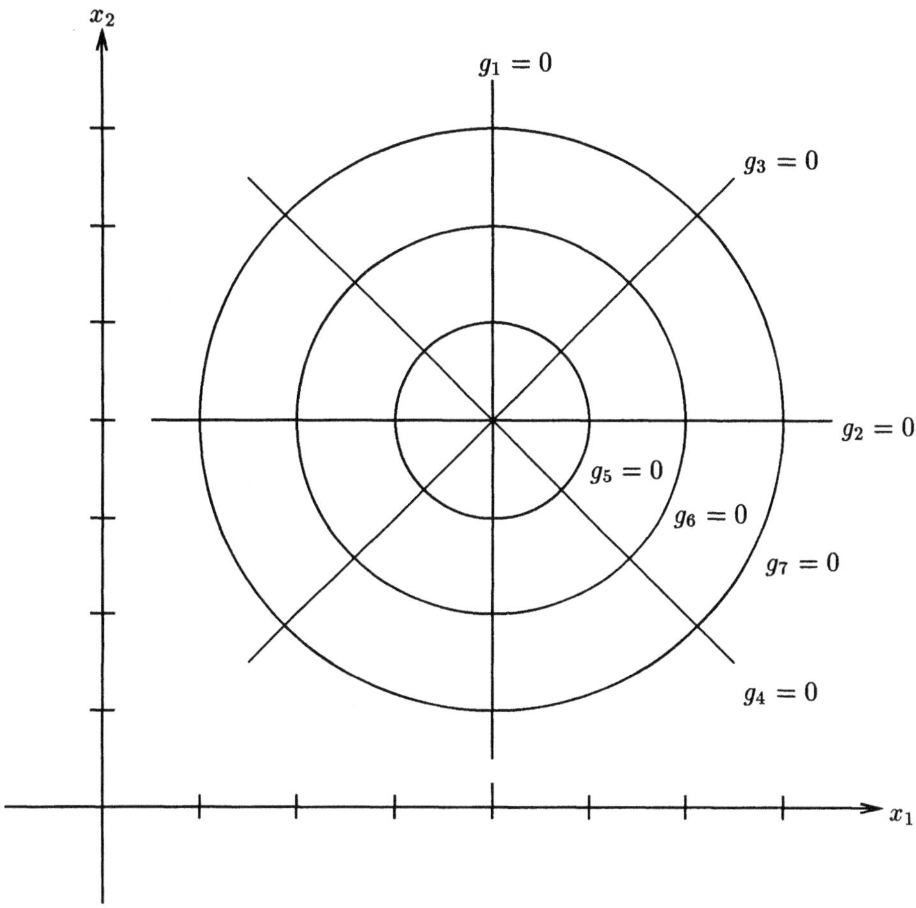

Figure 12.12 Decision boundaries corresponding to the discriminant functions in Example 3.

If there are cross products between variables of type $x_i x_j$, it becomes necessary to transform the coordinate system of the problem to a new coordinate system, such that a linear transformation in the new coordinate system is dependent upon a diagonal matrix. For this reason, the matrix $\mathbf{W} = [w_{ij}]$ must be reduced to a diagonal form. This is possible under the condition that the symmetric matrix $\mathbf{W} = [w_{ij}]$ is non-singular.

For that purpose, we can define a discriminant function equivalent to Equation (12.26), but in our usual matrix notation

$$g_i(\mathbf{x}) = \mathbf{x}^T \mathbf{W}_i \mathbf{x} + \mathbf{w}_i^T \mathbf{x} + w_{0i},$$

where

$$\begin{aligned} \mathbf{W}_i &= -\frac{1}{2}\mathbf{S}_i^{-1} \\ \mathbf{w}_i &= \mathbf{S}_i^{-1}\mathbf{m}_i \\ w_{0i} &= -\frac{1}{2}\mathbf{m}_i^T \mathbf{S}_i^{-1}\mathbf{m}_i - \frac{1}{2}\ln(\det(\mathbf{S}_i)) + \ln(P(\omega_i)). \end{aligned}$$

This discriminant function is based on the Bayes decision rule [37], and is usually referred to as a *Maximum Likelihood* classifier. In the Maximum Likelihood classification procedure the feature vectors in each class are assumed to be distributed about the mean vector in a Gaussian fashion. The shape of the Gaussian 'bump' is determined by a covariance matrix and its location by a mean vector. The term "Maximum Likelihood" refers to the method of obtaining estimates for the covariance matrix and the mean vector and not to the classification procedure itself. We will not give a full motivation for these issues, as they are outside the scope of this book. The reader is referred to [37]. An example of Maximum Likelihood classification is given in Chapter 13.4.7.

The estimate used for the mean vector $\mathbf{m}_i$, is the d-dimensional *sample mean*

$$\mathbf{m}_i = \frac{1}{n_i} \sum_{\mathbf{x} \in \omega_i} \mathbf{x}. \tag{12.27}$$

The estimate used for the covariance matrix is the *sample covariance matrix* $\mathbf{S}_i$

$$\mathbf{S}_i = \frac{1}{n_i - 1} \sum_{\mathbf{x} \in \omega_i} (\mathbf{x} - \mathbf{m}_i)(\mathbf{x} - \mathbf{m}_i)^T. \tag{12.28}$$

The last term in w_{0i}, $\ln(P(\omega_i))$, is the a priori probability as expressed in the Bayes model. The discriminant function $g_i(\mathbf{x})$ is calculated for each pixel and the pixel is said to belong to the class i with the largest $g_i(\mathbf{x})$. It can be shown that when $g_i(\mathbf{x})$ is maximum then also $P(\omega_i|\mathbf{x})$ is maximum and the Bayes decision rule is followed [37].

It is possible to continue to add terms of type $w_{ijk}x_ix_jx_k$; this leads to the class of polynomial discriminant functions. In practice it is difficult to use polynomial terms of higher degree. It becomes increasingly difficult to implement and interpret the complex hypersurfaces in multiple dimensions. A quadratic discriminant function contains $\hat{d} = \frac{1}{2}(d+1)(d+2)$ terms. If the dimensionality is high, the computation of many terms is required. The components of the weight vector $\mathbf{w}$ must be determined from the available data, and the number of samples must be considerably greater than the dimensionality or the number of degrees of freedom. It is apparent that a general series expansion of $g(\mathbf{x})$ would easily lead to a completely unrealistic situation for computation and classifier training data.

12.6 TRAINING OF LINEAR DISCRIMINANT CLASSIFIERS

The discriminant functions to use for a particular classification situation can be selected in a variety of ways. Sometimes a priori knowledge is available, which makes it possible to compute the important parameters with great precision. Or, qualitative behavior of the classes may be known, but it is necessary to tune the discriminants in order to achieve acceptable performance on actual patterns. This tuning process is performed by using a set of patterns that are representative of the patterns to classify. We now consider how we can produce the discriminant functions, $g(\mathbf{x})$, that are suitable for a particular classification situation.

The $g_i(\mathbf{x})$ for each class can be found using the following iterative method:

1. Start with some initial set of $g_i(\mathbf{x})$, one for each class.
2. Use an $\mathbf{x}$ which is already classified, such that we know that $\mathbf{x}$ belongs to class j.

3. Calculate $g_i(\mathbf{x})$ for every class i. If the discriminant functions are appropriately tuned, than the sample $\mathbf{x}$ belonging to class j should make $g_j(\mathbf{x}) > g_i(\mathbf{x}),\ i \neq j$.
4. If on the other hand $g_j(\mathbf{x}) < g_i(\mathbf{x}),\ i \neq j$, then the separating hyperplane has to be modified:

   IF $g_i(\mathbf{x}) > g_j(\mathbf{x}),\ i \neq j$ THEN

$$\begin{aligned} \mathbf{w}_i &:= \mathbf{w}_i - c_1\mathbf{x} \\ w_{0i} &:= w_{0i} - c_2 \\ \mathbf{w}_j &:= \mathbf{w}_j + c_1\mathbf{x} \\ w_{0j} &:= w_{0j} + c_2 \end{aligned}$$

It can be shown that a proper correction procedure is to modify the separating hyperplanes by a certain fraction of the misclassified vector $\mathbf{x}$ [107]. The variables c_1 and c_2 determine the speed of convergence, and may be varied during the iteration process, typically decreased toward zero.

It should be observed that adding new features may not necessarily improve the classifier performance on new samples. Rather, experience has shown that the addition of new correlated features may contribute to erratic performance on new samples. The reason for this is that the estimated covariance matrix will be nearly singular, resulting in numerical problems. For this reason a variety of methods have been developed to reduce the feature space under consideration in order to obtain a problem that is both better defined in a statistical sense and more manageable.

12.7 PERCEPTRONS AND FEEDFORWARD NEURAL NETWORKS

It is possible to restate what has been said in the previous sections on linear classifiers in a somewhat different language, that of *perceptrons* [116, 102]. The reason for doing this is that it allows us to make the transition to the modern theory of feedforward nonlinear neural networks, a class of very powerful classifiers.

A *simple perceptron* is a device that takes an input vector $\mathbf{x} = (x_1, x_2, \ldots, x_N)$ and produces an output vector $\mathbf{O} = (O_1, O_2, \ldots, O_M)$, by means of a mapping

$$O_i = \sigma\left(\sum_{k=1}^{N} w_{ik}\, x_k - w_{i0}\right) = \sigma\left(\sum_{k=0}^{N} w_{ik}\, x_k\right) \tag{12.29}$$

where the second equality holds if we include a "dummy" input component x_0 which is always -1. w_{ik} is an element of the *weight matrix* $\mathbf{w}$. σ is a monotonic function. Typical choices of σ are $\sigma(y) = y$, a sign function

$$\sigma(y) = \begin{cases} 1 & y \geq 0 \\ -1 & y < 0 \end{cases} \tag{12.30}$$

or a sigmoid function, e.g. $\tanh(\beta y)$ or $1/(1+\exp(-2\beta y))$. The machinery that produces a particular summation and function mapping is called a *unit*, and a simple perceptron thus consists of a number of units organized in a layer, see Figure 12.13, where we have included the "dummy" input x_0 though this is not customary.

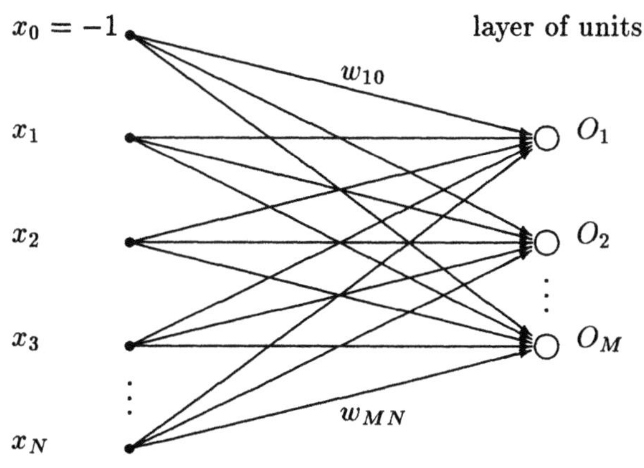

Figure 12.13 Simple perceptron with N input terminals and M output units. Note the "dummy" input terminal $x_0 = -1$, that is used for thresholding.

If we let $\mathbf{O} = O_1$ and $\sigma(y) = y$, the perceptron reduces to the linear two-category linear classifier. However, nothing prevents us from using nonlinear units.

We represent desired classification categories by output vectors $\hat{\mathbf{O}}(\mathbf{x})$, so that each input vector $\mathbf{x}$ is mapped to a corresponding output vector $\hat{\mathbf{O}}(\mathbf{x})$. We may, for instance, define three categories $\hat{\mathbf{O}}^1 = (1, 0, \ldots, 0)$, $\hat{\mathbf{O}}^2 = (0, 1, 0, \ldots, 0)$, $\hat{\mathbf{O}}^3 = (0, 0, 1, 0 \ldots, 0)$, and try to map input vectors onto these by adjusting the weights. To derive a learning algorithm for adjustment of the weights, we introduce a quadratic error function

$$E[\mathbf{w}] = \frac{1}{2} \sum_{i,p} (\hat{O}_i^p - O_i^p)^2 = \frac{1}{2} \sum_{i,p} (\hat{O}_i^p - \sigma(\sum_k w_{ik}\, x_k^p))^2 \tag{12.31}$$

for the training set $\{(\mathbf{x}^p, \hat{\mathbf{O}}^p)\}$, $p = 1 \ldots$ #examples. To minimize the error, we may use an optimization method, such as Newton's metod or the conjugate gradient method. The simplest choice is the classical gradient descent method

$$\Delta w_{ik} = -\epsilon \frac{\partial E}{\partial w_{ik}} = \epsilon \sum_p (\hat{O}_i^p - \sigma(y_i^p))\, \sigma'(y_i^p)\, x_k^p \tag{12.32}$$

where $y_i^p = \sum_k w_{ik} x_k^p$, and ϵ $(1 \geq \epsilon > 0)$ is the learning rate. This is usually referred to as the *Widrow-Hoff rule* [139]. We may use this rule incrementally, updating the weights after presentation of each input-output pair, so that

$$\Delta w_{ik} = \epsilon \Delta_i^p x_k^p \tag{12.33}$$

with $\Delta_i^p = (\hat{O}_i^p - \sigma(y_i^p))\, \sigma'(y_i^p)$. It is easy to see that if $\sigma(y) = y$ (linear units), this reduces to the learning rule for linear classifiers presented in the previous section. We see that a simple perceptron in this case basically is equivalent to a linear machine.

In a simple perceptron, we do not really gain anything from using nonlinear (continuous-valued) units, since this is equivalent to using linear units with output targets $\sigma^{-1}(\hat{O}_i^p)$ [53]. However, suppose now that we introduce an additional layer of units, as in Figure 12.14. (In the figure, there is in fact a coupling from the input threshold terminal x_0 to *all* units, although this has only been drawn for the first layer.) The new layer does not have direct access to the output values of the perceptron, and is referred to as a *hidden* layer. We know that each unit in the first layer works as a two-category linear classifier, generating a decision hyperplane in the feature space. What can we do with this new machine, a *multilayer* perceptron?

First, let us demonstrate that nothing is gained if we use linear units. In this case each hidden unit j generates an output $H_j = \sum_k w_{jk} x_k$, so that each

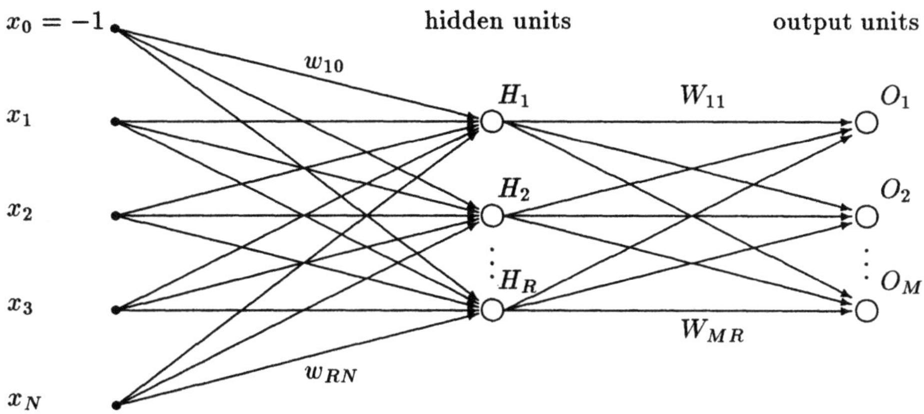

Figure 12.14 Two-layer perceptron with N input terminals, R hidden units and M output units. The "dummy" input x_0 feeds *all* units, but only the couplings to the first layer are shown.

output unit i produces

$$O_i = \sum_j W_{ij}\, H_j = \sum_j W_{ij} \sum_k w_{jk}\, x_k = \sum_k \left(\sum_j W_{ij}\, w_{jk} \right) x_k =$$

$$= \sum_k \tilde{W}_{ik}\, x_k \tag{12.34}$$

which could have been produced already in the first layer. Therefore, to produce more complex decision regions, we have to use nonlinear functions σ.

The decision region in Figure 12.15 can be produced with a network of sign functions or sigmoids, see Figure 12.16. In general, the decision regions of units in the second layer are *convex hulls* [93].

If we introduce yet another hidden layer, we see that an output unit receives convex hulls as inputs. By combining these one may construct decision regions of *arbitrary* shape. This means that no more than two layers are needed to have a *universal* classifier. Of course, the number of units in each hidden layer may have to be very large.

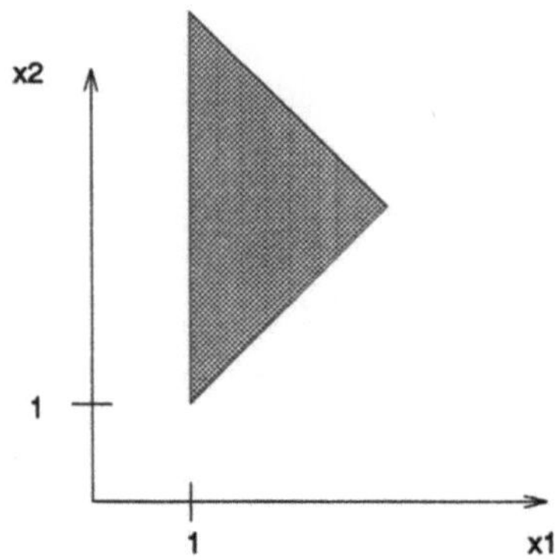

Figure 12.15 A convex decision region.

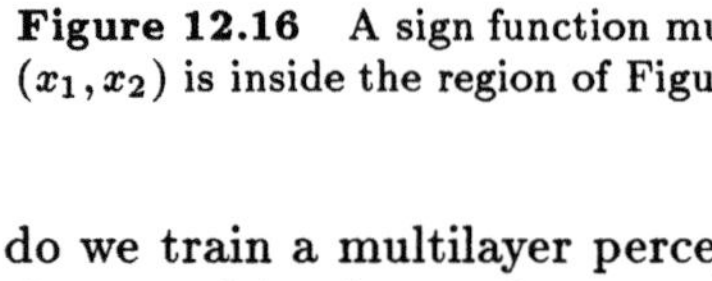

$$\mathbf{w}_1 = (1, 1, 0)$$
$$\mathbf{w}_2 = (-\sqrt{18}, -1/\sqrt{2}, -1/\sqrt{2})$$
$$\mathbf{w}_3 = (0, -1/\sqrt{2}, 1/\sqrt{2})$$
$$\mathbf{W}_1 = (2.5, 1.5, 1.5, 1.5)$$

Figure 12.16 A sign function multilayer perceptron that outputs a 1 only if (x_1, x_2) is inside the region of Figure 12.15.

How do we train a multilayer perceptron? We can use a generalization of the procedure used in the one-layer case [118]. Let us start with a two-layer net, and define an error function

$$E[\mathbf{w}, \mathbf{W}] = \frac{1}{2} \sum_{i,p} (\hat{O}_i^p - O_i^p(\{\mathbf{w}, \mathbf{W}\}))^2 =$$

$$\frac{1}{2} \sum_{i,p} (\hat{O}_i^p - \sigma(\sum_j W_{ij} \sigma(\sum_k w_{jk}\, x_k^p)))^2 \tag{12.35}$$

Gradient descent gives

$$\Delta W_{ij} = -\epsilon \frac{\partial E}{\partial W_{ij}} = \epsilon \sum_p (\hat{O}_i^p - O_i^p)\, \sigma'(y_i^p)\, H_j^p = \epsilon \sum_p \Delta_i^p H_j^p \tag{12.36}$$

for the hidden-to-output weights.

The adjustments of the input-to-hidden weights are somewhat more complex. The w_{jk}-weight is hidden inside the output $H_j^p = \sigma(\sum_k w_{jk} x_k^p)$ of the hidden unit j. We get

$$\Delta w_{jk} = -\epsilon \frac{\partial E}{\partial w_{jk}} = -\epsilon \sum_p \frac{\partial E}{\partial H_j^p} \frac{\partial H_j^p}{\partial w_{jk}} =$$

$$= \epsilon \sum_{p,i} (\hat{O}_i^p - O_i^p)\, \sigma'(y_i^p)\, W_{ij} \sigma'(y_j^p)\, x_k^p =$$

$$= \epsilon \sum_{p,i} \Delta_i^p W_{ij} \sigma'(y_j^p)\, x_k^p = \epsilon \sum_p \Delta_j^p x_k^p \tag{12.37}$$

with $\Delta_j^p = \sigma'(y_j^p) \sum_i W_{ij} \Delta_i^p$.

We see that the adjustments of the input-to-hidden layer weights are done by propagating the corrections of the hidden-to-output weights *backwards* through the network. The learning rule is called *error backpropagation.*

The procedure can be generalized to any number of layers. It will always look like

$$\Delta w_{out,in} = \epsilon \sum_p \Delta_{out}^p H_{in}^p \tag{12.38}$$

where Δ_{out}^p is either the output error if we update the hidden-to-output weights, or a backpropagated error, and H_{in}^p is either the input (if we update the input-to-hidden weights) or the activation of a unit in a preceding layer.

Usually the algorithm is used incrementally. The most common functions, σ, are sigmoids, but the requirement is that it be a differentiable function. As all gradient optimization methods, the algorithm may get locked in a *local minimum* of the error function. To overcome this, one may for instance add a small amount of noise either to the weight updates or to the input vectors. Local minima, however, usually do not cause great problems.

For more comprehensive presentations of the multilayer perceptron and other neural networks, see [119, 53].

12.8 CLUSTERING AND UNSUPERVISED CLASSIFICATION

There are a number of methods for carrying out classification without any a priori knowledge of size, shape, position or number of clusters in the feature space. Only the features are given. Such procedures are referred to as *unsupervised* classification procedures. These can be described as "self-learning". The algorithm learns and classifies simultaneously. The number of classes – the number of reasonably separated clusters in the feature space – is not known beforehand. We refer the interested reader to the pattern recognition literature [37, 35, 83], and restrict ourselves to an example.

12.8.1 A clustering algorithm

The following algorithm is based on the work of Mantaras [34] and uses a minimum distance classification procedure. The objective is to find a number of center points that characterize clusters as well as possible.

The feature vectors are fed into the algorithm sequentially, one at a time. Assume that we already have a number of center points, $\mathbf{m}_i$, $i = 1 \ldots M$ where M is the number of center points defined by the algorithm. The distances d_i from the new feature vector $\mathbf{x}_k$ to these center points are calculated and the shortest distance d_j is selected. If this distance is less than a certain predefined threshold T, then we say that the pixel that corresponds to the feature vector belongs to class j. We now want to adjust the center point $\mathbf{m}_i$ so that it takes into account $\mathbf{x}_k$ as well. This can be done with the following equation, where $\mathbf{m}_j$ becomes

$$\mathbf{m}_j^* = \mathbf{m}_j + \frac{K_0}{k}(\mathbf{x} - \mathbf{m}_j), \tag{12.39}$$

where k indicates that this is the k:th pixel that falls into this class.

If $d_j \geq T$, then the feature vector is not very close to any of the existing center points. In this case we increment the number of classes – clusters – by one and allow the center point for this new class be $\mathbf{x}$.

If T is small and anyone of the classes has a large variance, then this class might be divided into several subclasses. The result of the clustering consequently needs to be interpreted and each class in the image obtained should be mapped on a 'real' class. This may be a many-to-one mapping. The opposite, one-to-many mapping, is of course an indication of a failure.

13

TEXTURE ANALYSIS

13.1 INTRODUCTION

Texture is the term used to characterize the surface of a given object and it is unquestionably one of the main features used in computer vision and pattern recognition. Computer vision researchers have used measures of texture to segment scenes and discriminate between different objects. Psychophysicists have studied texture for knowledge about human low-level visual information processing. It has been assumed that a solution to the texture analysis problem will greatly advance the computer vision and pattern recognition field. But, in spite of its importance, texture lacks a precise definition.

Informally texture may be defined as an organized area phenomenon, composed of a large number of more or less ordered similar patterns or primitives, giving rise to a perception of surface homogeneity. These primitives or texture elements are often referred to as *texels*. It is possible to have texture within the primitives themselves and this nonhomogeneity is known as *microtexture*. The spatial organization or interaction between the primitives of the image texture may be random or may have a dependence on neighboring primitives. This definition of texture gives rise to texture properties that are local as well as global. A relevant description must, consequently, consider both local and global attributes of the texture.

Figure 13.1 (left) shows a texture with a relatively ordered array of nearly identical texels. If the texture is composed of identical, perfectly ordered texels it is called deterministic. Such a texture can be exactly described by the characteristics of the texels and by some placement rule defining the spatial organization of the texels. This way to represent a texture is called a *structural approach*.

Figure 13.1 **Left:** A fairly deterministic texture (cotton canvas). **Right:** A comparatively stochastic texture (grass lawn). Both images are taken from from Brodatz's album .

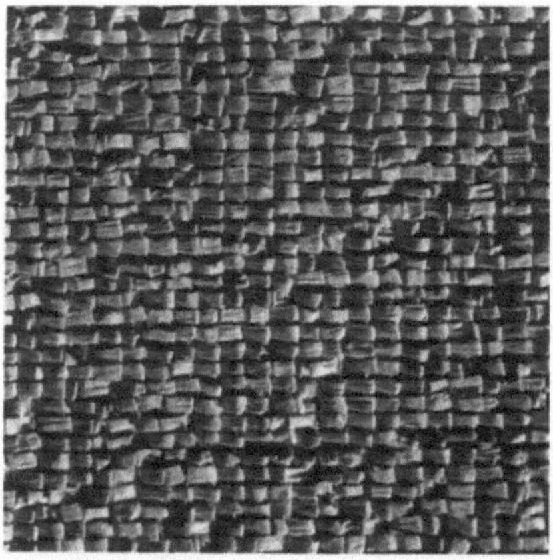

Figure 13.2 **Left:** A structural texture with variations in the texel description as well as in the placement rule (raffia). **Right:** A texture with high degree of randomness in the texel description and in the placement rule (straw). Both images are taken from from Brodatz's album .

Figure 13.1 (right) shows a more stochastic texture obeying some random laws. Observable textures are generally somewhere between deterministic and stochastic. They can be seen as deterministic textures with disturbed spatial organizations and with similar, but not identical, primitives. A typical example is shown in Figure 13.2 (left).

The album *Textures* by Phil Brodatz [24] is a photographic album with 112 textures that has become a standard reference in texture analysis. We shall refer to these textures as the *Brodatz textures.* The examples in Figure 13.1 and Figure 13.2 indicate the range of variation in real world textures.

13.2 ISSUES IN TEXTURE ANALYSIS

Texture analysis plays a fundamental role in classifying objects and outlining significant regions in an image. The main issues of concern in texture analysis may be described as follows.

13.2.1 Discrimination

Given a textured region in an image, to which of a finite number of classes does the sample belong? This problem is usually solved by extracting some characteristic parameters for each one of the given classes. These parameters are then used as input features for classification using techniques of statistical pattern recognition (see Chapter 12). One example of discrimination is given in 13.4.7.

13.2.2 Segmentation

Given a total scene in an image, how can the boundaries between regions of conspicuously different textures be extracted? The segmentation problem is often solved by first applying a feature extraction algorithm followed by a segmentation rule. The purpose of texture feature extraction is to map differences in spatial structures into differences in some representation. Segmentation rules then analyze the representation space in order to extract homogeneous regions. Sometimes there is a need for a method to select or reduce features if the number of features is too large [115]. Discrimination is performed as well, since the discrimination problem is embedded in the segmentation problem.

Feature extraction methods can classically be divided into feature-based, structural, and model-based ones. Examples of the two first methods are given in Sections 13.4 and 13.5. Structural methods look for detectable primitives, texels, and their corresponding placement rules. In feature-based methods, some characteristic features of the textures are extracted and regions with relatively constant features are searched. Model-based methods presume some underlying stochastic or fractal processes for textures and perform segmentation by comparing the parameters of these models.

Segmentation rules are often classified as either region-based or boundary-based, where the first attempts to detect feature homogeneity, while the second seeks feature inhomogeneity. Boundary-based methods often use an extension of the common edge detector. Instead of applying the edge detector to the image itself, it is applied to the texture feature field derived from the image. This operation extracts discontinuities in texture representation defined over some neighborhood; the result is sometimes called *texture gradient.*

13.3 HUMAN ASPECTS OF TEXTURE PERCEPTION

The main goal of any computational formalism for the texture problems above is the identification of easily computed and physically meaningful texture attributes that can be used for segmentation, discrimination, or description. Clues for the search for these texture attributes may be found in the mechanisms of early vision in humans. This may include physiological conformity as well as psychophysical plausibility in terms of perception.

13.3.1 Preattentive versus attentive vision

Investigations claim that human vision operates in two distinct modes, *preattentive vision* and *attentive vision* [63]. The human preattentive visual system performs a parallel, instantaneous perception over a large visual field and exhibits an ability to discriminate between certain types of patterns. This capacity is based on preattentive extraction and processing of some particular features. In psychophysical experiments on preattentive texture perception, the subject is required to perform a discrimination task after being presented with a pair of images for a duration of about 50 ms. A notable aspect of texture perception is that discrimination between textures is a much simpler task than identification of textures. Most psychophysical experiments on texture perception have focused on the preattentive aspect of perception. Experiments indicate that there may be a certain limited number of processing levels for preattentive vision. Julesz suggested that only variations up to second order can be discriminated for. Although counterexamples have later been found to this postulate applied to statistical moments, it may still be true that preattentive vision uses, say, two levels of processing in terms of still unknown primitives [63].

Attentive vision performs a serial search by *focal attention* in steps of approximately 50 ms. If the texture discrimination requires element-by-element scrutiny, this activity is directed by the focal attention which can be seen as a small "spotlight" moving serially over the visual array [63].

There are two different problems for psychological research [128]. One is to define which features or properties are the visual primitives of early vision. The other addresses how these primitives are combined to form preattentive perception. Models of texture segregation fall into at least two different classes which are described below [126].

13.3.2 Textons

In the first class of models, the segregation is based on differences in the first-order statistics of simple texture features, such as sizes and slopes of texture elements [99]. Julesz introduced the theory of *textons* and proposed that they are the fundamental features, responsible for the preattentive discrimination of textures [63]. These conspicuous features are elongated blobs like rectangles,

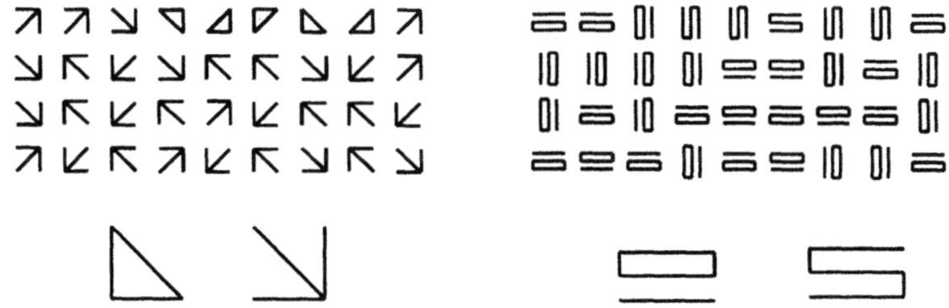

Figure 13.3 **Left:** Two regions which are possible to discriminate by preattentive vision. The elements differ in the number of line terminators (0 and 3). **Right:** The same regions are now impossible to discriminate preattentively because the two elements have the same number of line terminators (2 and 2).

ellipses or line segments, with specific properties including color, orientation, width, length, crossings of line segments and ends of lines (terminators). Texton detection may be consistent with the feature detection model of early visual processing that presumes the involvement the responses of the bar- and edge-sensitive receptive fields [22]. Only differences in the textons or in their density can be preattentively detected while the positional relationship between neighboring textons passes unnoticed.

Figure 13.3 demonstrates that line terminators are textons. The two elements in Figure 13.3 (left) are composed of three identical line segments with the same length, width, and orientation. The triangle-shaped element has no line terminators, while the other element has three line terminators. There is an apparent preattentive discrimination between the two regions. The elements in Figure 13.3 (right) are composed of three horizontal and two vertical line segments and both contain two line terminators. It is possible to perceive the positional relationship between these textons but this is done by the attentive visual system, not by the preattentive visual system. The preattentive system can only count the density of the textons but disregard their relative positions. The two regions are therefore impossible to discriminate without focal attention.

13.3.3 Spatial frequency channels

In the second model of preattentive segregation of textures the primitives are not geometric features but the output from operators with properties like those of the receptive fields of simple cells. Texture segregation is primarily based on differences in the outputs of these *spatial frequency channels.* Their responses are functions of the algebraic summation of the excitation and inhibition resulting from convolving a stimulus with a receptive field. Spatial frequency channels can be seen as parallel channels composed of local receptive fields that are distributed over the visual field and are similar in their sensitivity to spatial frequency and orientation [126]. Spatial frequency techniques for texture segmentation are based on image representations that indicate the frequency content in localized regions in the spatial domain. These methods are consequently able to achieve high resolution in both the spatial and frequency domains.

13.4 FEATURE-BASED TEXTURE ANALYSIS

We have seen in the preceding sections that local orientation and spatial frequency are the primitives in the spatial frequency model of preattentive vision. The apparent approach is to estimate these primitives by convolving the image with filter combinations tuned to these properties.

13.4.1 Estimation of local orientation using quadrature filters

Estimates of orientation and frequency, produced by combining the outputs from quadrature filter pairs with different orientations and frequency characteristics, are obvious elements in a computational approach for texture analysis. The quadrature filters are local in both the spatial and frequency domains and accordingly they achieve high resolution in both domains [69] [70].

We will recall major properties of the concept of quadrature filters and refer to Chapter 6 and 8 for a more complete treatment. The output $\mathbf{q}_k$ from quadrature filter k is a complex number where the magnitude $q_k = |\mathbf{q}_k|$ has phase invariant properties and the argument $\arg(\mathbf{q}_k)$ represents the local phase of the signal in the filter direction. Quadrature filters are polar separable and in this application they have a directional function

$$\begin{cases} D_k(\hat{\mathbf{u}}) = (\hat{\mathbf{u}} \cdot \hat{\mathbf{n}}_k)^2 = cos^2(\phi) & if \ \mathbf{u} \cdot \hat{\mathbf{n}}_k > 0 \\ D_k(\hat{\mathbf{u}}) = 0 & otherwise \end{cases} \tag{13.1}$$

where ϕ is the angle difference between the frequency vector $\mathbf{u}$ and the direction vector $\hat{\mathbf{n}}_k$ of the filter. The radial function $R(\rho)$ is chosen as a lognormal function described by

$$R(\rho) = e^{-\frac{4}{ln 2 B^2} ln^2(\rho/\rho_i)} \tag{13.2}$$

where B is the relative bandwidth, $\rho = |\mathbf{u}|$, and ρ_i is the center frequency of the filter.

The local orientation in two dimensions is estimated by a combination of the outputs from four lognormal quadrature filters. For texture segmentation the vector representation of local orientation is used and this is given by

$$\mathbf{z} = \sum_{k=1}^{4} q_k e^{i(k-1)\frac{\pi}{2}} \tag{13.3}$$

where q_k is the magnitude of the output $\mathbf{q}_k$ from filter k.

The structural content in the original image is transformed into a slowly varying vector field describing spatial orientation. This transform may be seen as a differentiation operation based on the variation in the image. An important aspect is that the vector field does not contain the high frequency components present in the texture but only a description of the structural properties of the texture and how these properties vary in the image [41].

13.4.2 Texture gradient

A differentiation operation can operate upon a vector field as well as on a scalar field. A second transformation upon the vector field will give high outputs only for spatial positions on boundary lines between regions with significant differences in the feature described by the vector field. In Chapter 11 the estimation of vector edges is denoted by

$$\mathbf{T} = \sum_{k} \left(\sum_{l} q_{kl}^2 \right)^{\frac{1}{2}} \mathbf{M}_k \tag{13.4}$$

where q_{kl} denotes the magnitude response of quadrature filter k obtained from filtering vector component l. $\mathbf{M}_k$ is a tensor associated with quadrature filter k. The tensor $\mathbf{T}$ represents the local variation in the vector field. In the case of two dimensions, a vector $\mathbf{v}$ describing these changes can be composed as

$$\mathbf{v} = \sum_{k=1}^{4} \left(\sum_{l=1}^{2} q_{kl}^2 \right)^{\frac{1}{2}} e^{i(k-1)\frac{\pi}{2}}. \tag{13.5}$$

This implies that the magnitude of $\mathbf{v}$ corresponds to the strength of the variations in the vector field and the orientation of $\mathbf{v}$ describes the direction of the variation. The result from this operation may be seen as a *texture gradient* and can be used for segmentation purposes as shown in the forthcoming examples.

13.4.3 Relaxation of vector fields

The vector representation of local orientation described above is in many cases very suitable for texture segmentation. However, in some situations there is a need of relaxation of a vector field. Vectors that point in more or less the same direction add up to a large value, while vectors with a lack of consistency, sum to a small vector as described in Chapter 9. The degree of stabilization in the field, caused by the relaxation, is directly coupled to a reduction of spatial resolution [70]. The problem of finding a suitable size for the relaxation operator is an important task, and the size depends primarily on the nonuniformity of the estimates and the size of the region. Let the orientation vector $\mathbf{z}$ be decomposed in the following way:

$$\mathbf{z} = c_z \; \hat{\mathbf{z}} \tag{13.6}$$

where $c_z = |\mathbf{z}|$ and $\hat{\mathbf{z}} = e^{i arg(\mathbf{z})}$. This means that c_z denotes the certainty of the estimate and $\hat{\mathbf{z}}$ is a unit vector describing the local orientation. The relaxation of the local orientation $\hat{\mathbf{z}}$ is implemented by normalized convolution of $\hat{\mathbf{z}}$, and a, the applicability function defined in Equation (9.13), in the following way:

$$\hat{\mathbf{z}}' = \frac{a \, * \, c_z \, \hat{\mathbf{z}}}{a \, * \, c_z} \tag{13.7}$$

The unit vector $\hat{\mathbf{z}}'$ represents the relaxed estimate of local orientation and the certainty of the relaxed orientation estimate, $c_{z'}$, is given by:

$$c_{z'} = (a \, * \, c_z)^{-1} \tag{13.8}$$

The scalar $c_{z'}$ can be interpreted as the length of the relaxed vector representation of local orientation. The applicability function a, defined in Equation (9.13), describes the locality of the relaxation operation; in the examples the parameters have been $\alpha = 0$ and $\beta = 2$. An illustration of this function appears in Figure 9.11. The size of the implemented applicability filter function a used in Section 13.4.4 and 13.4.7 is 9×9 pixels.

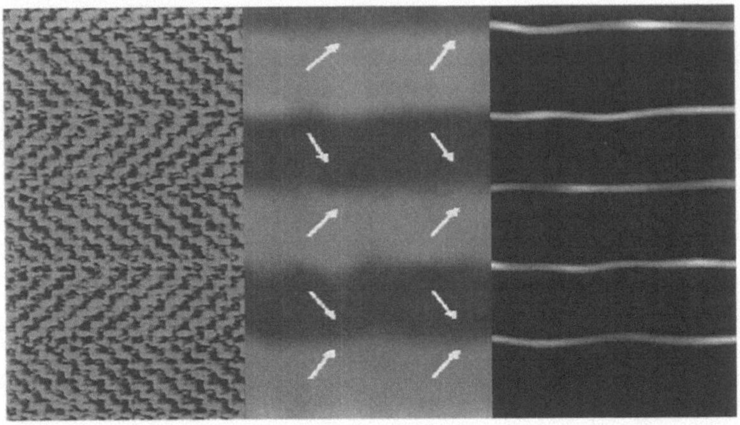

Figure 13.4 **Left:** Brodatz-texture herringbone weave. **Middle:** Orientation estimate described by the vector representation. **Right:** Segmentation by second order transform.

13.4.4 Example of segmentation by local orientation estimation

An example of segmentation based on differences in local orientation is given in Figure 13.4 in which the input image is the Brodatz-texture herringbone weave. Here, the vector field representing local orientation has been exposed to relaxation as described by Equation (13.7); the effect of this is an increased stability of the estimate within each region. The vector field with relaxed local orientation estimates is shown in Figure 13.4 (middle) where $\arg(\hat{\mathbf{z}}')$, i.e. the arguments describing different local orientations, are mapped into gray scale. Some vectors have been superimposed to make this clearer. The segmentation result from the magnitude of the second differentiation operation is presented in Figure 13.4 (right), where the texture gradient between regions with consistent orientations is emphasized.

13.4.5 Texture segmentation by local frequency estimation

The frequency estimation is performed as described in Chapter 8. In this case there are three sets of quadrature filters with different frequency weighting which implies that N is three. Each set contains two quadrature filters with orthogonal directional vectors $\hat{\mathbf{n}}_k$. Let $\mathbf{q}_{ki}$ be the complex output from the filter with directional vector $\hat{\mathbf{n}}_k$ and a frequency characteristic associated with filter set number i. By summing the magnitude of the filter responses from one filter set we obtain an orientation independent estimate of the signal energy in the corresponding frequency interval. This is due to the the choice of $\cos^2(\phi)$ as a directional function, as shown in Equation (13.1). Let the orientation-independent energy estimate from filter set i be denoted by

$$q_i = \sum_{k=1}^{2} q_{ki}. \tag{13.9}$$

where $q_{ki} = |\mathbf{q}_{ki}|$. The relative bandwidth B of the filters is in this case $2\sqrt{2}$ and the center frequencies are chosen to be an octave apart. The center frequencies for the three filter sets are $\pi/8, \pi/4$ and $\pi/2$. Equation (8.28) states that the estimated frequency ρ can be written as follows:

$$\rho = \rho_0 [\sum_{i=1}^{N-1} q_i]^{-1} \sum_{i=1}^{N-1} 2^{i+0.5} q_{i+1} \tag{13.10}$$

The certainty measure c_f of the estimated frequency is based on the variance σ^2 of the local spectrum, defined in Chapter 8 as

$$\sigma^2 = [\sum_{i=1}^{N-1} q_i^2]^{-1} \sum_{i=1}^{N-1} q_i^2 [2^{i+0.5} \frac{q_{i+1}}{q_i} - \rho]^2 \tag{13.11}$$

and

$$c_f = \frac{1}{1+\sigma^2}. \tag{13.12}$$

By combining ρ and c_f into the vector representation of frequency described in Section 1.6.2 we obtain a vector field where the certainty is the vector length and the frequency estimate is represented by the vector orientation. It is now possible to segment between image regions with different frequency characteristics by using the second-level differentiation operation, defined in Section 13.4.2, on the vector field.

13.4.6 Example of segmentation using local frequency

Figure 13.5 (top) is a synthesized image containing four different texture fields consisting of band-passed Gaussian noise with $\pi/2, \pi/3, \pi/5$, and $\pi/8$ as center frequencies. Frequency estimates for this image are shown in Figure 13.5 (lower left). The estimate of local frequency is displayed in gray scale. Some vectors are overlaid to show the vector representation of local frequency. The magnitude of the second differentiation operation is shown in Figure 13.5 (lower right), where the texture gradients between regions with homogeneous frequency estimates are emphasized.

13.4.7 Example of classification using local orientation and frequency

The example of Color image 6 illustrates the discrimination capability that use of local orientation and frequency gives in texture analysis. The estimates are extracted and relaxed by the methods described in this chapter and are then used for supervised classification. See Chapter 12 for a more complete description of classification. Color image 6 (top left) contains a mixture of the textures in Section 13.1; (top right) shows the training areas marked in color corresponding to the different texture classes. The vectors describing orientation and frequency in Color image 6 (middle right) and (left) are coded in color and it is apparent that there are significant differences in these features between the different textures.

The vector representations of orientation $\mathbf{z}$ and frequency $\mathbf{f}$ are split into their cartesian components z_1, z_2, f_1 and f_2. Each component is used as a feature in the classification procedure. There are, consequently, four different features for training and classification, and this means that the estimates of orientation and frequency, as well as their corresponding certainties are used.

Figure 13.5 **Top:** Original image containing colored noise. **Bottom left:** Frequency estimates. **Bottom right:** Segmentation by second order transform.

Projections of clusters from the training areas into two dimensional histograms, where the features used are basis vectors, are shown in Color image 6 (bottom right). Training data from different textures are marked with their corresponding colors; the bases in the histograms are (z_1, z_2), (f_1, f_2), (f_1, z_1) and (f_1, z_2) ordered in a clockwise sense starting with the top left projection. We can see that each cluster is well separated from the others in at least one projection. This implies that it should be possible to distinguish the different textures from each other. The result from the Maximum Likelihood classifier in Color image 6 (bottom left) shows that most errors in the classification are located on the boundaries between different textures. This is because the estimates of local orientation and frequency at these positions are based on more than one texture. The Maximum Likelihood classifier is described in Chapter 12.5.

13.5 STRUCTURAL APPROACH USING CERTAINTY GATING

In the framework of low-level image processing we can assume that we have access to estimates and certainties of local features, e.g. orientation, frequency and quadrature filter phase. These features have different significance in different spatial positions depending on the purpose of the next processing step as well as on the content of the image. If the significance of a feature can be extracted in some way, we will obtain a new certainty measure for that specific feature. This can be interpreted as a *region of significance* for that feature. This is illustrated in Figure 13.6, which shows an original image containing some objects. If the purpose is to perform a closer inspection of the centers or the boundaries of these objects, the regions of significance are those shown. A high value corresponds to high significance for that spatial position. The question is now how to extract this measure of importance. A feature which behaves well for this purpose is quadrature filter phase. The reason for this is that the phase shows a large degree of invariance to changes in scale and contrast. Its general properties are described in Chapter 7.

A combination of statements from different feature descriptors offers a possibility of extracting information about higher order features. This aggregation process can be seen as a "gating" procedure, where statements from different descriptors verify or reject a certain hypothesis at issue. The gating may be a "soft thresholding" depending on the importance of the local position for the actual task. When quadrature filter phase is used to perform this soft thresholding, the procedure is called *phase gating.*

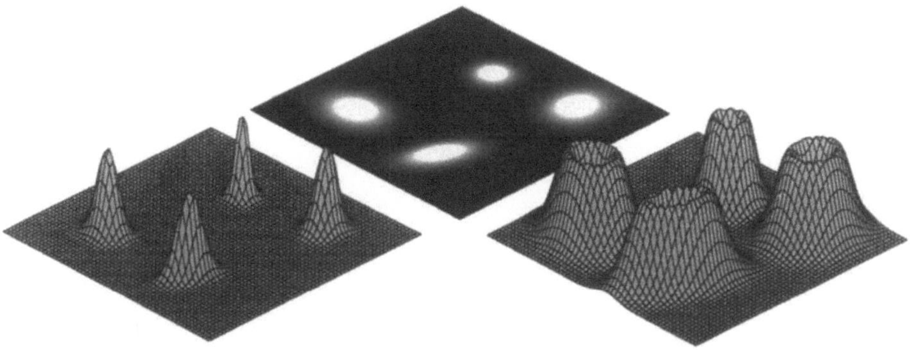

Figure 13.6 **Top:** Original image. **Left:** Regions of significance for object centers. **Right:** Regions of significance for object boundaries.

13.5.1 Phase gated frequency for description of size

In Section 13.1 structural textures are considered to be defined by texture elements that occur more or less repeatedly according to placement rules. In some texture related problems such as shape from texture the task can be to separate small texture elements from larger ones. This can be done by including the information from quadrature filter phase in the segmentation process.

Local spatial frequency is known to be a good size descriptor and this property is further improved when it is combined with the phase gating process. The purpose of this is to ensure that only neighborhoods where the spatial frequency has high relevance as a size descriptor are utilized in the texel description [47]. The phase describes the local relation of symmetry and antisymmetry in the neighborhood; it is computed by combining the outputs from quadrature filters with different orientations. The continuous vector description of phase in two dimensions is shown in Figure 7.3. In Chapter 1.10, this phase is represented by

$$\begin{cases} x_1 &= \lambda_1 \cos(\varphi)\sin(\theta) \\ x_2 &= \lambda_1 \sin(\varphi)\sin(\theta) \\ x_3 &= \lambda_1 \cos(\theta) \end{cases} \tag{13.13}$$

where

$$\begin{array}{lll} \lambda_1 & \in [0, \infty] & : \text{largest eigenvalue} \\ \varphi & \in [0, \pi[& : \text{orientation of largest eigenvector} \\ \theta & \in [-\pi, \pi[& : \text{phase.} \end{array}$$

The estimate of the size of an object should be invariant to the local orientation φ. Consequently, the phase used in the phase gating process must have this characteristic. The invariance can be achieved by using that

$$\begin{aligned} \sqrt{x_1^2 + x_2^2} &= \lambda_1 \sqrt{\cos(\varphi)^2 \sin(\theta)^2 + \sin(\varphi)^2 \sin(\theta)^2} \\ &= \lambda_1 |\sin(\theta)| = \lambda_1 \sin(|\theta|) \end{aligned} \tag{13.15}$$

in the following definition of the "rectified phase" $\mathbf{p}$.

$$\mathbf{p}(\lambda_1, \theta) = \lambda_1 e^{i|\theta|} = \lambda_1 [\cos(|\theta|) + i \sin(|\theta|)] = x_3 + i\sqrt{x_1^2 + x_2^2}. \tag{13.15}$$

This description of rectified phase, shown in Figure 13.7, disregards whether the local structure corresponds to an asymmetric signal with phase $\theta = \pi/2$ or $\theta = -\pi/2$, i.e. it shows invariance to the sign of an edge. Consequently, in the case of rectified phase, the problem discussed in Chapter 7.3 of defining the reference direction of $\hat{\mathbf{e}}_1$, the largest eigenvector, is no longer relevant. The certainty of the statement, i.e. the magnitude λ_1, corresponds to the signal energy.

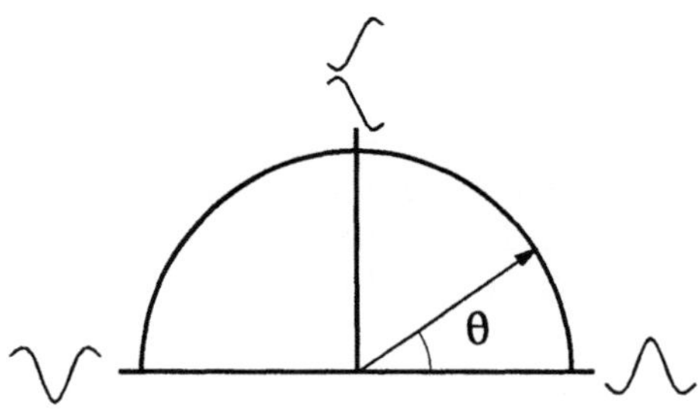

Figure 13.7 Illustration of rectified phase as a complex vector disregarding the sign of an edge.

The estimate of the size of an object should not be affected by how sharp the edges of the object are, thus, only areas in the local neighborhood with phase symmetry should be involved in the size description. A solution to this problem is to use a nonlinear function Γ to produce an updated certainty factor which describes the degree of local symmetry in combination with the certainties of the estimates of phase and frequency. This updated certainty factor marks the region of significance for frequency as a size descriptor.

The local frequency estimate ρ and its corresponding certainty c_f are given in Equations (13.10) and (13.12). Let c_g be the updated phase gated certainty factor describing the regions of significance for the frequency estimate ρ. It can be defined as

$$c_g = \Gamma(c_f, \lambda_1, \theta) \tag{13.16}$$

where Γ is a nonlinear function. It turns out that a suitable function Γ for the issues described above is defined as

$$\Gamma = \begin{cases} c_f \ \lambda_1 \ \cos(\theta)^\alpha & |\,\theta\,| \le \pi/2 \\ 0 & |\,\theta\,| > \pi/2. \end{cases} \tag{13.17}$$

The positive constant α defines how soft the threshold of local symmetry should be. If α is near zero, the function Γ will be approximately a rect(θ/π). On the other hand, if α is chosen equal to one, we will have a more smooth threshold. Larger values of α will give rise to a narrower gating region. Examples of this appear in Figure 13.8, where the certainties c_f and λ_1 are assumed to be equal to one. Equation (13.17) ensures that the certainty measure c_g is larger then zero only in areas where local frequency is a good size descriptor.

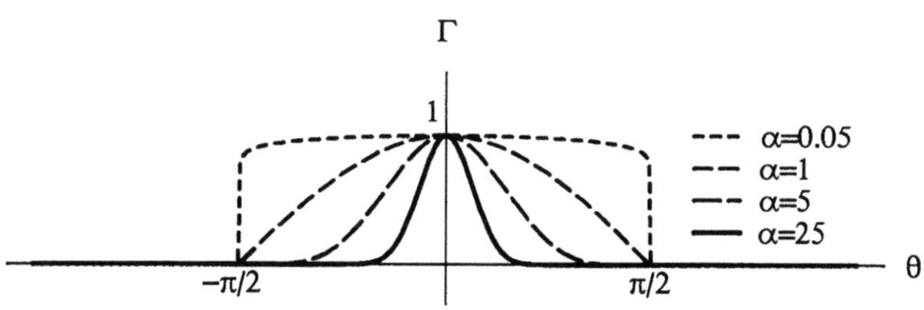

Figure 13.8 Γ as a function of θ for different values of α.

The gating process ensures that only symmetric areas, which are bright in the middle compared to the surroundings, will be regarded as interesting. High values of α will give rise to a rejection of components at all positions except those exactly on the center of a symmetric area. An example of this is shown in Figure 13.6 (left). By imposing other restrictions on the phase in Equation (13.17) it is possible to constrain the regions of significance into the boundaries of the objects where $\theta = \pi/2$ or into locally dark areas where $\theta = \pi$. The latter will lead to a phase gated frequency description of the background; this represents the placement rules describing spatial relations between the texels rather then their size.

13.5.2 Illustrative example of phase gating

The concept of phase gating is illustrated in Figure 13.9. The task is to segment between areas with different object sizes. The original image in Figure 13.9 (top left) contains objects whose radii are in the ratio 1.5:1. The rectified phase estimate is shown in Figure 13.9 (top right) where $|\theta|$ is mapped into gray scale. White corresponds to a light local symmetric input, i.e. an object which is bright compared to its surroundings ($\theta = 0$). Gray corresponds to an antisymmetric input, i.e. an edge ($|\theta| = \pi/2$). Phase vectors are superimposed to clarify the image content.

The phase gated frequency estimate is shown in Figure 13.9 (bottom left). In this example the parameter α in Equation (13.17) is set to one. The region of significance, i.e. the certainty c_g, is displayed in gray scale, with white implying high certainty and black implying low. The certainty c_g is zero everywhere except on the center regions of the objects. The frequency estimate ρ is displayed by the superimposed vectors whose lengths are equal to the corresponding certainties c_g. It is apparent that the difference in size between the objects is well transformed into a difference in the phase gated frequency estimate. It is now possible to segment between the areas with different object size. This can be done by applying a second differentiation operation to the phase gated frequency estimate. The magnitude of this operation is overlayed on the original image in Figure 13.9 (bottom right).

13.5.3 Differentiation of phase gated features

The phase gated segmentation algorithm described in the example above gives output only in areas where the differentiation operator spatially covers contrasting statements within regions of high significance. When the operator is in a region with only one object, it produces no segmentation. If the phase gating step is omitted and the segmentation is based only on the original fre-

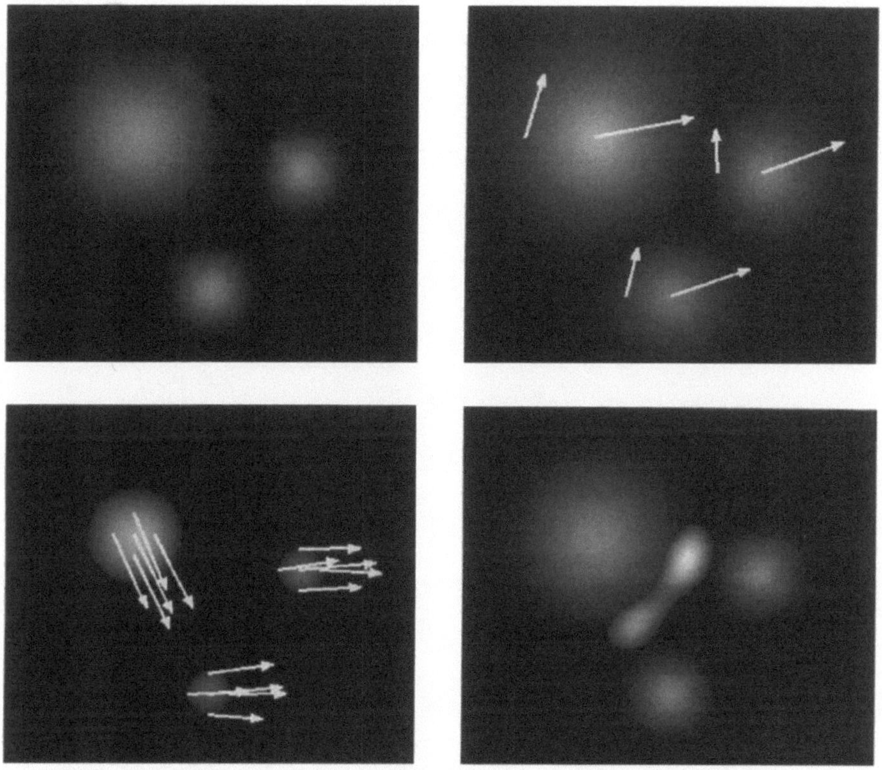

Figure 13.9 Top left: Original image. **Top right:** Rectified phase estimate with the vector representation. **Bottom left:** Regions of significance and corresponding phase gated frequency. **Bottom right:** Segmentation results superimposed on the original image.

quency estimate ρ and c_f, this would not be the case. An application where this behavior is desired is given below.

13.5.4 Example of phase gating for size segmentation

Figure 13.10 shows a more realistic texture example containing plastic pellets in two different sizes (from Brodatz' album [24]). The textures can be described in a structural manner by their texels and distribution laws. The goal is to segment between the two regions of different texel size. A frequency approach alone is insufficient for this task, as there are large variations in the distance between the texels within each region. This will give rise to significant variations in the frequency estimates within the regions as well as between the regions. Using the concept of phase gating, the problem can be solved in the same way as in the example shown in Figure 13.9. The parameter α is set to one.

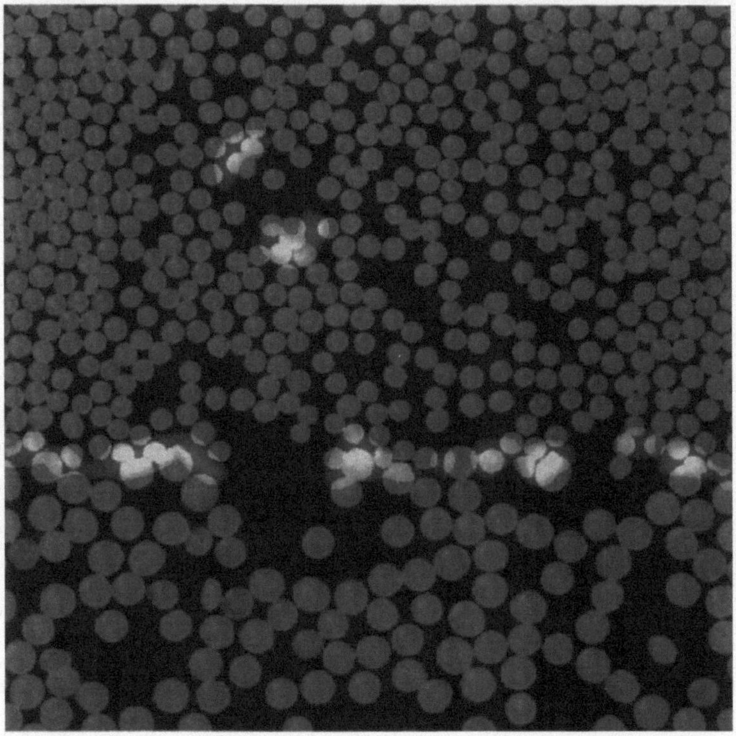

Figure 13.10 Segmentation of regions containing different texel sizes using phase gating on the frequency estimate.

The magnitude from the second differentiation operation on the phase gated frequency estimate is superimposed on the original image in Figure 13.10. There are two areas in the upper part of the image where the second differentiation operation indicates local differences in the sizes of the texels, and the reason for this is that pellets overlayed over each other will be seen as one larger pellet. The result from the segmentation algorithm is otherwise very satisfactory.

REFERENCES

[1] J. F. Abramatic and L. M. Silverman. Non-stationary linear restoration of noisy images. In *Proceedings IEEE Conf. on Decision and Control*, pages 92–99, Fort Lauderdale, FL., 1979.

[2] J. B. Allen and L. R. Rabiner. A unified approach to short-time fourier analysis and synthesis. In *Proc. IEEE*, volume 65:11, pages 1558–1564, November 1977.

[3] M. Andersson. *Controllable Multidimensional Filters in Low Level Computer Vision.* PhD thesis, Linköping University, Sweden, S–581 83 Linköping, Sweden, September 1992. Dissertation No 282, ISBN 91–7870–981–4.

[4] F. Attneave. Some informational aspects of visual perception. *Psychological Review*, 61, 1954.

[5] D. H. Ballard. *Vision, Brain, and Cooperative Computation*, chapter Cortical Connections and Parallel Processing: Structure and Function. MIT Press, 1987. M. A. Arbib and A. R. Hanson, Eds.

[6] D. H. Ballard. Animate vision. Technical Report 329, Computer Science Department, University of Rochester, Feb. 1990.

[7] D. H. Ballard and C. M. Brown. *Computer Vision.* Prentice-Hall, 1982.

[8] H. Bårman. *Hierarchical Curvature Estimation in Computer Vision.* PhD thesis, Linköping University, Sweden, S–581 83 Linköping, Sweden, September 1991. Dissertation No 253, ISBN 91–7870–797–8.

[9] H. Bårman, G. H. Granlund, and H. Knutsson. A new approach to curvature estimation and description. In *3rd International Conference on Image Processing and its Applications*, pages 54–58, Warwick, Great Britain, July 1989. IEE. ISBN 0 85296382 3 ISSN 0537–9989.

[10] H. Bårman, G. H. Granlund, and H. Knutsson. Tensor field filtering and curvature estimation. In *Proceedings of the SSAB Symposium on Image*

Analysis, pages 175–178, Linköping, Sweden, March 1990. SSAB. Report LiTH–ISY–I–1088, Linköping University, Sweden, 1990.

[11] J. Bigün. Recognition of local symmetries in gray value images by harmonic functions. In *The 9th International Conference on Pattern Recognition*, Rome, Italy, November 1988. Report LiTH–ISY–I–0893, Computer Vision Laboratory, Linköping University, Sweden, 1988.

[12] J. Bigün and G. H. Granlund. Optimal orientation detection of linear symmetry. In *Proceedings of the IEEE First International Conference on Computer Vision*, pages 433–438, London, Great Britain, June 1987. Report LiTH–ISY–I–0828, Computer Vision Laboratory, Linköping University, Sweden, 1986.

[13] J. Bigün, G. H. Granlund, and J. Wiklund. Multidimensional orientation estimation with applications to texture analysis and optical flow. *IEEE Transactions on Pattern Analysis and Machine Intelligence*, 13(8):775–790, August 1991. Report LiTH–ISY–I–1148, Linköping University, Sweden, 1990.

[14] J. Bigün, G. H. Granlund, and J. Wiklund. Multidimensional orientation: texture analysis and optical flow. *IEEE Transactions on Pattern Analysis and Machine Intelligence*, PAMI–13(8), August 1991.

[15] C. Blakemore, editor. *Vision: coding and efficiency.* Cambridge University Press, 1990.

[16] C. Blakemore and F. W. Campbell. On the existence of neurones in the human visual system selectively sensitive to the orientation and size of retinal images. *J. Physiol.*, 203:237–260, 1969.

[17] C. Blakemore and P. Sutton. Size adaptation: a new aftereffect. *Science*, 166:245–247, 1969.

[18] G. Blasdel and G. Salama. Voltage sensitive dyes reveal a modular organisation in monkey striate cortex. *Nature*, 321:579–585, 1986.

[19] T. V. P. Bliss and T. Lømo. Long lasting potentiation of synaptic transmission in the dentate area of the anaesthetized rabbit following stimulation of the perforant path. *J.Physiol.(London)*, 232:331–356, 1973.

[20] B. Boashash. Estimating and interpreting the instantaneous frequency of a signal - part 1: Fundamentals. *Proceedings of the IEEE*, 80(4), 1992.

[21] B. Boashash. Estimating and interpreting the instantaneous frequency of a signal - part 2: Algorithms and applications. *Proceedings of the IEEE*, 80(4), 1992.

[22] A. C. Bovik, M. Clark, and W. S. Geisler. Multichannel texture analysis using localized spatial filters. *IEEE Transactions on Pattern Analysis and Machine Intelligence*, 12(1):55–73, January 1990.

[23] R. Bracewell. *The Fourier Transform and its Applications*. McGraw-Hill, 2nd edition, 1986.

[24] P. Brodatz. *Textures*. Dover, New York, 1966.

[25] V. B. Brooks. *The Neural Basis of Motor Control*. Oxford University Press, 1986.

[26] V. Bruce and P. R. Green. *Visual Perception: physiology, psychology and ecology*. Lawrence Erlbaum Associates, second edition, 1990.

[27] P. J. Burt and E. H. Adelson. Merging images through pattern decomposition. In *Applications of digital image procesing VIII*. SPIE, 1985. vol. 575.

[28] J. Carson and T. Fry. Variable frequency electric circuit theory with application to the theory of frequency modulation. *Bell Syste Tech. J.*, 16:513–540, 1937.

[29] R. Chellappa and A. A. Sawchuk. Image enhancement, reconstruction and restoration. In *Volume 1: Digital Image Processing*, chapter 4 and 5, pages 426–695. IEEE Computer Society, 1985.

[30] S. Clippingdale and R. Wilson. Quad-tree image estimation: a new image model and its application to minimum mean square error image restoration. In *Proc. 5th Scand. Conf. on Image Anal.*, 1987.

[31] S. Coren and L. M. Ward. *Sensation & Perception*. Harcourt Brace Jovanovich, Publishers, San Diego, USA, 3rd edition, 1989. ISBN 0–15–579647–X.

[32] H. S. M. Coxeter. *Introduction to Geometry*. John Wiley & Sons, Inc., 1961.

[33] J. R. Cronly-Dillon, editor. *Vision and visual dysfunction*. The Macmillan Press, 1991. Series.

[34] R. Lopez de Mantaras and J. Aquilar-Martin. Self-learning pattern classification using a sequential clustering technique. *Pattern Recognition*, pages 271–277, 1985.

[35] P. A. Devijver and J. Kittler. *Pattern Recognition: A Statistical Approach.* Prentice-Hall, London, 1982.

[36] R. M. Dolezal, T. N. Mudge, J. L. Turney, and R. A. Volz. Determining the pose of an object. In O. D. Faugeras and Robert Kelley, editors, *Computer Vision for Robots*, pages 68–71, 1986. Proc. SPIE 595.

[37] R. O. Duda and P. E. Hart. *Pattern classification and scene analysis.* Wiley-Interscience, New York, 1973.

[38] D. J. Fleet. *Measurement of image velocity.* Kluwer Academic Publishers, 1992. ISBN 0–7923–9198–5.

[39] W. T. Freeman and E. H. Adelson. The design and use of steerable filters for image analysis, enhancement, and wavelet representation. Technical report, Vision and Modeling Group, Media Lab., MIT, Cambrige, September 1990.

[40] D. Gabor. Theory of communication. *Proc. Inst. Elec. Eng.*, 93(26):429–441, 1946.

[41] G. Granlund. Description of texture using the general operator approch. In *5th International Conference on Pattern Recognition*, pages 776–779, 1980.

[42] G. H. Granlund. In search of a general picture processing operator. *Computer Graphics and Image Processing*, 8(2):155–178, 1978.

[43] G. H. Granlund. Hierarchical image processing. In *Proceedings of SPIE Technical Conference*, Geneva, Switzerland, April 1983.

[44] G. H. Granlund, J. Arvidsson, and H. Knutsson. GOP, a paradigm in hierarchical image processing. In *Proceedings of The First IEEE Computer Society International Symposium on Medical Imaging and Image Interpretation, ISMI II'82*, Berlin, Federal Republic of Germany, October 1982.

[45] G. H. Granlund and H. Knutsson. Contrast of structured and homogenous representations. In O. J. Braddick and A. C. Sleigh, editors, *Physical and Biological Processing of Images*, pages 282–303. Springer Verlag, Berlin, 1983.

[46] T. Gustavsson and S. Nivall. Adaptive spatio-temporal filtering of ultrasound image sequences. In *Proceedings of the SSAB Conf. on Image Analysis*, pages 104–108, Gothenburg, Sweden, March 1989. SSAB.

[47] L. Haglund. *Adaptive Multidimensional Filtering.* PhD thesis, Linköping University, Sweden, S–581 83 Linköping, Sweden, October 1992. Dissertation No 284, ISBN 91–7870–988–1.

[48] L. Haglund, H. Knutsson, and G. H. Granlund. On phase representation of image information. In *The 6th Scandinavian Conference on Image Analysis*, pages 1082–1089, Oulu, Finland, June 1989.

[49] L. Haglund, H. Knutsson, and G. H. Granlund. Scale analysis using phase representation. In *The 6th Scandinavian Conference on Image Analysis*, pages 1118–1125, Oulu, Finland, June 1989.

[50] A. R. Hansen and E. M. Riseman. Constructing semantic models in the visual analysis of scenes. In *Proceedings Milwaukee Symp. Auto. & Contr. 4*, pages 97–102, 1976.

[51] M. Hedlund, G. H. Granlund, and H. Knutsson. A consistency operation for line and curve enhancement. In *The Computer Society Conference on PR&IP*, Anaheim, California, June 1982.

[52] P. Heggelund and K. Albus. Orientation selectivity of single cells in striate cortex of cat: The shape of orientation tuning curves. *Vision Research*, 18:1067–1071, 1978.

[53] J. Hertz, A. Krogh, and R. G. Palmer. *Introduction to the Theory of Neural Computation.* Addison-Wesley, 1991.

[54] W. C. Hoffman. The Lie algebra of visual perception. *Journal Math. Psychol.*, 3:65–98, 1966.

[55] J. J. Hopfield. Neural networks and physical systems with emergent collective computational capabilities. *Proceedings of the National Academy of Sciences*, 79:2554–2558, 1982.

[56] B. K. P. Horn. *Robot vision.* The MIT Press, 1986.

[57] D. H. Hubel. *Eye, Brain and Vision*, volume 22 of *Scientific American Library.* W. H. Freeman and Company, 1988. ISBN 0–7167–5020–1.

[58] D. H. Hubel and T. N. Wiesel. Receptive fields, binocular interaction and functional architecture in the cat's striate cortex. *J. Physiol.*, 160:106–154, 1962.

[59] D. H. Hubel and T. N. Wiesel. Receptive fields and functional architecture in two non–striate visual areas (18 and 19) of the cat. *J. Physiol.*, 28:229–289, 1965.

[60] D. H. Hubel and T. N. Wiesel. Receptive fields and functional architecture of monkey striate cortex. *J. Physiol.*, 195:215–243, 1968.

[61] D. H. Hubel and T. N. Wiesel. Brain mechanisms of vision. *Scientific American*, September 1978.

[62] D. H. Hubel, T. N. Wiesel, and S. LeVäy. Plasticity of ocular dominance columns in monkey striate cortex. *Philos. Trans. R. Soc.*, B 278:377–409, 1977.

[63] B. Julesz. Textons, the elements of texture perception, and their interactions. *Nature*, 290(12):91–97, March 1981.

[64] E. R. Kandel and J. H. Schwartz. Molecular biology of learning: modulation of transmitter release. *Science*, 218:433–443, 1982.

[65] H. Knutsson. *Filtering and Reconstruction in Image Processing.* PhD thesis, Linköping University, Sweden, 1982. Diss. No. 88.

[66] H. Knutsson. Producing a continuous and distance preserving 5-D vector representation of 3-D orientation. In *IEEE Computer Society Workshop on Computer Architecture for Pattern Analysis and Image Database Management - CAPAIDM*, pages 175–182, Miami Beach, Florida, November 1985. IEEE. Report LiTH–ISY–I–0843, Linköping University, Sweden, 1986.

[67] H. Knutsson. Representing local structure using tensors. In *The 6th Scandinavian Conference on Image Analysis*, pages 244–251, Oulu, Finland, June 1989. Report LiTH–ISY–I–1019, Computer Vision Laboratory, Linköping University, Sweden, 1989.

[68] H. Knutsson and H. Bårman. Robust orientation estimation in 2D, 3D and 4D using tensors. In *Proceedings of International Conference on Automation, Robotics and Computer Vision*, September 1992.

[69] H. Knutsson and G. H. Granlund. Fourier domain design of line and edge detectors. In *Proceedings of the 5th International Conference on Pattern Recognition*, Miami, Florida, December 1980.

[70] H. Knutsson and G. H. Granlund. Texture analysis using two-dimensional quadrature filters. In *IEEE Computer Society Workshop on Computer Architecture for Pattern Analysis and Image Database Management - CAPAIDM*, Pasadena, October 1983.

[71] H. Knutsson, L. Haglund, and G. Granlund. Adaptive filtering of image sequences and volumes. In *Proceedings of International Conference on Automation, Robotics and Computer Vision*, September 1992.

[72] H. Knutsson, L. Haglund, and G. H. Granlund. Tensor field controlled image sequence enhancement. In *Proceedings of the SSAB Symposium on Image Analysis*, pages 163–167, Linköping, Sweden, March 1990. SSAB. Report LiTH–ISY–I–1087, Linköping University, Sweden, 1990.

[73] H. Knutsson, M. Hedlund, and G. H. Granlund. Apparatus for determining the degree of consistency of a feature in a region of an image that is divided into descrete picture elements. Swedish patent 8502570-8 1986 (U.S. Patent 4.747.152, 1988), 1986.

[74] H. Knutsson, B. von Post, and G. H. Granlund. Optimization of arithmetic neighborhood operations for image processing. In *Proceedings of the First Scandinavian Conference on Image Analysis*, Linköping, Sweden, January 1980.

[75] H. Knutsson and C-F Westin. Normalized and differential convolution: Methods for interpolation and filtering of incomplete and uncertain data. In *Proceedings of IEEE Computer Society Conference on Computer Vision and Pattern Recognition*, New York City, USA, June 1993. IEEE.

[76] H. Knutsson, R. Wilson, and G. H. Granlund. Anisotropic non-stationary image estimation and its applications — part I: Restoration of noisy images. *IEEE Trans on Communications*, COM–31(3):388–397, March 1983. Report LiTH–ISY–I–0462, Linköping University, Sweden, 1981.

[77] H. E. Knutsson, P. Edholm, G. H. Granlund, and C. U. Petersson. Ectomography. A new radiographic reconstruction method. I Theory and error estimates. *IEEE Trans. on Biomedical Engineering*, BME–27(11):640–645, November 1980. Report LiTH–IST–I–0242.

[78] J. J. Koenderink. *Solid Shape.* The MIT Press, 1990.

[79] J. J. Koenderink and A. J. van Doorn. The singularities of the visual mapping. *Biol. Cybernetics*, 24:51–59, 1976.

[80] J. J. Koenderink and A. J. van Doorn. The structure of images. *Biological Cybernetics*, 50:363–370, 1984.

[81] W. Köhler and H. Wallach. Figural after–effect, an investigation of visual processes. In *Proc. Am. Philos. Soc.*, volume 88, pages 269–357, 1944.

[82] T. Kohonen. *Self-organization and associative memory.* Springer–Verlag, Berlin, third edition, 1989.

[83] P. R. Krishnaiah and L. N. Kanal. Classification, pattern recognition, and reduction of dimensionality. In *Handbook of Statistics.* North-Holland, Amsterdam, 1982.

[84] E. H. Land. The retinex theory of colour vision. *Scientific American*, 237:108–128, 1977.

[85] S. R. Lehky and T. J. Sejnowski. Neural model of stereoacuity and depth interpolation based on a distributed representation of stereo acuity. *Journal of Neuroscience*, 10:2281–2299, 1990.

[86] R. Lenz. *Group Theoretical Methods in Image Processing.* Springer-Verlag, 1990.

[87] D. Letalick, I. Renhorn, and 0. Steinvall. Noise sources in laser radar systems. *Applied Optics*, 28:2657–2665, 1989.

[88] S. LeVay, D. H. Hubel, and T. N. Wiesel. The pattern of ocular dominance columns in macaque visual cortex revealed by a reduced silver stain. *J. Comp. Neurol.*, 159:559–576, 1975.

[89] M. W. Levine and J. M. Shefner. *Fundamentals of sensation and perception.* Addison–Wesley, 1981.

[90] L. M. Lifshitz. Image segmentation via multiresolution extrema following. Tech. Report 87-012, University of North Carolina, 1987.

[91] J. Limb, C. Rubinstein, and J. Thompson. Digital coding of color video signals - a review. *IEEE Trans. on Communication*, 25(11):530–533, November 1977.

[92] R. Linsker. Development of feature-analyzing cells and their columnar organization in a layered self-adaptive network. In Rodney M. L. Cotteril, editor, *Computer Simulation in Brain Science*, chapter 27, pages 416–431. Cambridge University Press, 1988.

[93] R. P. Lippmann. An introduction to computing with neural nets. *ASSP Magazine*, pages 4–22, April 1987.

[94] M. Livingstone and D. Hubel. Segregation of form, color, movement, and depth: Anatomy, physiology, and perception. *Science*, 240:740–749, 1988.

[95] D. M. MacKay. *Information, Mechanism and Meaning.* M.I.T. Press, Cambridge, Massachusetts and London, England, 1969.

[96] P. R. MacLeish, H. Knutsson, and J. H. Stern. The control of the rod outer segment conductance by cyclic-gmp and divalent cations. *Photobiochemistry and Photobiophysics*, 13:359–372, 1986.

[97] B. Maclennan. Gabor representations of spatiotemporal visual images. Technical Report CS-91-144, Computer Science Department, University of Tennesse, September 1981.

[98] K. V. Mardia. *Statistics of Directional Data.* Academic Press, 1972.

[99] D. Marr. Early processing of visual information. AI Memo, 340, MIT, Cambridge, Mass., 1975.

[100] C. McCollough. Colour adaption of edge-detectors in the human visual system. *Science*, 149:1115–1116, 1965.

[101] R. Milanese. Focus of attention in human vision: a survey. Technical Report 90.03, Computing Science Center, University of Geneva, Geneva, August 1990.

[102] M. Minsky and S. Papert. *Perceptrons.* M.I.T. Press, Cambridge, Mass., 1969.

[103] M. Mishkin, L. G. Ungerleider, and K. A. Macko. Object vision and spatial vision: Two cortical pathways. *Trends in Neurosciences*, 6:414–417, 1983.

[104] J. A. Movshon, I. D. Thompson, and D. J. Tolhurst. Spatial summation in the receptive fields of simple cells in the cat's striate cortex. *J. Physiol.*, 283:53–77, 1978.

[105] W. J. H. Nauta and M. Feirtag. *Fundamental Neuroanatomy.* W.H. Freeman and Company, New York, 1986.

[106] I. Newton. *Opticks.* Dover Publications, New York, fourth edition, 1952. Reprint.

[107] N. J. Nilsson. *Learning Machines.* McGraw-Hill, 1965.

[108] G. Oesterberg. Topography of the layer of rods and cones in the human retina. *Acta Ophtalmol. Suppl.*, 6:1–103, 1935.

[109] T. Pavlidis. *Structural Pattern Recognition.* Springer-Verlag, New York, 1977.

[110] D. I. Perett, E. T. Rolls, and W. Caan. Visual neurones responsive to faces in the monkey temporal cortex. *Experimental Brain Research*, 47:329–342, 1982.

[111] P. Perona. Steerable-scalable kernels for edge detection and junction analysis. In *Proceedings ECCV Conf. on Computer Vision*, pages 3–18, 1992.

[112] A. Peters and E. G. Jones, editors. *Cerebral Cortex*, volume 3, Visual Cortex. Plenum, New York, 1985.

[113] W. H. Press, B. P. Flannery, S. A. Teukolsky, and W. T. Vetterling. *Numerical Recipes*. Cambridge University Press, 1986.

[114] S. Ramón y Cajal. *Histologie du système nerveux*, volume II. C.S.I.C, Madrid, 1955.

[115] T. R. Reed and J. M. H. du Buf. A review of recent texture segmentation and feature extraction techniques. *Computer Vision, Graphics, and Image Processing: Image Understanding*, 57(3):359–372, May 1993.

[116] F. Rosenblatt. *Principles of Neurodynamics: Perceptrons and the theory of brain mechanisms*. Spartan Books, Washington, D.C., 1962.

[117] A. Rosenfeld, R. Hummel, and S. W. Zucker. Scene labeling by relaxation operations. *IEEE Trans.s on Systems, Man and Cybernetics*, 6:420–422, 1976.

[118] D. E. Rumelhart, G. E. Hinton, and R. J. Williams. Learning representations by back-propagating errors. *Nature*, 323:533–536, 1986.

[119] D. E. Rumelhart, J. L. McClelland, and the PDP Research Group. *Parallel Distributed Processing: Explorations in the Microstructure of Cognition*. MIT Press, Cambridge, 1986. 2 vols.

[120] T. D. Sanger. Stereo disparity computation using gabor filters. *Biological cybernetics*, 59:405–418, 1988.

[121] P. H. Schiller, B. L. Finlay, and S. F. Volman. Quantitative studies of single cells in monkey striate cortex. II. Orientation specificity and ocular dominance. *J. Neurophysiol.*, 39:1320–1333, 1976.

[122] P. H. Schiller and N. K. Logothetis. The color–opponent and broad–band channels of the primate visual system. *Trends in Neurosciences*, 13:392–398, 1990.

[123] J. Serra. *Image Analysis and Mathematical Morphology.* Academic Press, 1982.

[124] D. L. Sparks and J. S. Nelson. Sensory and motor maps in the mammalian superior colliculus. *Trends in Neurosciences*, 10:312–317, 1987.

[125] J. H. Stern, H. Knutsson, and P. R. MacLeish. Divalent cations directly affect the conductance of excised patches of rod photoreceptor membrane. *Science*, 236, June 1987.

[126] A. Sutter, J. Beck, and N. Graham. Contrast and spatial variables in texture segregation: Testing a simple spatial-frequency channels model. *Perception and Psychophysics*, 46(4):312–332, 1989.

[127] S. L. Tanimoto and T. Pavlidis. A hierarchical data structure for picture processing. *Computer Graphics and Image Processing*, 2:104–119, June 1975.

[128] A. Treisman. Preattentive processing in vision. *Computer Vision, Graphics, and Image Processing*, 31:156–177, 1985.

[129] L. Uhr. Layered 'recognition cone' networks that preprocess, classify and describe. In *Proceedings Conf. on Two-Dimensional Image Processing*, 1971.

[130] B. van der Pol. The fundamental principles of frequency modulation. *Proceedings of the IEEE*, 93:153–158, 1946.

[131] Esprit basic research action 3038, vision as process, final report. Project document, April 1992.

[132] H. Wässle, B. B. Boycott, and J. Roehrenbeck. Horizontal cells in the monkey retina. Cone connections and dendritic network. *Eur. J. Neurosci.*, pages 421–435, 1989.

[133] U. Wehmeier, D. Dong, C. Koch, and D. Van Essen. Modeling the mammalian visual system. In C. Koch and I. Segev, editors, *Methods in neuronal modelling: from synapses to networks.* MIT Press, 1989.

[134] C-J Westelius. Preattentive gaze control for robot vision, June 1992. Thesis No. 322, ISBN 91–7870–961–X.

[135] C-J Westelius, H. Knutsson, and G. H. Granlund. Focus of attention control. In *Proceedings of the 7th Scandinavian Conference on Image Analysis*, pages 667–674, Aalborg, Denmark, August 1991. Pattern Recognition Society of Denmark.

[136] G. Westheimer. Spatial sense of the eye. *Investigative Ophtalmology and Visual Science*, 18:893–912, 1979.

[137] C-F Westin. Feature extraction based on a tensor image description, September 1991. Thesis No. 288, ISBN 91–7870–815–X.

[138] C-F Westin. *A Tensor Framework for Multidimensional Signal Processing.* PhD thesis, Linköping University, Sweden, S–581 83 Linköping, Sweden, 1994. Dissertation No 348, ISBN 91–7871–421–4.

[139] B. Widrow and M. E. Hoff. Adaptive switching circuits. In *1960 IRE WESCON Convention Record*, pages 96–104, New York, 1960.

[140] J. Wiklund, L. Haglund, H. Knutsson, and G. H. Granlund. Time sequence analysis using multi-resolution spatio-temporal filters. In *The 3rd International Workshop on Time-Varying Image Processing and Moving Object Recognition*, pages 258–265, Florence, Italy, May 1989. Invited Paper. Report LiTH–ISY–I–1014, Computer Vision Laboratory, Linköping University, Sweden, 1989.

[141] H. R. Wilson. Development in of spatiotemporal mechanisms in infant vision. *Vision research*, 28(5):611–628, 1988.

[142] R. Wilson, A. D. Calway, and E. R. S. Pearson. A Generalized Wavelet Transform for Fourier Analysis: The Multiresolution Fourier Transform and Its Application to Image and Audio Signal Analysis. *IEEE Trans. on Information Theory*, 38(2):674–690, 1992.

[143] R. Wilson and G. H. Granlund. The uncertainty principle in image processing. *IEEE Transactions on Pattern Analysis and Machine Intelligence*, PAMI–6(6), November 1984. Report LiTH–ISY–I–0576, Computer Vision Laboratory, Linköping University, Sweden, 1983.

[144] R. Wilson and H. Knutsson. Uncertainty and inference in the visual system. *IEEE Transactions on Systems, Man and Cybernetics*, 18(2), March/April 1988.

[145] R. Wilson and H. Knutsson. A multiresolution stereopsis algorithm based on the Gabor representation. In *3rd International Conference on Image Processing and Its Applications*, pages 19–22, Warwick, Great Britain, July 1989. IEE. ISBN 0 85296382 3 ISSN 0537–9989.

[146] R. Wilson and M. Spann. *Image segmentation and uncertainty.* Research Studies Press, 1988.

[147] R. Wilson and M. Spann. A new approach to clustering. *Pattern Recognition*, 23:1413–1425, 1990.

[148] A. Witkin. Scale-space filtering. In *8th Int. Joint Conf. Artificial Intelligence*, pages 1019–1022, Karlsruhe, 1983.

[149] R. H. Wurtz and M. E. Goldberg, editors. *The neurobiology of saccadic eye movements.* Elsevier, New York, 1989.

[150] S. Zeki. The visual image in mind and brain. In *Mind and Brain: Readings from Scientific American Magazine.* W. H. Freeman and Company, New York, 1993. ISBN 0–7167–2376–X.

[151] S. Zeki and S. Shipp. The functional logic of cortical connections. *Nature*, 335:311–317, 1988.

INDEX

Abstraction hierarchy, 7
Active Vision, 1
Adaption, 85–86
Adaptive filtering, 309, 311
Affine coordinate changes, 136
Aliasing, 185
Amacrine cell, 52, 56–57
Analytic function, 167
Analytic signal, 164, 167
  application of, 169
  in multiple dimensions, 172
  in one dimension, 165
Angular coordinates
  in the Fourier domain, 124
  in the spatial domain, 124
Anti-Hermitian function, 141
Argument representation, 300
Associative, 26
Asymptotic behaviour, 143
Attentive vision, 402
Averaging, 304
Axon, 43
Backpropagation, 395
Bessel function, 160
Binary image, 98
Bipolar cell, 52, 54
Blind spot, 49
Blob, 91
Bottom-up, 33
Box classification, 371
Brodatz texture, 400–401
Cartesian separability, 153
Cascaded operations, 111
Central symmetry, 23
Cerebral cortex, 69
Channel representation, 88
Chromaticity coordinates, 63
Chromaticity diagram, 65
Circular symmetries, 350
Class membership, 37
Classification, 367
Clustering, 396
Color constancy, 62
Color image, 99
Color space
  RGB, 64
  XYZ, 64
Color vision, 59
Color-coding cells, 61
Compact representation, 106
Compatibility, 26, 37, 112
Complex cell, 71–72, 76, 81
Complex convolution, 102
Complex representation, 101
Cone, 49
Consistency operation, 356
Consistency, 37
Continuous Fourier transform, 117
Control tensor, 313
Controllable filter, 310
Conventions, 123
Convolution, 13, 135
  and multiplication, 136
Cortex
  cerebral, 69
  inferior temporal, 95
  layers, 70
  neocortex, 70
  posterior parietal, 95
  primary visual, 68, 70, 91

secondary visual, 68, 91
striate, 68, 70, 91
Cross-orientation inhibition, 83
Curvature estimation, 361
Dendrite, 43
Depolarization, 46
DFT, 117, 192
definition of, 193
inverse, 194
versus Fourier transform, 194
Direction selective cells, 76, 82
Directional coordinates, 123
Discrete Fourier transform, 117, 192
definition of, 193
inverse, 194
Discriminant functions, 371
Disparity, 88
Double-angle, 19
Dual elements, 240, 322
Dual tensors, 240, 322
End-stopped cells, 76, 83
Energy function, 147
Equivariance, 226
Equivariant, 298
Even functions, 141
Feature, 368
Feedback, 93
Filter implementations, 80
Filter synthesis, 312
Focal attention, 403
Fourier domain, 132
angular coordinates, 124
radial coordinates, 124
Fourier slice theorem, 151
Fourier transform, 131
and derivatives, 139
and discrete functions, 180
and moments, 139
basic theorems for, 139
continuous, 117
definition of, 131
dimensionality of, 117
discrete, 117, 192
global, 119
inverse, 134
linearity, 139
of periodic functions, 186
of truncated functions, 187
one-dimensional, 132
other definitions of, 133
pairs, 177
preserves even and oddness, 143
relative symmetry of, 135
separability of, 153
statistical calculations with, 174
three-dimensional, 133
two-dimensional, 133
versus DFT, 194
weighted, 121
local, 119
Fovea, 49
Frame, 4
Frequency domain, 132
Frequency estimation, 279
Frequency function, 132
Frequency variable, 132
Frequency
local, 259
Function
analytic, 167
energy, 147
even, 141
Hermitian, 141
odd, 141
periodic, 186
rank of, 156
sampled, 180
simple, 156
truncated, 187
Gabor function, 280
Ganglion cell, 52, 61
W, 57, 90
X, 57, 67, 90

Y, 57, 67, 90
Global Fourier transform, 119
Graded signals, 48
Group theory, 21
Habituation, 48
Hankel transform, 160
Heaviside function, 126
Hering's color theory, 61
Hermitian function, 141
Hierarchy, 3
Higher order fields, 343
Hilbert transform, 165
Horizontal cell, 52, 54, 56
Hypercomplex cell, 71
Hyperpolarization, 50
Image enhancement, 309
Image primitive, 7
Image restoration, 309
Impulse function, 127
multi-dimensional, 128
one-dimensional, 127
Inferior temporal cortex, 95
Instantaneous amplitude, 171
Instantaneous frequency, 171, 279
Instantaneous phase, 171
Invariance, 2, 226
Invariant, 299
Inverse Fourier transform, 134
Ion channel, 46
Ion pump, 45
Kanitza triangle, 93
Kernel optimization, 199
Lateral geniculate nucleus, 43, 67
Lateral posterior nucleus, 69
Learning, 48
LGN, 43, 67
Local dominance, 106
Local Fourier transform, 119
Local one-dimensionality, 18, 106
Local orientation, 220
Locality, 3
Lognormal filter, 282
Long-term potentiation, 48
Magno system, 67, 88
Matrix, 38
McCollough effect, 87
Mean, 146
Memory, 48
Microtexture, 399
Minimum distance, 378
Model selectivity, 356
Modulation transfer function, 84
Moments, 139
Monochrome, 98
Multi-dimensionality, 99
Multilayer perceptron, 392
Myelin, 46
Neocortex, 70
Nerve impulse, 45
Neural network, 390
Neural transmitters, 47
Neuron, 43
Nodes of Ranvier, 46
Normalized averaging, 306
Normalized convolution, 306
Notations, 123
Nucleus
lateral geniculate, 43, 67
lateral posterior, 69
pulvinar, 69
superior colliculus, 43, 68, 89
Object recognition, 95
Ocular dominance columns, 78
Odd functions, 141
Off-center cell, 53
On-center cell, 53
Opponent-color theory, 61
Optic chiasm, 42
Optic nerve, 42
Optic radiations, 67
Optic tectum, 43, 68
Optic tract, 42
Optimal, 150
Orientation columns, 77

Orientation estimation, 219
Orientation selectivity, 72
Orientation tensor, 227
Orientation vector, 243
Parabolic symmetries, 354
Parseval's theorem, 140
Parvo system, 67, 88
Perceptron, 390
Periodic function, 186
Phase gating, 413
Phase invariance, 231
Phase, 169
  local, 259
Pherical coordinates, 124
Photopic vision, 59
Poisson's summation formula, 183
Polar coordinates, 124
Population coding, 88–89
Posterior parietal cortex, 95
Power theorem, 140
Preattentive vision, 2, 402
Primary visual cortex, 43, 68, 70, 91
Projection slice theorem, 151
Pulvinar nucleus, 69
Purkinje shift, 59
Pyramid, 5
Quadrature filter, 172, 230, 262
Quadrature, 18
Radial coordinate
  in the Fourier domain, 124
  in the spatial domain, 124
Rank of function, 156
Receptive field, 52, 84
Receptor, 49, 59
  cone, 49
  density, 49
  pigment, 49
  rod, 49
Rectangular function, 125
Rectified phase, 414
Reentrant paths, 93
Reference function, 200
Refractory time, 46
Regions of significance, 412
Relaxation, 37
Resolution pyramids, 266
Retina, 42, 49
Rhodopsin, 49
Rod, 49
Rotational symmetries, 350
Rotational symmetry, 23
Saccadic eye movements, 69
Sample covariance matrix, 388
Sample function, 128
Sample mean, 388
Sampled functions, 180
Sampling process, 180
Sampling theorem, 184
Sampling
  locality aspects of, 190
Scalar image, 98
Scalar product, 123
Scalar representation, 300
Scale hierarchy, 5
Scale-space, 6, 266
Scotopic vision, 59
Segmentation, 401, 406, 416
Separability, 153
  Cartesian, 153
  spherical, 162
Separators, 114
Shah function, 128
Shah function
  multidimensional, 130
Shift-variant filter, 311
Sign function, 126
Simple cell, 71–72, 76, 80
Simple function, 23, 156, 224, 344
Simple perceptron, 390
Sinc function, 125
Size description, 413
Spatial domain, 132
  angular coordinates, 124

radial coordinate, 124
Spatial frequency channels, 404
Spatial frequency, 27
Spatial function, 132
Spatial variable, 132
Spatial vision, 95
Spectrally opponent processes, 60
Spectrum variance estimation, 289
Sperical symmetry, 159
Spherical separability, 162
Standard observer, 63
Steerable filter, 310
Step function, 126
Stereopsis, 88
Striate cortex, 68, 70, 91
Structural texture, 399, 412
Sub-sampling, 7
Superior colliculus, 43, 68, 89
Symmetries, 350, 354
Symmetry operation, 20
Synapse, 43, 47
Tectum, 43, 68
Tensor edge, 345
Tensor representation, 301
Tensor, 38, 222
Texels, 399
Textons, 403
Texture, 399
  Brodatz texture, 401
  discrimination, 401
  feature-based, 405
  gradient, 402, 406
  microtexture, 399
  orientation, 405
  preattentive segmentation, 402
  segmentation, 401, 406, 416
  spatial frequency channels, 404
  structural, 399, 412
  textons, 403
Tomography device, 152
Top-down, 33
Topographic map, 76, 90
Training, 389
Transform, 131
Transmitter substance, 47
Transposition, 123
Truncated function, 187
Uncertainty principle, 119, 146
  in multiple dimensions, 151
Uncertainty relation, 146
Unsupervised classification, 396
Variance, 146
Vector edge, 345
Vector representation, 301
Vector, 13
Velocity estimation, 219, 253
Visual cortex, 68–69, 91
  primary, 43
Visual pathways, 67
  geniculostriate, 67
  retinotectal, 68
  tectopulvinar, 69
W cell, 57, 90
Wavelet, 281
Weighted Fourier transform, 121
Weighting function, 201
Wide range frequency estimation, 288
Widrow-Hoff rule, 392
Wiener filtering, 310–311
Window, 12
X cell, 57, 61, 67, 90
Y cell, 57, 67, 90
Young-Helmholz model, 59

MIX
Papier aus verantwortungsvollen Quellen
Paper from responsible sources
FSC® C105338

If you have any concerns about our products,
you can contact us on
ProductSafety@springernature.com

In case Publisher is established outside the EU,
the EU authorized representative is:
Springer Nature Customer Service Center GmbH
Europaplatz 3, 69115 Heidelberg, Germany

Printed by Libri Plureos GmbH
in Hamburg, Germany